METHODS IN MOLECULAR BIOLOGY

For further volumes:
http://www.springer.com/series/7651

Alpha-Synuclein

Methods and Protocols

Edited by

Tim Bartels

Dementia Research Institute, University College London, London, UK

Humana Press

Editor
Tim Bartels
Dementia Research Institute
University College London
London, UK

ISSN 1064-3745 ISSN 1940-6029 (electronic)
Methods in Molecular Biology
ISBN 978-1-4939-9123-5 ISBN 978-1-4939-9124-2 (eBook)
https://doi.org/10.1007/978-1-4939-9124-2

Library of Congress Control Number: 2019930436

This Humana Press imprint is published by the registered company Springer Science+Business Media, LLC, part of Springer Nature.
The registered company address is: 233 Spring Street, New York, NY 10013, U.S.A.

Preface

This book is a lab manual targeted at researchers working on topics related to alpha-synuclein. This small protein is the object of intense investigation in a plethora of fields for the last two decades. The reason lies in its still puzzling biochemical properties and hard to define physiological function.

In biophysics, its properties as an "intrinsically disordered protein" have led to many interesting results in the field of protein folding, while its amyloidogenic aggregation has shed light on protein oligomerization. In biochemistry, its specific lipid-binding capabilities and multiple proposed interactors have been broadly studied. In molecular biology, its transient interaction with biological membranes and impact on vesicle trafficking have led to important insights in cellular function, while its misfolding has shed light on a number of mechanisms important for cell survival. Last but not least, major efforts to understand its physiological role have stemmed from biomedicine. Alpha-synuclein seems to be a central point of pathogenesis in Parkinson's disease, Lewy body dementia, and multiple system atrophy, diseases therefore commonly grouped as "synucleinopathies."

Hence, this book is meant to supply the reader with detailed protocols that use a variety of in vitro and in vivo techniques and assays that are applicable in reductionist systems as well as in human samples. The methods try to encompass a broad spectrum of functional aspects of alpha-synuclein biology, from lipid interaction and misfolding to functional deficits in cells and mice. While obviously not being able to cover all techniques used in this ever-expanding field, the experts that have written the chapters and I hope to enable the reader to get started in the most common methods if interested in a scientific question related to alpha-synuclein biology.

London, UK *Tim Bartels*

Contents

Contributors

JAVIER ALEGRE-ABARRATEGUI • *Division of Brain Sciences, Faculty of Medicine, Imperial College London, London, UK*
NORA BENGOA-VERGNIORY • *Department of Physiology, Anatomy and Genetics, Oxford Parkinson's Disease Centre, University of Oxford, Oxford, UK*
PAYEL BHATTACHARJEE • *Department of Psychiatry and Neurochemistry, Institute of Neuroscience and Physiology, The Sahlgrenska Academy at the University of Gothenburg, Mölndal, Sweden*
INÊS CALDEIRA BRÁS • *Department of Experimental Neurodegeneration, Center for Biostructural Imaging of Neurodegeneration, University Medical Center Goettingen, Goettingen, Germany*
THERESA BRAUN • *Department of Chemistry and Konstanz Research School Chemical Biology, University of Konstanz, Konstanz, Germany*
GERHARD H. BRAUS • *Department of Molecular Microbiology and Genetics, Institute of Microbiology and Genetics, Georg-August-Universität Göttingen, Göttingen, Germany; Center for Nanoscale Microscopy and Molecular Physiology of the Brain, Goettingen, Germany*
ANN BRINKMALM • *Department of Psychiatry and Neurochemistry, Institute of Neuroscience and Physiology, The Sahlgrenska Academy at the University of Gothenburg, Mölndal, Sweden*
STEVEN A. CARR • *Broad Institute of MIT and Harvard, Cambridge, MA, USA*
JULIA CATTANI • *Department of Chemistry and Konstanz Research School Chemical Biology, University of Konstanz, Konstanz, Germany*
CHEE YEUN CHUNG • *Yumanity Therapeutics, Cambridge, MA, USA*
LUIS CONCHA-MARAMBIO • *Amprion Inc., San Diego, CA, USA*
TAPOJYOTI DAS • *Weill Cornell Medicine, New York, NY, USA*
ULF DETTMER • *Ann Romney Center for Neurologic Diseases, Brigham and Women's Hospital and Harvard Medical School, Boston, MA, USA*
MALTE DRESCHER • *Department of Chemistry and Konstanz Research School Chemical Biology, University of Konstanz, Konstanz, Germany*
TARA A. EASTWOOD • *Molecular Biology, School of Biosciences, University of Kent, Canterbury, Kent, UK*
DAVID ELIEZER • *Weill Cornell Medicine, New York, NY, USA*
SARANNA FANNING • *Ann Romney Center for Neurologic Diseases, Brigham and Women's Hospital and Harvard Medical School, Boston, MA, USA*
ALLAN CHRIS M. FERREON • *Department of Pharmacology and Chemical Biology, Baylor College of Medicine, Houston, TX, USA*
JOSEPHINE C. FERREON • *Department of Pharmacology and Chemical Biology, Baylor College of Medicine, Houston, TX, USA*
RON GATHAGAN • *Center for Neurodegenerative Disease Research, Institute on Aging, Department of Pathology and Laboratory Medicine, University of Pennsylvania Perelman School of Medicine, Philadelphia, PA, USA*
THIBAUT IMBERDIS • *Ann Romney Center for Neurologic Diseases, Brigham and Women's Hospital and Harvard Medical School, Boston, MA, USA*

CHI WANG IP • *Department of Neurology, University Hospital of Würzburg, Würzburg, Germany*
AKUA A. KARIKARI • *Department of Neurology, University Hospital of Würzburg, Würzburg, Germany*
VICTORIA KEHM • *Center for Neurodegenerative Disease Research, Institute on Aging, Department of Pathology and Laboratory Medicine, University of Pennsylvania Perelman School of Medicine, Philadelphia, PA, USA*
VIKRAM KHURANA • *Ann Romney Center for Neurologic Disease, Department of Neurology, Brigham and Women's Hospital and Harvard Medical School, Boston, MA, USA; Broad Institute of MIT and Harvard, Cambridge, MA, USA; Harvard Stem Cell Institute, Cambridge, MA, USA*
MANDY KOOPMAN • *University of Groningen, University Medical Center Groningen, European Research Institute for the Biology of Aging, Groningen, The Netherlands*
JAMES B. KOPRICH • *University Health Network, Krembil Research Institute, Toronto Western Hospital, Toronto, ON, Canada*
NIELS KRUSE • *Institute of Neuropathology, University Medical Center Göttingen, Göttingen, Germany*
VIRGINIA M.-Y. LEE • *Center for Neurodegenerative Disease Research, Institute on Aging, Department of Pathology and Laboratory Medicine, University of Pennsylvania Perelman School of Medicine, Philadelphia, PA, USA*
SUSAN N. LEIGHT • *Center for Neurodegenerative Disease Research, Institute on Aging, Department of Pathology and Laboratory Medicine, University of Pennsylvania Perelman School of Medicine, Philadelphia, PA, USA*
KELVIN C. LUK • *Center for Neurodegenerative Disease Research, Institute on Aging, Department of Pathology and Laboratory Medicine, University of Pennsylvania Perelman School of Medicine, Philadelphia, PA, USA*
ERIC S. LUTH • *Department of Biology, Simmons University, Boston, MA, USA*
BRIT MOLLENHAUER • *Paracelsus-Elena-Klinik, Kassel, Germany; Department of Neurology, University Medical Center Göttingen, Göttingen, Germany*
MAHDI MUHAMMAD MOOSA • *Department of Pharmacology and Chemical Biology, Baylor College of Medicine, Houston, TX, USA*
DANIEL P. MULVIHILL • *Molecular Biology, School of Biosciences, University of Kent, Canterbury, Kent, UK*
ANDREW NEWMAN • *Ann Romney Center for Neurologic Diseases, Brigham and Women's Hospital and Harvard Medical School, Boston, MA, USA*
ELLEN A. A. NOLLEN • *University of Groningen, University Medical Center Groningen, European Research Institute for the Biology of Aging, Groningen, The Netherlands*
ANNIKA ÖHRFELT • *Department of Psychiatry and Neurochemistry, Institute of Neuroscience and Physiology, The Sahlgrenska Academy at the University of Gothenburg, Mölndal, Sweden*
TIAGO F. OUTEIRO • *Department of Experimental Neurodegeneration, Center for Biostructural Imaging of Neurodegeneration, University Medical Center Goettingen, Goettingen, Germany; Center for Nanoscale Microscopy and Molecular Physiology of the Brain, Goettingen, Germany; Max Planck Institute for Experimental Medicine, Goettingen, Germany*
PATRICK OVANDO-ROCHE • *Ann Romney Center for Neurologic Disease, Department of Neurology, Brigham and Women's Hospital and Harvard Medical School, Boston, MA, USA*

DUSHYANT S. PATEL • *Ann Romney Center for Neurologic Diseases, Brigham and Women's Hospital, Harvard Medical School, Boston, MA, USA*
BLAGOVESTA POPOVA • *Department of Molecular Microbiology and Genetics, Institute of Microbiology and Genetics, Georg-August-Universität Göttingen, Göttingen, Germany; Center for Nanoscale Microscopy and Molecular Physiology of the Brain, Goettingen, Germany*
ALEX E. POWERS • *Ann Romney Center for Neurologic Diseases, Brigham and Women's Hospital, Harvard Medical School, Boston, MA, USA*
NAGENDRAN RAMALINGAM • *Ann Romney Center for Neurologic Diseases, Brigham and Women's Hospital and Harvard Medical School, Boston, MA, USA*
ROSALIND F. ROBERTS • *Montreal Neurological Institute, McGill University, Montréal, QC, Canada*
MATTEO ROVERE • *Ann Romney Center for Neurologic Diseases, Brigham and Women's Hospital and Harvard Medical School, Boston, MA, USA*
JOHN B. SANDERSON • *The Warren Alpert Medical School of Brown University, Providence, RI, USA*
RICARDO SANZ • *Ann Romney Center for Neurologic Disease, Department of Neurology, Brigham and Women's Hospital and Harvard Medical School, Boston, MA, USA*
RENÉE I. SEINSTRA • *University of Groningen, University Medical Center Groningen, European Research Institute for the Biology of Aging, Groningen, The Netherlands*
MOHAMMAD SHAHNAWAZ • *Mitchell Center for Alzheimer's Disease and Related Brain Disorders, Department of Neurology, University of Texas Houston Medical School, Houston, TX, USA*
CLAUDIO SOTO • *Amprion Inc., San Diego, CA, USA; Mitchell Center for Alzheimer's Disease and Related Brain Disorders, Department of Neurology, University of Texas Houston Medical School, Houston, TX, USA*
IRINA G. STAVROVSKAYA • *Department of Neurology, Columbia University Medical Center, New York, NY, USA*
ARON SZABO • *Institute of Genetics, Biological Research Centre of the Hungarian Academy of Sciences, Szeged, Hungary*
GEORGE K. TOFARIS • *Nuffield Department of Clinical Neurosciences, University of Oxford, John Radcliffe Hospital, Oxford, UK*
JOHN Q. TROJANOWSKI • *Center for Neurodegenerative Disease Research, Institute on Aging, Department of Pathology and Laboratory Medicine, University of Pennsylvania Perelman School of Medicine, Philadelphia, PA, USA*
NAMRATA D. UDESHI • *Broad Institute of MIT and Harvard, Cambridge, MA, USA*
KARINA J. VARGAS • *Department of Neurology, Yale University, New Haven, CT, USA; Program in Cellular Neuroscience, Neurodegeneration and Repair, Yale University, New Haven, CT, USA; Eugene Bell Center for Regenerative Biology and Tissue Engineering, Marine Biology Laboratory, Woods Hole, MA, USA*
LAURA A. VOLPICELLI-DALEY • *Department of Neurology, Center for Neurodegeneration and Experimental Therapeutics, University of Alabama at Birmingham, Birmingham, AL, USA*

HENRIK ZETTERBERG • *Department of Psychiatry and Neurochemistry, Institute of Neuroscience and Physiology, The Sahlgrenska Academy at the University of Gothenburg, Mölndal, Sweden; Clinical Neurochemistry Laboratory, Sahlgrenska University Hospital, Mölndal, Sweden; Department of Molecular Neuroscience, UCL Institute of Neurology, Queen Square, London, UK; UK Dementia Research Institute at UCL, London, UK*
BIN ZHANG • *Center for Neurodegenerative Disease Research, Institute on Aging, Department of Pathology and Laboratory Medicine, University of Pennsylvania Perelman School of Medicine, Philadelphia, PA, USA*

Chapter 1

Assays for Neuronal Defects Caused by Early Formation of α-Synuclein Inclusions in Primary Cultured Neurons

Laura A. Volpicelli-Daley

Abstract

Parkinson's disease (PD) and dementia with Lewy bodies (DLB) are characterized by intracellular inclusions composed mostly of α-synuclein (Baba et al., Am J Pathol 152:879–884, 1998). How inclusion formation impacts neuronal function prior to death is key to understanding disease progression and identifying therapeutic windows. In the α-synuclein fibril model, exposure of primary neurons to α-synuclein fibrils induces endogenously expressed α-synuclein to form inclusions which closely resembles pathologic mechanisms in humans with PD and DLB (Volpicelli-Daley et al., Neuron 72, 57–71, 2011). In this model, at 7 days after exposure of neurons to fibrils, when there is no neuron death, inclusions in the axon selectively impair axonal transport of endosomes carrying the TrkB receptor and LC3-positive autophagosomes (Volpicelli-Daley et al., Mol Biol Cell 25:4010–4023, 2014). In addition, the frequency and amplitude of spontaneous Ca^{2+} transients are reduced in neurons 7 days after fibril exposure. Here we discuss protocols for plating primary hippocampal neurons, generating fibrils and measuring axonal transport and Ca^{2+} transients. These assays provide additional assays of neurotoxicity allowing researchers to determine if a therapeutic intervention can prevent neuronal defects before intractable neurodegeneration.

Key words α-Synuclein, Fibril, Parkinson's disease, Primary neurons, Axonal transport, Ca^{2+} imaging, Lewy body, Lewy neurite, Inclusion

1 Introduction

Neuronal inclusions composed of α-synuclein characterize Parkinson's disease (PD) and dementia with Lewy bodies (DLB) [1]. Most studies utilizing models of PD and synucleinopathies study cell death as an endpoint to determine if a therapeutic target prevents neurodegeneration. However, it is critical to determine if putative therapeutics can prevent or reverse neuronal defects before intractable neurodegeneration. The α-synuclein fibril model allows researchers to study inclusion formation from their initial formation [2]. In this model, exposure of neurons to α-synuclein fibrils formed from recombinant α-synuclein induces normal endogenous α-synuclein to form inclusions that biochemically and morphologically closely

Tim Bartels (ed.), *Alpha-Synuclein: Methods and Protocols*, Methods in Molecular Biology, vol. 1948,
https://doi.org/10.1007/978-1-4939-9124-2_1,

resemble those found in neurons of PD and DLB brains; the inclusions are insoluble, hyperphosphorylated, ubiquitinated, and filamentous by electron microscopy. When added to primary cultured neurons, inclusions are apparent 3–4 days after addition of fibrils, are abundant by 7–10 days after fibril addition, and cause cell death by 14 days after fibril exposure. An additional advantage of this model is that it is the only model in which inclusion formation can be induced in primary neuron cultures. Primary neurons and live-cell microscopy provide the spatial and temporal resolution for the visualization and quantification of potential neuron defects. Finally, in contrast to in vivo models, primary neuron cultures allow multiple compounds to be tested at different concentrations without the blood–brain barrier. Thus, putative drugs can be screened for whether they prevent α-synuclein inclusion formation and reduce neuronal dysfunction before moving on to more time-consuming and costly in vivo models.

We have found that formation of α-synuclein inclusions selectively impairs axonal transport of endosomes and autophagosomes at a time point well before neuron death [3]. We have also found that early formation of α-synuclein inclusions reduces the frequency and amplitude of Ca^{2+} transients. These early effects of α-synuclein accumulations may predict points of intervention in the neurodegenerative process. Here we describe methods for inducing inclusion formation in primary neurons with a modified protocol for generating α-synuclein fibrils that is more reliable and robust. We describe methods for imaging and quantifying axonal transport of endosomes and autophagosomes. Finally, we describe methods for imaging calcium transients in primary neurons by transducing them with AAV-synapsin-GCAMP6 [4]. Overall, these assays provide additional "toxicity" endpoints for researchers to assess the potential of a therapeutic target in preventing neuron dysfunction.

2 Materials

2.1 Primary Neuron Culture

1. Pregnant mice or rats carrying embryonic day E16 to E18 embryos.
2. 35 mm MatTek dishes with No. 1.5 coverslip.
3. Borate buffer: 50 mM Boric acid (Sigma, B6768) in ddH_2O; adjust pH to 8.5 and filter sterilize.
4. Poly-D-lysine (Sigma, cat no. P0899, www.sigmaaldrich.com) at 2 mg/mL in borate buffer: Filter sterilize. Make 2 mL aliquots in polystyrene tubes and freeze at −20 °C (*see* **Note 1**).
5. Hibernate E (*see* **Note 2**).
6. Plating media: Neurobasal media, 1× B27 from a 50× stock, 1× Glutamax from a 100× stock, 50 U/mL penicillin and

50 μg/mL streptomycin, 10% fetal bovine serum. Filter sterilize and store at 4 °C.

7. Neuronal media: Neurobasal media, 1× B27 from a 50× stock, 1× Glutamax from a 100× stock. We do not use antibiotics in our neuronal media because they interfere with the transfections. Filter sterilize and store at 4 °C (*see* **Note 3**).
8. HBSS: To calcium- and magnesium-free HBSS, add 6 mL of 1 M Hepes pH 7, 3 mL of 50 U/mL penicillin and 50 μg/mL streptomycin, 6 mL of 100 mM pyruvic acid. Bring to 500 mL with the remainder of HBSS, and add 100 mL of water (final volume 600 mL). Filter sterilize and store at 4 °C.
9. Papain solution: Make fresh before starting dissection. For tissue from 6–8 pups, make 10 mL. Add papain to HBSS (above) to a final concentration of 20 U/mL, 2 mg L-cysteine, and 22 μL of 500 mM EDTA, pH 8.0. Incubate with shaking at room temperature to allow papain to dissolve.
10. DNase: Dissolve DNase to 10 U/mL (Sigma-Aldrich, DN25) in HBSS (above), aliquot, and store at −20 °C.
11. 40 mm Nylon mesh cell strainer.

2.2 Mouse α-Synuclein Fibrils

1. Mouse α-synuclein monomer: We recommend purifying untagged monomeric mouse α-synuclein; see Mulvihill et al. in this issue. It is preferable to store monomeric α-synuclein at a concentration of >5 mg/mL because of possible spontaneous aggregates [5, 6]. These aggregates will be lost to the pellet after high-speed centrifugation, slightly reducing the protein concentration. Aliquot monomer to avoid freeze/thaw cycles and store at −80 °C. Store α-synuclein monomer in 10 mM Tris, pH 7.5, and 50 mM NaCl. Human α-synuclein monomer is also acceptable for primary neuron experiments (it is less efficient for in vivo studies). Monomer should have low levels of endotoxin. We use the Pierce High Capacity endotoxin removal resin to remove endotoxin and the Pierce LAL endotoxin kit. We typically have about 1.0 ET units/mg of protein or 20.0 ET unit/mL.
2. 8 M Guanidinium chloride (GnCl): Can be stored at −20 °C.
3. 1 M KCl.
4. 1 M Tris–HCl, pH 7.5.
5. 4–20% Gradient gels.
6. 2× Laemmli buffer: 126 mM Tris–HCL (pH 6.8), 4% SDS (vol/vol), 20% glycerol (vol/vol), 0.002% wt/vol bromophenol blue.

7. Coomassie stain: 0.2% Coomassie Brilliant Blue R250 (wt/vol), 50% methanol (vol/vol), dissolve dye, then add 10% acetic acid (vol/vol), and bring to final volume with water.
8. Coomassie destain: 50% Methanol (vol/vol), dissolve dye, then add 10% acetic acid (vol/vol), and bring to final volume with water.
9. Spectrophotometer for measuring protein concentration at A280: We prefer the Nanodrop to minimize volume and ease of use.
10. Thermomixer with block for 1.5 mL tubes placed inside standard lab incubator set at 37 °C.
11. Benchtop ultracentrifuge with tubes for 100,000 × *g* spins.
12. Sonic Dismembrator System with 0.16 in. microtip and 0.25 in. tip (Fisher Scientific cat no. FB120110, www.fishersci.com).
13. Squirt bottle with 10% SDS.

2.3 Axonal Transport

1. Plasmids: TrkB-GFP, Addgene 32500, GFP-LC3, Addgene 21073.
2. Lipofectamine LTX (*see* **Note 3**).
3. DMEM.
4. Phosphate-buffered saline (PBS).
5. Imaging buffer: 136 mM NaCl, 2.5 mM KCl, 2 mM $CaCl_2$, 1.3 mM $MgCl_2$, 10 mM glucose, and 10 mM 4-(2-hydroxyethyl)-1-piperazineethanesulfonic acid.
6. BDNF: 100 μg/mL Reconstituted in endotoxin-free, sterile water with 0.1% BSA, aliquoted, and stored at −80 °C.
7. 4% Paraformaldehyde in PBS: We recommend diluting 16% paraformaldehyde in aqueous solution (from Electron Microscopy Sciences) in PBS.
8. Phospho-α-synuclein antibody: EP1536Y. This antibody recognizes α-synuclein inclusions with minimal nonspecific background.
9. 3% Bovine serum albumin in PBS.
10. Goat anti-rabbit Alexa 555.
11. 10% Sodium azide.
12. Inverted epifluorescent microscope with temperature-controlled incubator, 60× oil objective, equipped with CoolSNAP HQ^2 camera or equivalent.
13. Fiji/Image J.

2.4 Ca^{2+} Imaging

1. AAV9.Syn.GCamp6m.WPRE.SV40 (University of Pennsylvania School of Medicine Vector Core): Make 10–20 μL aliquots and store in the −80 °C freezer.
2. 10% Bleach.
3. Imaging buffer: 136 mM NaCl, 2.5 mM KCl, 2 mM $CaCl_2$, 1.3 mM $MgCl_2$, 10 mM glucose, and 10 mM 4-(2-hydroxyethyl)-1-piperazineethanesulfonic acid.
4. Inverted epifluorescent microscope with temperature-controlled incubator, 60× oil objective, equipped with CoolSNAP HQ^2 camera or equivalent.

3 Methods

3.1 Primary Neuron Cultures

1. Coat MatTek dishes with 0.1 mg/mL poly-D-lys for 1.5 h (or overnight) at 37 °C. Do not coat for longer than 16 h. Rinse 2× with water followed by neuronal plating media. Store in 37 °C tissue culture incubator in plating media for no longer than 1 week.
2. Calculate the amount of plating and neuronal media that will be needed. Twenty mL of plating media will be needed for rinses. For plating, 2 mL of plating and neuronal media per dish. Place plating and neuronal media into separate tissue culture flasks, label clearly, and place in CO_2 cell culture incubator overnight. Pre-equilibrating the media with CO_2 is a critical step.
3. Make HBSS with papain the day of the dissection. Do not make in advance.
4. Dissect the hippocampi as previously demonstrated [7] using Hibernate E.
5. Place dissected hippocampi in the 15 mL sterile tube with 10 mL of Hibernate E.
6. Move tube to laminar flow hood for cell culture. Remove all but 1 mL of Hibernate E. Add 10 mL of HBSS. Remove all but 1 mL of HBSS. Repeat one more time. Leave 1 mL HBSS with dissected hippocampi. Filter sterilize the 10 mL HBSS+papain with 0.2 μm syringe filter. Add to tube with dissected hippocampi.
7. Set tube with hippocampi in HBSS/papain on its side inside cell culture incubator at 37 °C. Make sure that hippocampi are inside solution. Incubate for a minimum of 30 min and a maximum of 90 min.
8. Place tube with HBSS/papain/hippocampi upright in laminar flow hood. The tissue will be sticky and it will take a few minutes for hippocampi to float toward the bottom of the

tube. In the meantime, make 10 mL of plating media with 50 μL of DNase. Carefully remove as much HBSS/papain from hippocampi. Again, the tissue is sticky; take care not to pipet the tissue into the pipette. We use a 1 mL filtered pipette for this. Leave a little HBSS/papain behind. Add the 10 mL of plating media/DNase. Allow the hippocampi to float back down to the bottom of the tube. The tissue will no longer be sticky. Remove all but 1 mL of plating media/DNase. Add 10 mL of plating media (no DNase). Allow the hippocampi to float back down to the bottom of the tube. Remove all but 1 mL of plating media. Add 10 mL of HBSS (no DNase). Remove all but 1 mL of HBSS. Repeat one more time, and leave 1 mL of HBSS in tube.

9. Triturate tissue. We use a 1 mL pipetman with a filter tip. Pipet up and down 10–15 times (up and down is one time). Be careful not to introduce bubbles. Do not overtriturate.
10. Add 5 mL of HBSS. Filter over 40 mm nylon mesh cell strainer to remove clumps of undigested tissue.
11. Perform cell count.
12. Plate neurons on MatTek dishes at 5.2×10^4 cells per cm^2 (500,000 cells per 35 mm MatTek dish).
13. Anywhere between 1 h later to 18 h later, aspirate plating media and replace with 2 mL of neuronal media. Do not incubate in plating media any longer than 18 h to avoid excessive proliferation of glia or fibroblast cells.

3.2 Making Fibrils

1. Rapidly thaw purified α-synuclein monomer in 1.5 mL microcentrifuge tube in 37 °C water bath until small ice cube is visible. Place on ice until completely thawed.
2. Centrifuge at 100,000 × *g* at 4 °C for 60 min to pellet any aggregated material. Remove the supernatant (*see* **Note 4**).
3. Calculate protein concentration using an A280 measurement. Use Beer's law A = cdε. ε for synuclein is 5960/M/cm for human synuclein, and 7450 M/cm for mouse synuclein. Use molecular weight of 14,460 g/mol for α-synuclein.
4. Dilute supernatant from **step 2** in PBS in new, sterile 1.5 mL microcentrifuge tube to a volume of 500 μL and final concentration of 5 mg/mL in 50 mM Tris, pH 7.5, and 150 mM KCl.
5. Shake in thermomixer set at 37 ° C at 700 RPM. Place the thermomixer inside a 37° incubator; this prevents evaporation.
6. Shake for 7 days. At the end of 7 days of shaking, the fibrils should be turbid/cloudy. If it is not cloudy, there will not be enough fibrillar α-synuclein fibrils to seed inclusion formation.

7. Spin the fibrils at 20,000 × *g* in a benchtop centrifuge for 30 min. Resuspend pellet in 300 μL 50 mM Tris, pH 7.5, and 150 mM KCl. To dissociate the fibrils into monomer, take 5 μL and incubate with 5 μL 8 M GnCl for 1 h at room temperature (can be left overnight). Measure protein concentration using A280 measurement as in **step 3**. Use 5 μL buffer and 5 μL 8 M GnCl as a blank. The mix with the GnCl is a ½ dilution of the sample, so the concentration measured should be multiplied by 2.
8. Add 50 mM Tris, pH 7.5, and 150 mM KCl to the 300 μL of fibrils so they are a final concentration of 5 mg/mL.
9. To ensure that the fibrils will induce inclusion formation in neurons, we perform a sedimentation assay. Spin 20 μL at 100,000 × *g*. Remove supernatant to another tube, and add equal volume 2× laemmli buffer. Resuspend pellet in 20 μL 20–50 mM Tris, pH 7.5, and 150 mM KCl. Spin again 20 μL at 100,000 × *g* and resuspend in 20 μL 20–50 mM Tris, pH 7.5, and 150 mM KCl. Add equal volume of 2× laemmli. Run 10 μL of supernatant and pellet on 4–20% gradient gel [5].

   There should be more α-synuclein in the pellet than the supernatant.
10. For primary neuron experiments, it is acceptable to aliquot fibrils into 10–20 μL aliquots and store in the back of the −80 °C freezer. Freezing and thawing can result in disaggregation of the fibrils (*see* **Note 5**).

3.3 Axonal Transport

1. At DIV4, transfect neurons (*see* **Note 6**): Prewarm DMEM to 37 °C. For two 35 mm MatTek dishes, in microcentrifuge tube 1, mix 200 μL of DMEM with 16 μL of Lipofectamine LTX. In microcentrifuge tube 2, mix 200 μL of DMEM with 6 μg of DNA and 6 μL of Plus reagent. Pipet up and down to mix, and let sit at room temperature for 5 min. Mix tube 1 and tube 2 by pipetting up and down. Add 100 μL of mixture to one MatTek dish (*see* **Note 7**).
2. At DIV 7, add sonicated fibrils [5]. As a control, we use an equal volume of PBS. We sonicate immediately before adding the fibrils to the media.

   PERFORM SONICATION STEP IN BSL2 LAMINAR FLOW HOOD. FOLLOW BSL2+ PROCEDURES (GOGGLES, LAB COAT, PROTECTIVE DISPOSABLE LAB SLEEVES, SHOE COVERINGS, DOUBLE GLOVE). FURTHER SAFETY INSTRUCTIONS ARE FOUND IN THE PUBLICATION FROM DR. MELKI'S GROUP [8].

   Add fibrils to sterile PBS to a final concentration of 0.1 mg/mL, e.g., 10 μL in 490 μL PBS. For our sonicator (listed above), we use 30% power, 30 s total sonication time, 1 s on, 1 s off. It is critical to generate 50 nm fibrils. We recommend examining for sufficient sonication using transmission EM (*see* Fig. 3 [6]). This is not necessary for every experiment, but at least initially to ensure that the fibrils are broken into 50 nm fibrils.

For a 35 mm MatTek dish with 2 mL of media, we add 40 μL of fibrils to 2 mL.

3. At 7 days after adding fibrils (DIV 14) perform axonal transport assays.

The researcher performing the live-cell imaging should be blinded to experimental conditions.

Work one MatTek dish at a time. Remove media. Add 2 mL of imaging buffer. Perform imaging on inverted microscope at 32 °C. Use a 60× oil immersion lens. Allow oil to equilibrate to 32 °C which will help maintain focus.

Identify axons [9, 10]. Axons are thin and lack spines. We typically identify an axon by identifying the tip. We then image 25–50 μm away from the tip (Fig. 1) (the pattern and kinetics of vesicle traffic are different at the tip of the axon). We choose axons in which mobile vesicles can be identified. We also choose axons in which the GFP-tagged organelle marker is not highly expressed as the puncta tend not to be mobile in neurons with high expression of the marker.

For the TrkB-GFP experiments, we image one set of MatTek dishes (control and fibril) without BDNF. Axonal transport of TrkB-GFP in neurons with inclusions is further impaired upon treatment with BDNF. We image another set of MatTek dishes in the presence of 50 ng/mL BDNF. For this, we dilute BDNF in imaging buffer. We remove the media and add the imaging buffer/BDNF to the dishes. Neurons are incubated for 30 min before imaging. Imaging is performed with the imaging buffer/BDNF.

Perform live imaging. Adjust the percent output of the LED and the exposure time to a minimum value that allows visualization of the GFP marker yet prevents photobleaching. For example, in our microscope system, we capture images with a 50% 488 LED and exposure time of 100 ms. We adjust the LUT so that the GFP puncta can be seen.

Capture a video every 300 ms for 3 min. We watch as the video is captured to ensure that the focus is maintained throughout the entire times.

Draw an arrow to indicate retrograde transport which is away from the tip of the growth cone (Fig. 1).

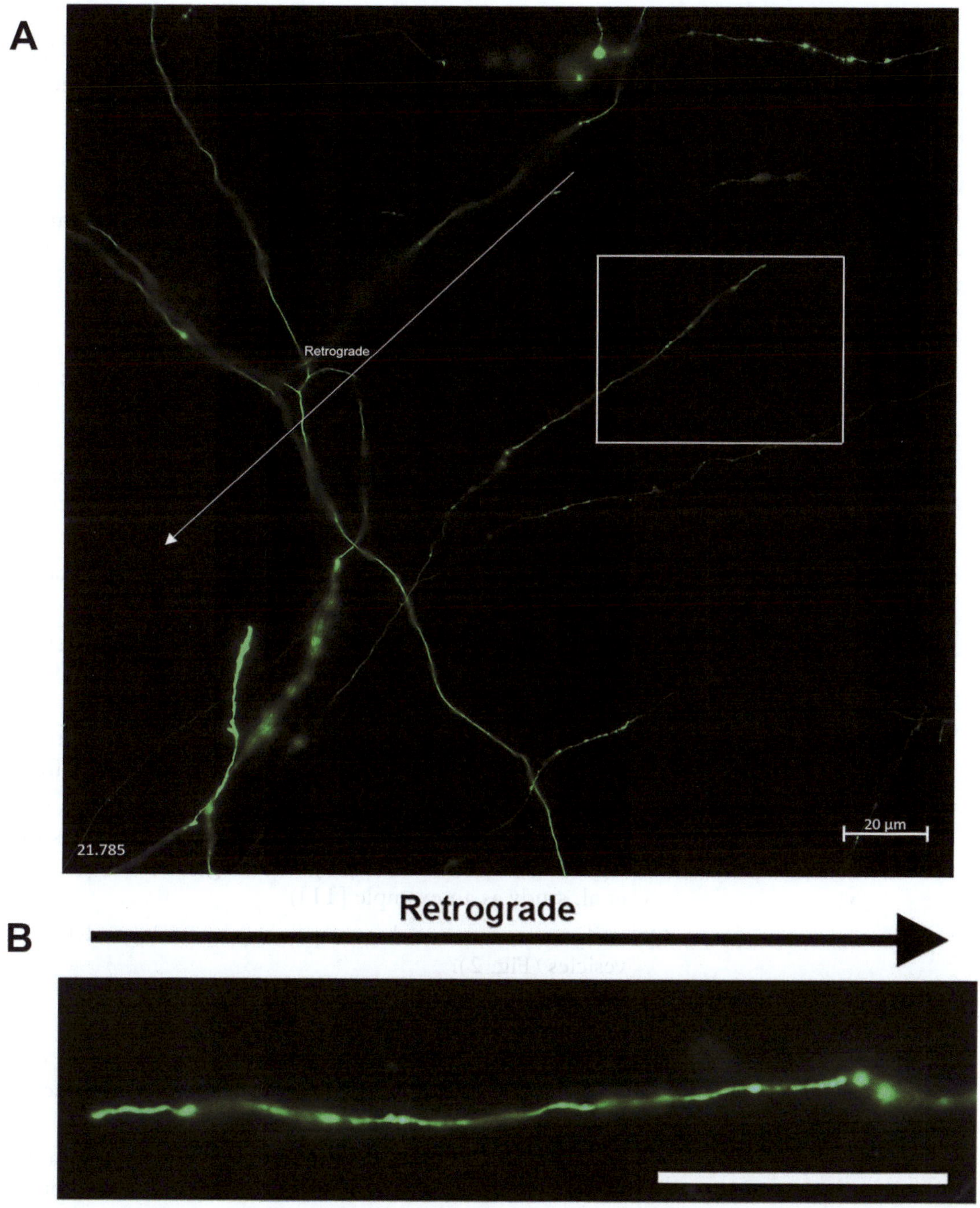

Fig. 1 (**a**) Screenshot from our imaging with a 60× objective of primary hippocampal neurons transfected with GFP-LC3. (**b**) The boxed area from A was cropped and rotated using Fiji. The arrow indicates the direction of retrograde axonal transport away from the tip of the axon

Insert scale bar (20 μm) and time stamp.

Export movie as .avi. Alternatively export each frame as individual .tif file. In Fiji, open each tif file and in the menu go to Image-Stack-Images to Stack, and then save as an .avi file.

Quantify images.

Open video in Fiji. Make sure that the movie is 8bit file (Image-Type-8bit). Crop the image to selectively image the axon of interest (Fig. 1). We orient the video so retrograde transport is to the left and anterograde transport is to the right.

Set the scale: Draw a straight line along the length of the scale bar. Go to Analyze-Set Scale. In known distance box, enter 20. In Unit of Length box, enter um.

Go to Image-Properties and under frame interval enter the seconds between each from (for example, 0.5 s between each frame).

Use FijiTrack mate- under Plug-ins, and find Tracking-Manual Tracking with Trackmate.

Follow the instructions at https://imagej.net/Manual_tracking_with_TrackMate to track individual spots/GFP-positive puncta.

Click the Analysis button. Go to the window "Links in tracks statistics." One of the columns will be velocity from frame to frame.

Data can be analyzed by binning the number of particles that travel 0.5, 1, 0, and 2.0 μm/s (*see* Fig. 4B of the Kalmer et al. study as an example [11]).

Quantify velocities, number of pauses, reversals, and mobile vesicles (Fig. 2).

3.4 Ca^{2+} Imaging

1. On DIV 7, add fibrils as described in Subheading 3.3.
2. On DIV 8, add AAV-Synapsin-GCAMP6 at 1:1000 to each MatTek dish (2 μL in 2 mL). Have a beaker with 10% bleach to decontaminate pipet tips.
3. Six days later (DIV 14) perform imaging. Work one MatTek dish at a time. Remove media, and add 2 mL of imaging buffer to the dish.

Add 2 mL of imaging buffer. Perform imaging on inverted microscope at 32 °C. Use a 10× objective.

Decontaminate media with virus and any pipette tips or supplies that came in contact with the virus with 10% bleach.

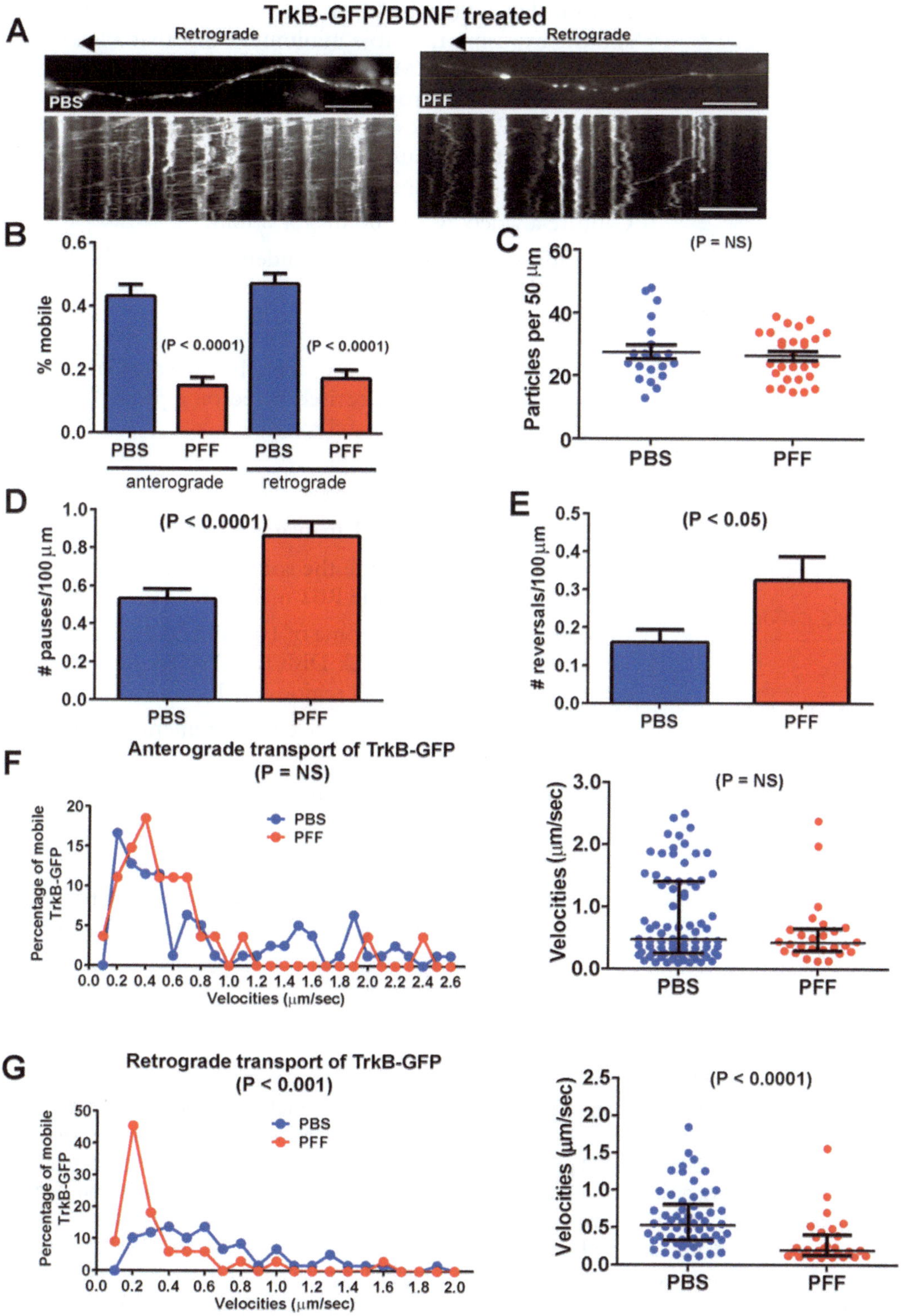

Fig. 2 Example of axonal transport of TrkB-GFP in neurons with and without exposure to fibrils. Reproduced from Volpicelli-Daley et al. 2014 *Mol Biol Cell* **25**, 4010–4023. Primary hippocampal neurons were transfected

4. Perform live imaging. Adjust the percent output of the LED and the exposure time to a minimum value that allows visualization of the GCAMP6 yet prevents photobleaching. For example, in our microscope system, we capture images with a 50% 488 LED and exposure time of 100 ms. We adjust the LUT so that the neurons expressing GCAMP6 can be seen (Fig. X).

   Capture a video every 300 ms for 5 min.

   We capture videos for six independent fields for each MatTek dish.

   Export video as .avi.

5. Use Fiji to quantify Ca^{2+} transients in the soma of individual neurons. This method is revised based on a protocol provided at http://akiramuto.net/archives/148.
   Open movie using Fiji.

   On the menu, go to Image-Type-8bit.

   Go to Analyze-Tools-RO1 manager.

   Select the circle icon, circle the soma of an individual neuron, and click "Add" in the R01 manager.

   Repeat this; we circle the soma of 10 individual neurons in each field from each MatTek Dish (total of 60 neurons). Click "Show all" to show all regions circled.

   On the menu, go to Analyze-Set Measurement; check only the Mean gray value box.

   On R01 manager, go to More-Multi measure; click ok.

   The results window will show image intensities for every frame for each neuron selected.

   Go to select all and copy and paste into Excel. Line plots can be generated for the mean fluorescence intensity over time (Fig. 3).

Fig. 2 (continued) with TrkB-GFP and imaged 7 days after PBS or PFF addition. Neurons were treated with BDNF for 30 min prior to imaging and BDNF was included in the imaging media. (**a**) The top panels show representative images. Scale bar = 10 μm. (**b**) Of the mobile particles, the percentages of anterograde and retrograde particles was quantified. (**c**) There was not a significant difference in the mean number of GFP-Rab7 particles per 50 μm of axonal membrane. (**d**) Neurons with α-synuclein inclusions showed a significant increase in (**d**) the number of pauses and (**e**) the number of reversals. (**f**) The velocities were binned with 10 cut points. The right panel presents the median and interquartile ranges of the anterograde velocities of the mobile TrkB-GFP particles in BDNF-treated neurons. (**g**) For retrograde transport, TrkB-GFP velocities were binned with 10 cut points The scatter plot on the right shows a striking decrease in the velocities of mobile vesicles

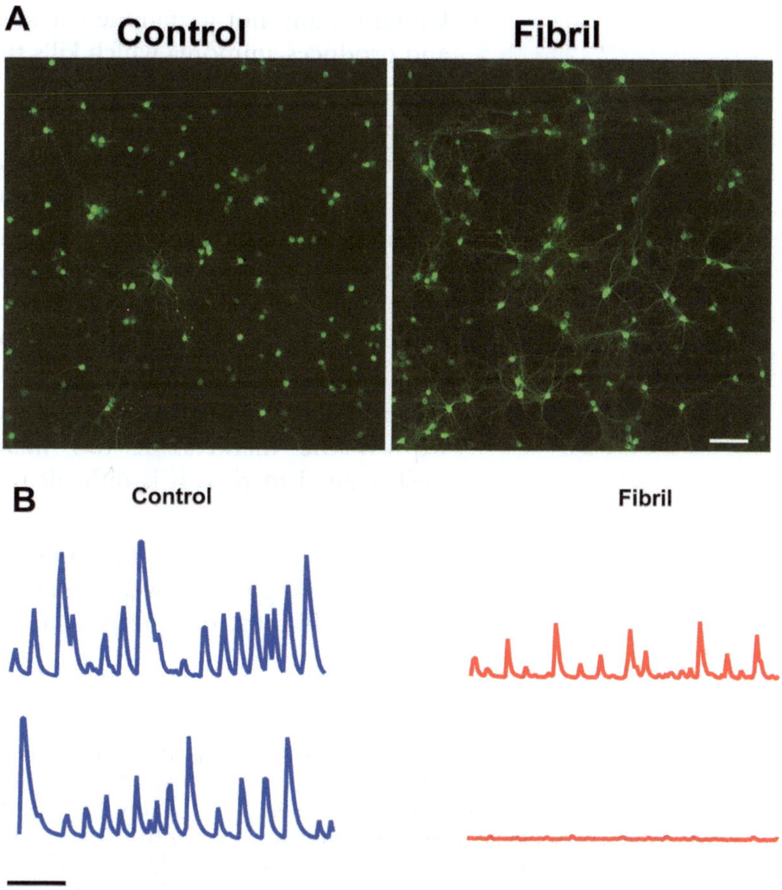

Fig. 3 Primary hippocampal neurons from wild-type mice were exposed to fibrils (or PBS) at DIV 7. On DIV 8, primary neurons were transduced with AAV9-Synapsin1-GCamp6 to visualize Ca^{2+} transients selectively in neurons. On DIV 14 (7 days postinduction of inclusions with fibrils), neurons were imaged live. Images were captured every 300 ms for 5 min at 37 °C. (**a**) Images of neurons expressing AAV9-Synapsin1-GCamp6. Scale bar = 100 μm. (**b**) Representative traces of Ca^{2+} transients over time in control neurons and neurons with inclusions

4 Notes

1. Poly-D-lysine: We have tried substrates and vendor other than poly-D-lysine from Sigma (P0899 specifically). This is the only one that is not toxic in our hands. This is also the only substrate in which we do not have clumps of neurons growing together; the neurons in general grow as individual, nicely separated cells. It is also critical to dissolve the poly-D-lysine in borate buffer which helps the poly-D-lysine go into solution. If it is dissolved in water, again the neurons tend to grow in clumps.
2. We find that the neurons appear the healthiest when we dissect in Hibernate E.

3. Glutamax and not glutamine is essential. Glutamine degrades and produces ammonia which kills the neurons.
4. We find that Lipofectamine LTX produces the least toxicity. Other colleagues that study axonal transport have told us that other transfection reagents alter axonal transport on their own, whereas Lipofectamine LTX does not.
5. α-Synuclein readily aggregates. Keeping the monomer at 4 °C helps prevent aggregation. But if monomer is being used as a control in the assays it is critical to centrifuge the monomer (20,000 × *g*, 15 min at 4 °C) and only use the supernatant.
6. The advantage of transfecting neurons with the organelle makers such as GFP-LC3 or TrkB-GFP is that few neurons express the markers. If too many neurons express the GFP-tagged marker, it is difficult to identify and image individual axons.
7. It is critical not to store the fibrils at 4 °C or −20 °C because they will disaggregate. Even storage at −80 °C with the freeze/thawing will cause disaggregation.

References

1. Baba M, Nakajo S, Tu PH, Tomita T, Nakaya K, Lee VM, Trojanowski JQ, Iwatsubo T (1998) Aggregation of alpha-synuclein in Lewy bodies of sporadic Parkinson's disease and dementia with Lewy bodies. Am J Pathol 152:879–884
2. Volpicelli-Daley LA, Luk KC, Patel TP, Tanik SA, Riddle DM, Stieber A, Meaney DF, Trojanowski JQ, Lee VM (2011) Exogenous alpha-synuclein fibrils induce Lewy body pathology leading to synaptic dysfunction and neuron death. Neuron 72:57–71
3. Volpicelli-Daley LA, Gamble KL, Schultheiss CE, Riddle DM, West AB, Lee VM (2014) Formation of alpha-synuclein Lewy neurite-like aggregates in axons impedes the transport of distinct endosomes. Mol Biol Cell 25:4010–4023
4. Chen TW, Wardill TJ, Sun Y, Pulver SR, Renninger SL, Baohan A, Schreiter ER, Kerr RA, Orger MB, Jayaraman V, Looger LL, Svoboda K, Kim DS (2013) Ultrasensitive fluorescent proteins for imaging neuronal activity. Nature 499:295–300
5. Volpicelli-Daley LA, Luk KC, Lee VM (2014) Addition of exogenous alpha-synuclein pre-formed fibrils to primary neuronal cultures to seed recruitment of endogenous alpha-synuclein to Lewy body and Lewy neurite-like aggregates. Nat Protoc 9:2135–2146
6. Polinski NK, Volpicelli-Daley LA, Sortwell CE, Luk KC, Cremades N, Gottler LM, Froula J, Duffy MF, Lee VM, Martinez TN, Dave KD (2018) Best practices for generating and using alpha-synuclein pre-formed fibrils to model Parkinson's disease in rodents. J Parkinsons Dis 8(2):303–322
7. Seibenhener ML, Wooten MW (2012) Isolation and culture of hippocampal neurons from prenatal mice. J Vis Exp. https://doi.org/10.3791/3634
8. Bousset L, Brundin P, Bockmann A, Meier B, Melki R (2016) An efficient procedure for removal and inactivation of alpha-synuclein assemblies from laboratory materials. J Parkinsons Dis 6:143–151
9. Roy S, Winton MJ, Black MM, Trojanowski JQ, Lee VM (2008) Cytoskeletal requirements in axonal transport of slow component-b. J Neurosci 28:5248–5256
10. Dotti CG, Sullivan CA, Banker GA (1988) The establishment of polarity by hippocampal neurons in culture. J Neurosci 8:1454–1468
11. Kalmar B, Innes A, Wanisch K, Kolaszynska AK, Pandraud A, Kelly G, Abramov AY, Reilly MM, Schiavo G, Greensmith L (2017) Mitochondrial deficits and abnormal mitochondrial retrograde axonal transport play a role in the pathogenesis of mutant Hsp27-induced Charcot Marie Tooth Disease. Hum Mol Genet 26:3313–3326

Chapter 2

Immunohistochemical Detection of Alpha-Synuclein in Unfixed Human Brain Tissue

John B. Sanderson

Abstract

In both research and diagnostics, immunohistochemistry is an essential method for assessing pathology in neurodegenerative diseases. Typically, at autopsy, one hemisphere of the brain is formalin fixed for sectioning and histochemical analysis, while the other hemisphere is flash frozen for biochemical studies. While this process ensures the generation of high-quality slides, it is important to assess the pathology of tissue near pieces used for biochemical analyses, since pathology burden can be unevenly distributed throughout the brain (Fig. 1). Here, we describe a method for assessing alpha-synuclein (αSyn) pathology in unfixed human brain tissue using a novel monoclonal antibody.

Key words Alpha-synuclein, Immunohistochemistry, Human tissue, Frozen, Fresh

1 Introduction

While chemically modified antibodies had previously been used to establish structural properties of immunoglobulin and interrogate the mechanisms of the antibody-antigen interactions, the concept of using antibody derivatives to detect antigenic material originated in 1930, when Reiner and colleagues conjugated antipneumococcus I and II antibodies to atoxyl-azo in a study published in *Science* [1]. Four years later, Dr. John Marrack and colleagues generated antisera to typhoid and cholera, conjugated the antibodies using diazotized benzidine-azo-R-salt, and visualized the appropriate organisms in a study published in *Nature* [2]. In 1941, three investigators from the Department of Bacteriology and Immunology and the Chemical Library at Harvard Medical School, led by Dr. Albert Coons, prepared a β-anthryl-carbamido derivative of an antipneumococcus III antibody. The group observes the characteristic blue fluorescence of the conjugate only in the presence of pneumococcus III, but not antipneumococcus II. In the report, published in the *Proceedings of the Society for Experimental Biology and Medicine*, they stated that they hoped to develop

Tim Bartels (ed.), *Alpha-Synuclein: Methods and Protocols*, Methods in Molecular Biology, vol. 1948,
https://doi.org/10.1007/978-1-4939-9124-2_2,

"a method by which antigenic substances could be revealed in mammalian tissues" [3]; the idea became the foundation of modern immunohistochemistry.

Several years after Dr. Coons and his colleagues published their study, Drs. Morris Karnovsky and Richard Graham, also both of Harvard Medical School, published a series of studies demonstrating the feasibility of enzyme-linked detection of cellular structures [4–6]. At the same time, Dr. Stratis Avrameas, of the Pasteur Institute in Paris, and Dr. Paul Nakane, of the University of Michigan, independently reported two methods of antibody-enzyme conjugation [7, 8]. Both ultimately settled on horseradish peroxidase (HRP) as the enzyme conjugate due to its stability and sensitivity. Immunoenzyme-labeled structures did not fade with visualization as immunofluorescence-labeled structures did. Additionally, tissue autofluorescence does not interfere with immunoenzyme-labeled structures, and enzyme-labeled structures can be visualized at the electron microscopy level [9]. These conjugates remain essential in current immunohistochemical techniques visualized using bright-field and electron microscopy. In the years since these important discoveries, the technique has been refined, but the HRP methodology remains the standard for immunohistochemical staining.

In the field of neurodegeneration, *postmortem* immunohistochemical staining is often necessary for definitive diagnosis. In synucleinopathies, such as Parkinson's disease and dementia with Lewy bodies, uneven pathological burden within the brain of an individual patient can pose a challenge to biochemical studies, as it is difficult to assess whether negative results provide information about the disease process or if they are caused by low pathological burden in the studied area (Fig. 1). To assess the pathological burden of brain regions adjacent to non-fixed pieces used for biochemical testing, a protocol for staining this material is essential.

2 Materials

2.1 Tissue Fixation

1. Tissue cassette.
2. 4% Formaldehyde in phosphate-buffered saline (PBS) (*see* **Note 1**).
3. Ethanol-resistant marker or pencil.

2.2 Tissue Paraffinization and Embedding

1. ThermoFisher Excelsior™ AS Tissue Processor.
2. Leica EG 1150 Tissue Embedding Center.
3. Paraffin wax for histology.

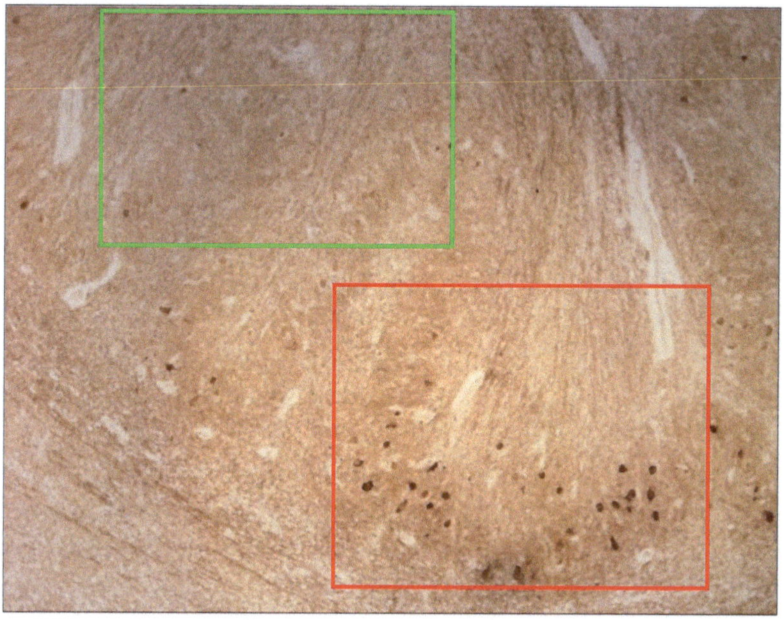

Fig. 1 2F12 immunohistochemical staining of the striatum from the brain of a Parkinson's disease patient reveals uneven distribution of αSyn-positive pathological lesions. Red box illustrates area of heavy pathology burden, while nearby green area encompasses an area of relatively little pathology

2.3 Tissue Sectioning

1. Jung AG Heidelberg Microtome.
2. Leica HI1210 Water bath for paraffin sections.
3. Glass microscope slides (25 mm × 75 mm).
4. 37 °C Incubator.

2.4 Deparaffinization

All reagents can be stored at room temperature.

1. Histo-Clear II (cat. #: HS-202, National Diagnostics, Atlanta, GA).
2. 100% Ethanol.
3. 95% Ethanol (prepared from 100% ethanol and ultrapure water).
4. Ultrapure water.
5. Staining dish system with glass rack and cover, 20-slide capacity (such as Wheaton cat. #: 900200, Millville, NJ).

2.5 Antigen Retrieval and Staining

1. 500 mL 10 mM citrate buffer (*see* **Note 2**).
2. Staining dish system with glass rack and cover.
3. Microwave.
4. Nonfat powdered milk.

5. Phosphate-buffered saline (PBS).
6. PBS with 0.05% TWEEN® 20 (volume:volume) (PBS-T).
7. Anti-αSyn monoclonal antibody, clone 2F12 (cat. #: MABN1817, Millipore Sigma, Burlington, MA).
8. Goat anti-mouse IgG, human ads-biotin (cat. #: 1030-08, Southern Biotech, Birmingham, AL).
9. 3,3′-Diaminobenzidine (DAB) peroxidase (HRP) substrate kit (with nickel) (cat. # SK-4100, Vector Laboratories, Burlingame, CA).
10. Vectastain avidin-biotin-peroxidase complex (ABC) HRP kit (peroxidase, standard) (cat. #: PK-4000, Vector Laboratories, Burlingame, CA).
11. Acid alcohol (*see* **Note 3**).
12. Richard-Allan Scientific™ Signature Series Hematoxylin 7211 (cat. #: 7211, Thermo Fisher Scientific, Waltham, MA).
13. Xylene.
14. Mounting medium (such as cat. #: SP15, Fisher Scientific, Hampton, NH).
15. Polyethylene transfer pipette.
16. Slide-staining system (such as cat. #: Z670146, Millipore Sigma, St. Louis, MO).
17. Rectangular cover glasses (such as cat. #: 12-545-M, Fisher Scientific, Hampton, NH).
18. Kimwipes.

3 Methods

3.1 Tissue Fixation

Small, frozen tissue pieces (~300 mg) should be placed in a cassette labeled with sample information using a pencil or an ethanol-resistant marker. Incubate the cassette at 4 °C in 4% formaldehyde prepared in phosphate-buffered saline (pH: 7.4) for 72–96 h (*see* **Note 4**).

3.2 Tissue Paraffinization and Sectioning

Once the tissue has been sufficiently fixed, place it in the sample basket of the tissue processor. Ensure that reagents are loaded appropriately and that run parameters are set according to Table 1.

After tissue processing and paraffinization is complete, embed the tissue into paraffin blocks prior to sectioning using a tissue-embedding center with paraffin kept above its melting point of 57 °C. Place the paraffinized tissue piece in the bottom of a block mold. Using forceps, press the tissue piece against the bottom of the block mold and inject molten paraffin until the tissue piece is fully submerged. Keeping the tissue pressed against the bottom of

Table 1
Tissue processing and paraffinization parameters

Reagent	Time (h)	Vacuum	Temp. (°C)
75% Ethanol	1:00	On	Ambient
90% Ethanol	1:00	On	Ambient
95% Ethanol	1:30	On	Ambient
100% Ethanol	2:00	On	Ambient
100% Ethanol	2:00	On	Ambient
100% Ethanol	2:00	On	Ambient
Xylenes	1:00	On	Ambient
Xylenes	1:00	On	Ambient
Xylenes	1:00	On	Ambient
Paraffin	0:45	On	60
Paraffin	0:45	On	60
Paraffin	0:45	On	60

the mold, transfer the mold to the cooling zone of the embedding center and allow the paraffin to harden. Once this is complete, fill the rest of the mold with molten paraffin, place the bottom piece of the tissue cassette in the mold with the bottom facing the tissue, and transfer to −20 °C for 15–20 min.

Once the paraffin has hardened completely, remove the block and the labeled cassette piece from the mold. Bring the water bath to 42 °C and set the microtome to 6 μM. Place the block in the block holder. Adjust the block holder to ensure even sectioning (*see* **Note 5**). Once the block is in place, warm the paraffin surface using gauze pads dipped in the 42 °C water bath. Begin sectioning by turning the drive wheel (*see* **Note 6**). Using forceps or wooden skewers, transfer the tissue sections to the water bath, taking care to lie them flat on the surface of the water. Once the sections have expanded and flattened in the water bath, gently push the seams apart using forceps. Use a single microscope slide to remove each desired section from the water bath, ensuring central placement before triggering the final adhesion by removing the slide and section from the water. Label the slides using pencil. Once desired sections have been mounted, transfer slides to glass rack and incubate at 37 °C overnight to dry.

3.3 Deparaffinization, Antigen Retrieval, and Staining

Fill two staining dishes per reagent with about 150 mL of each of the following reagents: Histo-Clear II, 100% ethanol, 95% ethanol, and ultrapure water. In that sequence, transfer the slides through this series, allowing the slides to incubate for three minutes at each station.

Once the slides have been deparaffinized and rehydrated, transfer them to a staining dish containing 10 mM citrate buffer (pH: 6.0). After microwaving for 7.5 min, replenish the evaporated citrate buffer and microwave for another 7.5 min (*see* **Note 7**). After microwaving, allow the slides and buffer to cool to ambient temperature, or about 25 °C.

After the slides have cooled, discard the citrate buffer and replace with 2% (weight:volume) nonfat dry milk in PBS-T. Incubate the slides in this blocking solution for 2 h at ambient temperature. During this incubation, prepare 150 mL of primary antibody solution by adding 2F12 anti-αSyn monoclonal antibody to 150 mL of 2% nonfat milk in PBS-T to a final concentration of 45 ng/mL (from a 1 mg/mL antibody stock, this is a 1:15,000 dilution). When the 2 h blocking incubation is complete, transfer the slides to this antibody solution, and incubate stationary for 1 h at ambient temperature.

Prepare 400 μL per slide of the secondary antibody solution during the primary antibody incubation by adding biotinylated anti-mouse IgG antibody to a final concentration of 2.5 μg/mL in 2% nonfat milk in PBS-T. Following the primary antibody incubation, rinse the slides three times in fresh PBS-T. Prepare the slide-staining system by filling the bottom with about 0.5 cm water and paper towels to ensure that moisture is retained during incubations. Carefully transfer the slides to the rails in the staining system, placing them with at least 0.5 cm space between adjacent slides. Pipette 300 μL of secondary antibody solution onto each slide, making sure to completely cover each tissue section. Cover the slide-staining system and incubate for 2 h at ambient temperature.

Once the secondary antibody incubation has begun, prepare 400 μL ABC HRP reagent mixture per slide by adding a single drop each of Reagent A and Reagent B per 2.5 mL PBS. After the secondary antibody incubation is complete, transfer the slides back to the staining rack. Rinse slides twice with PBS-T and once with PBS. Transfer the slides back to the slide-staining system and pipette on 300 μL of ABCHRP reagent mixture per slide, again ensuring that each tissue section is completely covered. Place the cover on the tray and incubate for 30 min at ambient temperature.

While the slides are incubating in the ABC HRP solution, prepare 400 μL DAB HRP substrate reagent per slide by adding 1 drop of buffer stock solution, 2 drops of DAB stock solution, and 2 drops of hydrogen peroxide solution per 2.5 mL of ultrapure water. Once the ABC HRP incubation is complete, transfer the slides back to the rack and dish and rinse three times in PBS. Return the slides to the rails of the slide-staining system and pipette 300 μL DAB HRP substrate solution per slide. As the HRP reaction occurs, the stained tissue will begin to turn brown. For light staining, carefully transfer the slide to fresh PBS in a staining dish to rinse off reagents and stop the reaction as soon as the brown color appears. For darker stains, allow the reaction to progress as desired. The reaction should be complete after 5 min of incubation at ambient temperature (*see* **Note 8**).

Once the HRP reaction has been stopped, rinse the slides three times in ultrapure water. Then, if a nuclear counterstain is desired, submerge the slides in hematoxylin for 3–8 s and transfer the slides to a bath of acid alcohol, again submerging for 3–8 s. Rinse slides three times in fresh ultrapure water.

To dehydrate stained slides, dip slides twice in 95% ethanol, twice in 100% ethanol, and twice in xylenes. Add two drops of mounting medium per slide on top of the tissue section. Carefully lay a glass coverslip across the slide, slowly lowering it into the medium to prevent bubbles from forming. Allow mounted slides to dry overnight at ambient temperature (*see* **Note 9**). Slides can be visualized using bright-field microscopy.

4 Notes

1. To prepare 1 L of 4% formaldehyde solution in PBS, first bring 800 μL of ultrapure water to 60 °C using a hot plate. When the water has come to temperature, add 40 g of paraformaldehyde and stir with a magnetic stir bar. Add 2 N NaOH dropwise with a Pasteur pipette. Once the solution turns clear, bring the solution to 1 L with 5× PBS. Allow the solution to cool to room temperature (25 °C), and then adjust the pH of the solution to 7.2 with HCl. Filter the solution using a 22 μM vacuum filter. Store at 4 °C for up to 2 weeks.
2. Prepare a 10× stock by adding 29.4 g trisodium citrate dihydrate to 600 mL of ultrapure water, and adjust pH to 6.0 using hydrochloric acid and sodium hydroxide. Bring volume to 1 L with ultrapure water. This stock can be stored indefinitely at room temperature.
3. To prepare, add concentrated hydrochloric acid to 50% isopropanol to 1% final concentration (volume:volume).
4. Time of tissue fixation must be optimized depending on the size of the brain slice to be stained. In this protocol, brain pieces weighing approximately 300 mg with a maximum thickness of approximately 5 mm were used.
5. Before sectioning, adjusting the block holder to the appropriate angle to ensure slices of even thickness requires extensive fine-tuning. Before the first turn of the drive wheel, be sure that the blade is close to the plane of the block face by giving a

gentle half-turn and seeing the relative position of blade and block. Adjust accordingly.

6. During sectioning, keep the block face warm and moist with the occasional application of water from the 42 °C bath with a gauze pad.
7. It is important not to let the level of the buffer dip below the level of the tissue on the slide.
8. Determining the proper amount of time for DAB development also takes optimization and careful observation of the color change of the tissue. This can occur quickly, and inattention can lead to stains that are too dark to be useful.
9. Excess mounting medium can be removed from dried mounted slides using a razor blade and Kimwipes dipped in Histo-clear.

Acknowledgments

The author would like to thank Dr. Mel Feany, Dr. Liqun Wang, and Kit Sin Tuen from the Feany Laboratory at the Ann Romney Center for Neurologic Disease at Brigham and Women's Hospital for assistance with developing the protocol. The author would also like to acknowledge the Pathology Core at the Dana-Farber Cancer Institute for their work in optimizing the 2F12 antibody concentration.

References

1. Reiner L (1930) On the chemical alteration of purified antibody-proteins. Science 72 (1871):483–484
2. Marrack J (1934) Nature of antibodies. Nature 133:292–293
3. Coons AH, Creech HJ, Jones RN (1941) Immunological properties of an antibody containing a fluorescent group. Exp Biol Med 47 (2):200–202
4. Graham RC, Lundholm U, Karnovsky M (1965) Cytochemical demonstration of peroxidase activity with 3-amino-9-ethylcarbazole. J Histochem Cytochem 13:150–152
5. Graham RC, Karnovsky MJ (1965) The histochemical demonstration of uricase activity. J Histochem Cytochem 13(6):448–453
6. Graham RC, Karnovsky MJ (1965) The histochemical demonstration of monoamine oxidase activity by coupled peroxidatic oxidation. J Histochem Cytochem 13(7):604–605
7. Avrameas S, Uriel J (1966) Method of antigen and antibody labelling with enzymes and its immunodiffusion application. C R Acad Sci Hebd Seances Acad Sci D 262(24):2543–2545. French
8. Nakane PK, Pierce GB (1966) Enzyme-labeled antibodies: preparation and application for the localization of antigens. J Histochem Cytochem 14(12):929–931
9. Childs, G (2014) History of immunohistochemistry. In: McManus LM, Mitchell RN (ed) Pathobiology of human disease, 1. Academic Press, Cambridge, MA

Chapter 3

Kinetic Measurements of Endocytosis and Exocytosis in Cultured Neurons

Karina J. Vargas

Abstract

The measurement of synaptic vesicle recycling in live neurons transfected with vesicular glutamate transporter fused to pHluorin (vGLUT-pHluorin) allows us to study exocytosis and endocytosis in neurons. When neurons are transfected with this protein we can measure the rate of vesicles fusing and internalizing from the membrane using live total internal reflection fluorescence (TIRF) imaging. Here, we describe transfection, culturing, and imaging of wild-type and αβγ-synuclein knockout hippocampal neurons. This technique can be used to evaluate the effect of different phenotypes and treatments in the physiology of synaptic vesicle recycling in cultured neurons.

Key words Neuron culturing, Imaging, Endocytosis, Exocytosis, Synaptic vesicle recycling, vGLUT-pHluorin, Synuclein, Parkinson's disease

1 Introduction

In the late 1990s, a fusion protein between pHluorin, a pH-sensitive version of the green fluorescent protein (GFP), and synaptobrevin was developed [1, 2]. First, pHluorin was strategically cloned to face the lumen of a synaptic vesicle within the sequence of synaptobrevin (synaptopHluorin) [1]. Later, it was cloned into the sequence of several other synaptic vesicle proteins as synaptophysin (sypHy) [3] and synaptotagmin (pHluorin-syt) [4, 5]. One of the most commonly used proteins for this chimera is vesicular glutamate transporter (vGLUT-pHluorin) (Fig. 1a) [6]. Vesicular glutamate transporter (vGLUT) is a transmembrane protein expressed in synaptic vesicles. It functions as a transporter for glutamate from the cytosol to the lumen of the vesicle [7]. The key property of pHluorin is its pH sensitivity. At neutral pH, this protein is fluorescent, but when the media is acidic the fluorescence is quenched. In neurons, the lumen of synaptic vesicle is acidic and the extracellular media is neutral. As a result, we can monitor the release of synaptic vesicles using this fusion. At steady state, just a

Tim Bartels (ed.), *Alpha-Synuclein: Methods and Protocols*, Methods in Molecular Biology, vol. 1948,
https://doi.org/10.1007/978-1-4939-9124-2_3,

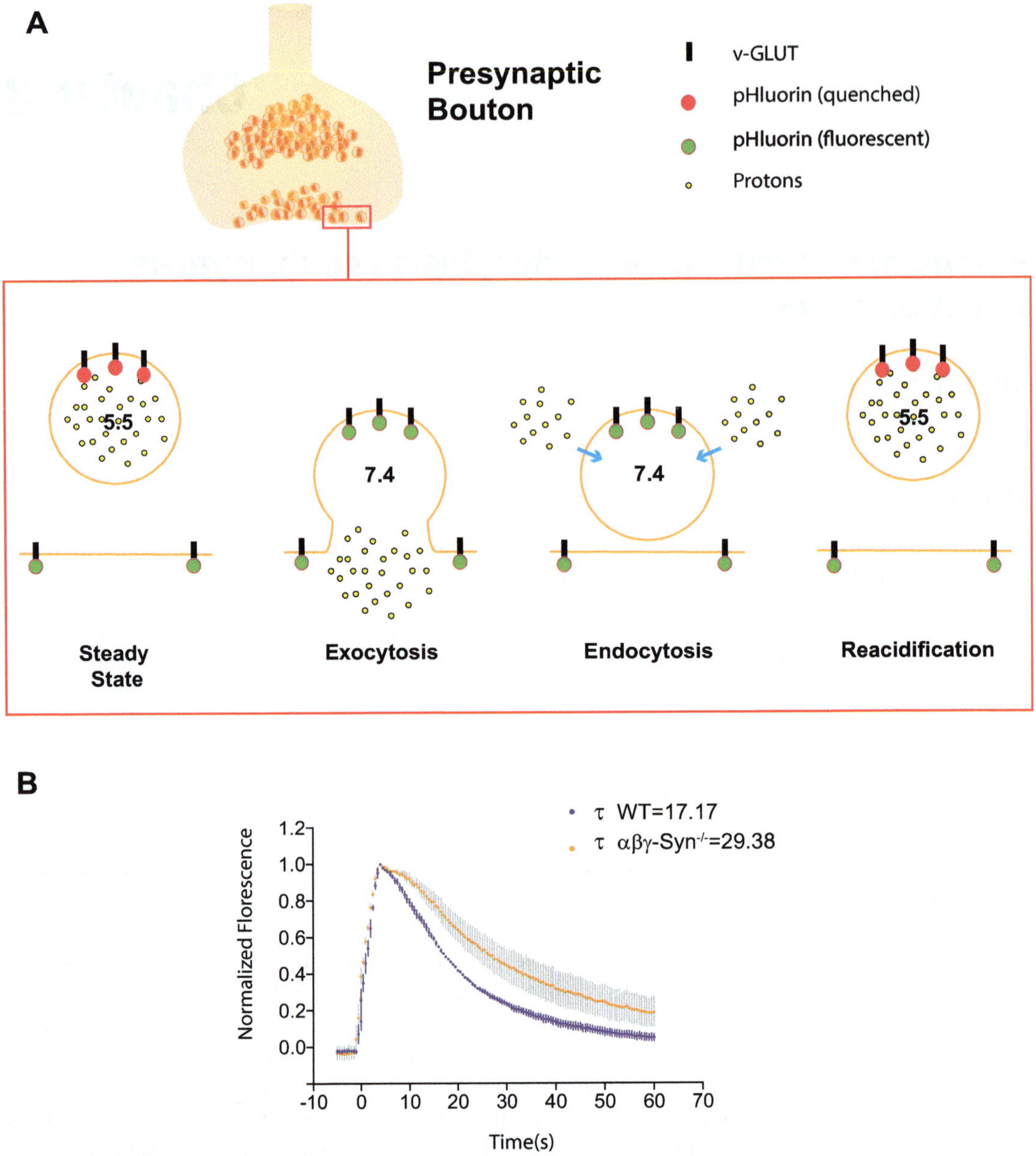

Fig. 1 Measuring endocytosis in vGLUT-pHluorin transfected neurons. (**a**) Cartoon of vGLUT-pHluorin mechanism. Upper left, presynaptic bouton showing the distribution of synaptic vesicles as well as the localization of the membrane zoomed in. Upper right, symbols representing different molecules. Red frame: pHluorin at steady state is facing the lumen of the vesicle, then in contact with the acidic media. At this pH, the emission of the protein is quenched. When the synaptic vesicle is fused to the membrane (exocytosis), the pHluorin faces the extracellular media and the protons diffuse. The pH change from acid to neutral and pHluorin becomes fluorescent, increasing the fluorescence over time. The membrane is endocytosed and the pHluorin with it, now facing the luminal space of the membrane once again. After a few seconds, the lumen of the vesicle is refilled with protons and the pH reaches the acidic level that quenches pHluorin, decreasing the fluorescence over time. (**b**) Traces showing the increase in fluorescence over time when stimulating the neurons (20 Hz, 100AP). Wild-type (purple trace) and αβγ-synuclein knockout (orange trace) traces show a difference in the decay after the stimulation. αβγ-Synuclein knockout hippocampal neurons are slower in decreasing the fluorescence over time. Therefore, the endocytosis in these neurons is impaired

few synaptic vesicles are released and the vGLUT-pHluorin is quenched (Fig. 1a, steady state). When the neurons are stimulated, synaptic vesicles fuse to the membrane and their lumen faces the extracellular media. As a consequence, vGLUT-pHluorin becomes fluorescent and acts as an indicator of synaptic vesicle exocytosis rate (Fig. 1a, exocytosis). When the stimulation is stopped, synaptic vesicles will internalize and vGLUT-pHluorin will face the luminal space again (Fig. 1a, endocytosis). After several seconds, the proton pump will acidify this compartment, quenching the fluorescence (Fig. 1a, reacidification). If the reacidification rate is constant, the decrease in the fluorescence will act as an indicator of synaptic vesicle endocytosis. The use of this tool has been a great advance in the study of synaptic vesicle recycling in neurons. It allows to study the kinetics of synaptic vesicle recycling and dissect its different components. Other techniques have been used in the past to study synaptic vesicle recycling. Electron microscopy-based measurements of horseradish peroxidase (HRP) uptake by vesicle membranes have been used successfully to study synaptic vesicle recycling and characterize proteins associated to this process [8–10]. However, this technique only shows snapshots of the membrane internalization without any information on the details of endocytosis kinetics. In addition, exocytosis cannot be clearly separated from endocytosis; this technique shows only the net result between exocytosis and endocytosis. Techniques measuring the uptake of *N*-(3-triethylammoniumpropyl)-4-(4-(dibutylamino) styryl) pyridinium dibromide, also known as FM™ 1-43, have been successfully used in the study of synaptic vesicle recycling as well. Different stimulation protocols in synapses from different species were studied resulting in the characterization of different synaptic vesicle pools and endocytosis modes [10, 11]. But there are some disadvantages due mainly to the high background signal, making the sensitivity of this method and the performance very low.

Using the assay described here, we were able to image the fusion and internalization of synaptic vesicles using TIRF microscopy, and quantify the kinetics of endocytosis and exocytosis of synaptic vesicles separately. As a result, we were able to assess the effect of the lack of function of synuclein in the synaptic vesicle recycling of hippocampal neurons. This protocol can also be used to evaluate the effect of different insults and acute perturbation of the neurons as well as overexpression of proteins.

2 Materials

2.1 Neuronal Dissection

1. Wild-type and αβγ-synuclein knockout pups, day 0 or 1 postnatal.
2. Pasteur pipet siliconization: 2% v/v Silicone oil in petroleum ether and cotton-plugged Pasteur pipets.

3. Coverslip preparation: Glass coverslips (diameter 24 mm, thickness 0.09–0.12 mm), xylenes, acetone, 95% and 70% ethanol, and filter paper.
4. HBS buffer: Hanks' balanced salt. Supplement to 0.35 mg/mL $NaHCO_3$, 1 mM Hepes pH 7.2–7.5. Sterile filter 0.2 μm and store at 4 °C.
5. Papain solution: 2% v/v Papain, 1 mM $CaCl_2$, 0.5 mM EDTA, in HBS buffer. Sterile filter 0.2 μm. Store in the incubator at 37 °C for 20 min.
6. Dissection tools: Scissors, curved-tip tweezer, two straight-tip tweezers, small and big spatula, microscissors, dissection microscope.
7. Medium: 5 mg/mL Glucose, 0.2 mg/mL $NaHCO_3$, 0.1 mg/mL bovine transferrin in MEM (no phenol red or L-glutamine). Sterile filter at 0.2 μm and store at 4 °C.
8. Plating medium: 10% v/v FBS (heat inactivated), 2 mM L-glutamine, 8 μg/mL insulin in medium. Sterile filter at 0.2 μm and store at 4 °C. This solution is stable for 1 month.
9. Growth medium: 5% v/v FBS (heat inactivated), 0.5 mM L-glutamine, 2% v/v B-27 supplement in medium. Prepare growth media with and without 4 μM Ara-C. Sterile filter 0.2 μm and store at 4 °C.
10. Matrigel solution: 2% v/v Matrigel® in sterile medium. Do not filter Matrigel or the Matrigel solution. Matrigel is stored as aliquots at −20 °C to prevent repeated freeze-thawing. Thaw one aliquot overnight at 4 °C and keep cold prior to dilution.

2.2 Neuronal Transfection

1. Amaxa® neuron transfection system.
2. High-quality DNA plasmid preparation: vGLUT-pHluorin cloned into pCAGGS.
3. Opti-MEM™.

2.3 vGLUT-pHluorin Imaging

1. Modified Tyrode buffer: 25 mM Hepes · NaOH pH 7.4, 119 mM NaCl, 2.5 mM KCl, 2 mM $MgCl_2$, 2 mM $CaCl_2$, 30 mM glucose (*see* **Note 1**).
2. Inhibitors: 50 μM AP-5 and 10 μM CNQX in all imaging buffers. Store stocks as single-use aliquots at −20 °C.
3. 50 mM NH_4Cl.
4. 1 μM Bafilomycin A.
5. Chamlide® EC perfusion-type electric stimulation magnetic chamber.

6. Perfusion system.
7. Total internal reflection fluorescence (TIRF) microscope.

2.4 Data Analysis

1. ImageJ Software.

3 Methods

3.1 Pasteur Pipet Siliconization

1. Remove the cotton plugs from the Pasteur pipets, and store the cotton plugs for later in a covered and clean beaker.
2. Submerge the Pasteur pipets in a beaker filled with siliconized petroleum ether; incubate overnight in chemical fume hood.
3. After siliconization, pour the solution back into a screw-cap bottle for storage and reuse.
4. Rinse the siliconized Pasteur pipets with running distilled water either overnight or from morning until afternoon; check that smell has subsided.
5. Dry the siliconized Pasteur pipets in a beaker with paper at the bottom for about 1 week. Keep free of dust.
6. Plug the ends of the siliconized Pasteur pipets with stored cotton and autoclave to sterilize.

3.2 Coverslip Treatment

1. Place the coverslips into a glass container with lid.
2. Rinse once with xylenes. All rinses are for at least 30 min or longer on a shaker.
3. Rinse twice with acetone.
4. Rinse once with 95% ethanol.
5. Rinse once with 70% ethanol.
6. Separate coverslips and place them on filter paper in a cell culture laminar flow hood.
7. Dry overnight.
8. Collect in a round Pyrex glass dish with a lid and autoclave to sterilize.

3.3 Coverslip Coating

1. Cool down the coverslips to room temperature.
2. Put one coverslip in each well of a 12-well culture plate.
3. Drop 100 μL of Matrigel on top of each coverslip.
4. Incubate the plate with the coverslips at 37 °C in a cell incubator.

3.4 Hippocampal Dissection

1. Take all the wild-type and αβγ-synuclein triple-knockout pups you will use for the procedure and place them in a dish.
2. Take one pup and cut off the head using the scissors. Place the head into a dish.

3. Place the forceps through the eyes and use the scalpel to make a cut in a rostral-dorsal direction. Use the tweezers to peel the skin until the entire top of the skull is visible.
4. Using small scissors, cut the skull entering from the opening in the neck and finishing on the top of the head, always cutting in between the skull and the brain. Peel the skull using forceps until the entire brain is visible.
5. Scoop the brain into a new dish.
6. Repeat **steps 2–5** for all the pups before proceeding.
7. Carry on next steps under the dissection microscope.
8. Grab the cerebellum with forceps. Separate the brain hemispheres using the scalpel to cut the brain along the longitudinal fissure. Make a space between both hemispheres and cut them from the choroid plexus.
9. Pull off the meninges from the isolated hemispheres.
10. Looking from the medial side of the separated hemispheres, you will be able to see the hippocampus, a banana-shaped density close to the base of the hemispheres.
11. Using fine-curved microscissors, cut out the hippocampus and place them all into a tube with cold HBS.

3.5 Neuron Disaggregation

1. Pour syringe-filtered papain solution onto the hippocampi and incubate at 37 °C in the incubator for 20 min.
2. While the tissue is incubating, clean dissection area, flame polish coated Pasteur pipet to make the tip smaller, and prepare the hemocytometer for cell counting.
3. Remove the enzyme solution and wash the tissue two times with 5 mL warm plating medium.
4. Add 2 mL plating medium and break down the tissue by careful trituration using a siliconized Pasteur pipet (passing through 20–40 times).
5. Pass the cell suspension through a 70 μm cell strainer into a 50 mL conical tube.
6. Pipet the cell suspension up and down and transfer 10 μL to a hemocytometer. Count and calculate the total number of neurons yielded.
7. Pellet neurons at 4000 × *g* for 5 min at 4 °C.
8. While waiting, pipet off the excess Matrigel from the coverslips.

3.6 Neuron Transfection

1. Prepare one microcentrifuge tube with 3 μg of vGLUT-pHluorin DNA for wild-type and one for αβγ-synuclein triple-knockout neurons.

2. Resuspend 6×10^6 neurons with 100 μL Amaxa® transfection solution.
3. Transfer 100 μL of the neuron resuspension to a microcentrifuge tube containing vGLUT-pHluorin DNA.
4. Mix carefully and transfer to the cuvette avoiding formation of bubbles.
5. Use the appropriate neuron nucleofection protocol.
6. Transfer to 1.5 mL of warm plating media.
7. Add 100 μL onto every coverslip.
8. Incubate at 37 °C for 1 h.
9. Add 2 mL of plating medium to each well.
10. After 24 h change the neuronal medium to 2 mL of growth medium.
11. After 24 h add 2 mL of 4 μM Ara-C growth medium.
12. Culture the neurons for at least 14 days and proceed to the next steps (*see* **Note 2**).

3.7 vGLUT-pHluorin Imaging

1. Fill the tubes of the perfusion system: one with Tyrode buffer and another with 50 mM ammonium chloride in Tyrode buffer.
2. Drain all the solution inside the tubing to ensure that there are no air bubbles.
3. Place a coverslip in the Chamlide® chamber. Put the magnetic lid on and add 500 μL of Tyrode buffer.
4. Ensure that the electrodes are in solution.
5. Place a drop of oil on the 60× lens, avoiding bubble formation.
6. Place the chamber in the heating plate, and fix the tubing for perfusion and suction to the chamber.
7. Open the perfusion and the suction. Wash the neurons with Tyrode buffer for 1 min (be sure you hear the suction noise).
8. Start imaging.
9. Image for 2 min and stimulate with 100AP at 20 Hz.
10. Wash with Tyrode buffer for 5 min.
11. Wash with 1 μM bafilomycin in Tyrode buffer for 30 s.
12. Stimulate at 100AP, 20 Hz.
13. Wash with 1 μM bafilomycin in Tyrode buffer for 5 min.
14. Wash with 50 mM ammonium chloride in Tyrode buffer for 1 min.
15. Wash with Tyrode buffer for 2 min.
16. Stop imaging.
17. Save video as multi-tiff file.

3.8 vGLUT-pHluorin Analysis

1. Open the video in ImageJ.
2. Examine the response of the synapses.
3. Draw a region of interest (ROI) around the synapses that responded to the stimulation (at least 100).
4. Multi-measure the fluorescence in every frame.
5. Export the measurement file.
6. Calculate the fluorescence over time compared to the initial fluorescence (Fig. 2(1), F0) for each ROI. Subtract the initial fluorescence from every frame and normalize against the initial fluorescence: $\Delta F = (F - F_0)/F_0$.
7. To normalize against expression of vGLUT-pHluorin, divide the obtained fluorescence in every frame by the ammonium chloride stimulation peak value $\Delta F/F_{max}$ (Fig. 2(5)).
8. Then, average the $\Delta F/F_{max}$ obtained for each ROI in the same frame. Do this for every frame to obtain the average fluorescence over the entire time course.
9. You will observe a first peak (Fig. 2(2)), stimulation, and a second peak (Fig. 2(4)), stimulation plus bafilomycin.

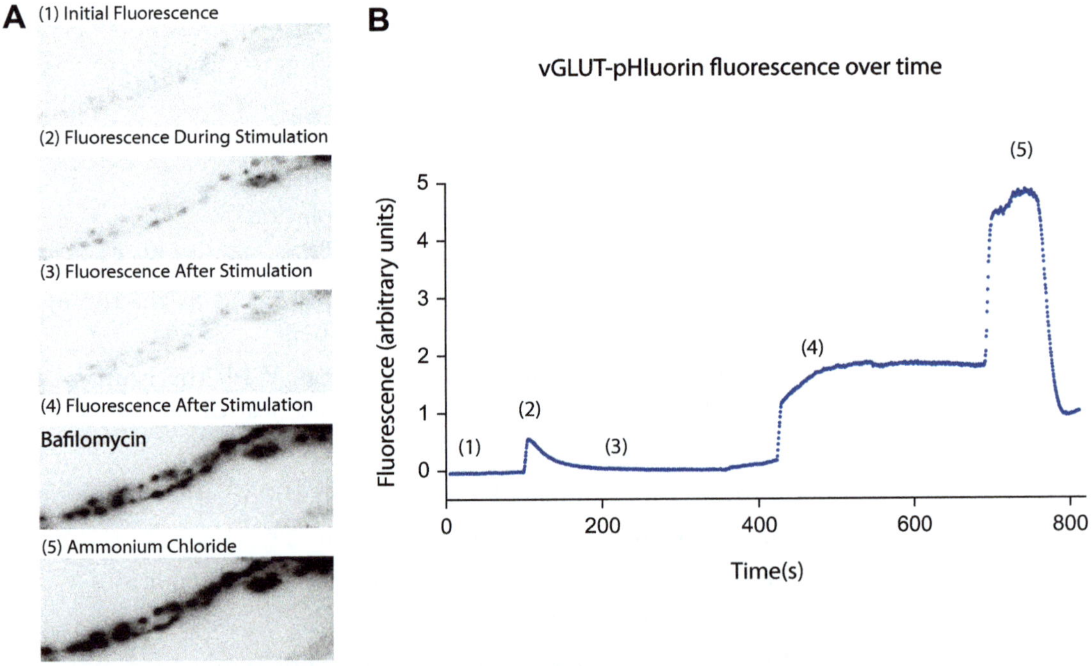

Fig. 2 vGLUT-pHluorin behavior upon bafilomycin treatment. (**a**) Snapshots of vGLUT-pHluorin at different time points. (1) Initial background fluorescence, (2) increase in the fluorescence when neurons are stimulated, (3) fluorescence 2 min after stimulation, (4) fluorescence after 2 more min when neurons are stimulated in the presence of 1 μM bafilomycin A, and (5) fluorescence after incubation with 50 mM ammonium chloride showing the total expression of v-GLUT-pHluorin. (**b**) Blue trace, example of a wild-type neuron, fluorescence quantification over the time showing the snapshots location

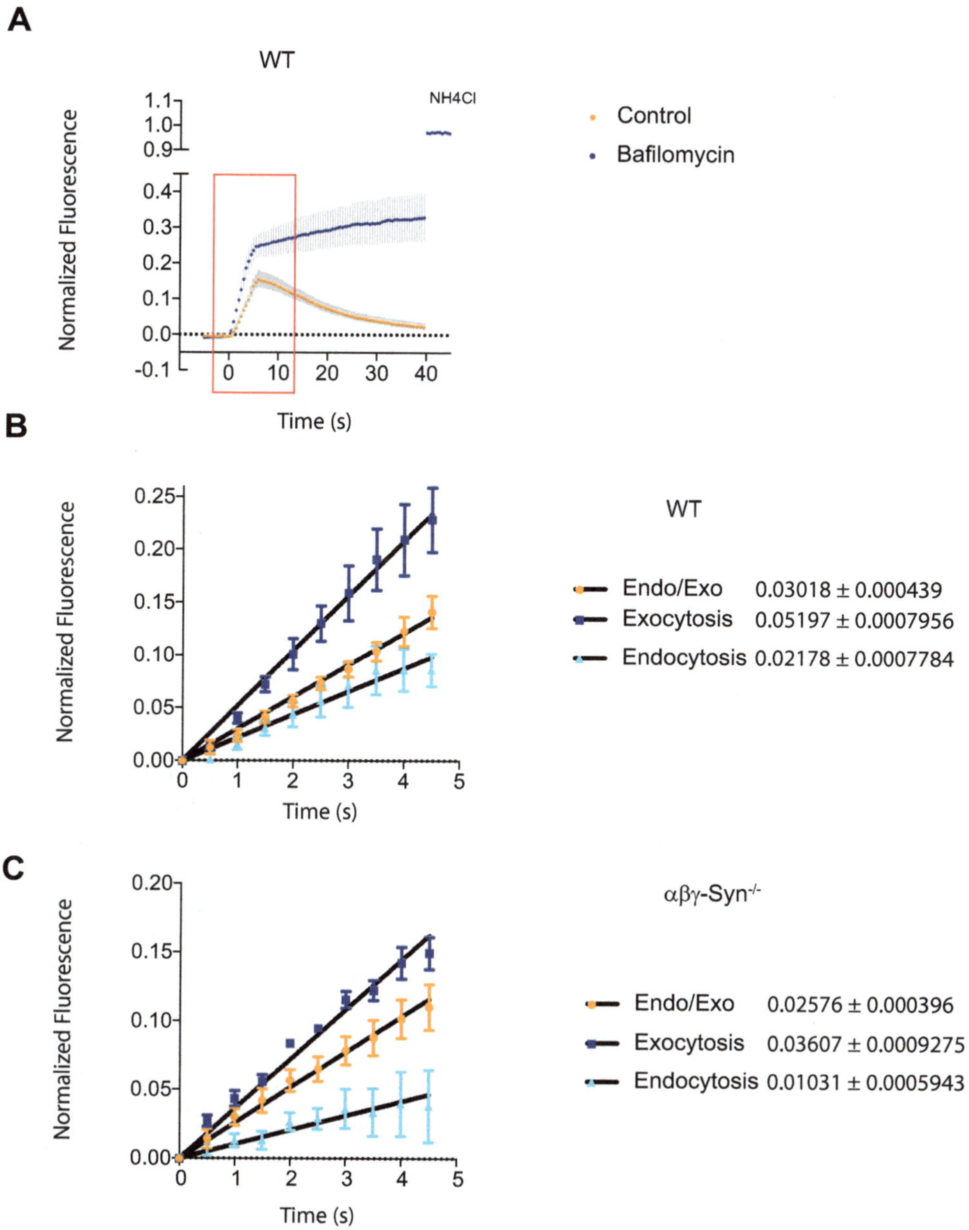

Fig. 3 Analysis of bafilomycin experiment. (**a**) Stimulation versus stimulation in the presence of bafilomycin A. Increase in fluorescence upon field stimulation (orange trace) overlapped to same stimulation protocol in the presence of bafilomycin (blue trace) for wild-type neurons ($n = 3$). Normalization of the data to levels of vGLUT-pHluorin expression (NH_4Cl). (**b**) Calculation of endocytosis constant for wild-type and (**c**) αβγ-Synuclein triple-knockout neurons. Orange trace shows the increase in fluorescence upon stimulation in wild-type neurons, net balance between exocytosis and endocytosis. Blue trace shows the increase in fluorescence when stimulating in the presence of bafilomycin A, exocytosis. Light blue trace is a calculated curve after subtracting endo/exo values from exocytosis values. Calculated constants are indicated for each genotype

10. Calculate the endocytosis and exocytosis constant upon stimulation. Calculate the slope of the fluorescence increase upon stimulation for both peaks (Fig. 3). The first slope has two components due to the contribution of exocytosis and endocytosis that happens simultaneously when stimulating. The second slope shows just the exocytosis because in the presence of bafilomycin, synaptic vesicles can not reacidify.
11. Subtract the exocytosis curve (blue trace) from the two component curve (orange trace) (Fig. 3b, c).
12. Calculate the slope of the new curve (light blue trace), where endocytosis is the only component (Fig. 3b, c).
13. Calculate the fluorescence decay constant after stimulation. First, isolate the data of the first stimulation. Normalize to the maximum value of the first peak, and calculate the fluorescence decay constant after stimulation, tau (τ) (Fig. 1b).

4 Notes

1. The pH and osmolality of the imaging buffers are critical for the survival of the neurons through the experiment. Tyrode buffer has to be prepared and adjusted to pH 7.4 just before use. The osmolality of the solution is also important for the long-term imaging of neurons (245 mOsm). When one of these values is not correctly adjusted we can expect abnormal events. In some cases, the neurons will change their morphology and die quickly. They can also not respond to any kind of stimulation. And in some other cases they will die in the course of the experiment; this will show as a lack of response for a second stimulus or a rundown in the fluorescence when bafilomycin is applied.
2. The quality of the culture is critical for a successful imaging. Wild-type and αβγ-synuclein knockout hippocampal neurons have to be cultured simultaneously to be imaged the same day. Wild-type neurons have a consistent and standard developmental pattern. After a couple of days, they start to grow neurites. After 7 days the neurites start to connect. Under white light microscope observation, the somas look round and tight, dendrites thicken, and the axonal cone is observed clearly. In a healthy culture, dendrites and axon are continuous and the membrane looks turgid and shiny. Nevertheless, the electrophysiological maturity of the culture is reached only at 14 days in vitro. The density is not lost over time and the connectivity increases. For the experimental culture, in this case αβ-γ-synuclein knockout neurons, some neuronal death due to neurodegeneration is expected.

Acknowledgments

The Pew Foundation Fellowship for Latin Americans supported this work.

References

1. Miesenbock G, De Angelis DA, Rothman JE (1998) Visualizing secretin and synaptic transmission with pH-sensitive green fluorescent proteins. Nature 394:192–195
2. Sankaranarayanan S, De Angelis DA, Rothman JE, Ryan TA (2000) The use of pHluorin for optical measurements of presynaptic activity. Biophys J 79:2199–2208
3. Granseth B, Odermatt B, Royle SJ, Lagnado L (2006) Clathrin-mediated endocytosis is the dominant mechanism of vesicle retrieval at hippocampal synapses. Neuron 51:773–786
4. Diril MK, Wienisch M, Jung N, Klingauf J, Haucke V (2006) Stonin2 is an AP-2-dependent endocytic sorting adaptor for synaptotagmin internalization and recycling. Dev Cell 10:233–244
5. Dean C, Dunning FM, Liu H, Bomba-Warczak E, Martens H, Bharat V et al (2012) Axonal and dendritic synaptotagmin isoforms revealed by a pHluorin-syt functional screen. Mol Biol Cell 23:1715–1727
6. Voglmaier SM, Kam K, Yang H, Fortin DL, Hua Z, Nicoll RA et al (2006) Distinct endocytic pathways control the rate and extent of synaptic vesicle protein recycling. Neuron 51:71–84
7. Bellocchio EE, Reimer RJ, Fremeau RT Jr, Edwards RH (2000) Uptake of glutamate into synaptic vesicles by an inorganic phosphate transporter. Science 289:957–960
8. Heuser JE, Reese TS (1973) Evidence for recycling of synaptic vesicle membrane during transmitter release at the frog neuromuscular junction. J Cell Biol 57:315–344
9. Takei K, Mundigl O, De Camilli DL (1996) The synaptic vesicle recycling: a single budding step involving clathrin and dynamin. J Cell Biol 133:1237–1250
10. Cheung G, Jupp OJ, Cousin MA (2010) Activity-dependent bulk endocytosis and clathrin-dependent endocytosis replenish specific synaptic vesicle pools in central nerve terminals. J Neurosci 30(24):8151–8161
11. Gaffield MA, Romberg CF, Betz WJ (2011) Live Imaging of Bulk Endocytosis in Frog Motor Nerve Terminals using FM Dyes. J Neurophysiol 106:599–607

Chapter 4

Detection of Misfolded α-Synuclein Aggregates in Cerebrospinal Fluid by the Protein Misfolding Cyclic Amplification Platform

Luis Concha-Marambio, Mohammad Shahnawaz, and Claudio Soto

Abstract

Here, we describe a detailed protocol to set up the αS-PMCA assay using αS synthetic aggregates in buffer and to accurately detect endogenous αS aggregates from human CSF samples. Given the amplificative nature of the technique, minute amounts of misfolded protein aggregates circulating in human bodily fluids can be multiplied and thereafter detected by more conventional methods, such as immune assays or fluorescence. Following these principles, αS-PMCA was standardized for the highly sensitive and specific detection of αS misfolded aggregates in cerebrospinal fluid (CSF) of patients with synucleinopathies.

Key words α-Synuclein, Parkinson's, Synucleinopathies, PMCA

1 Introduction

The protein misfolding cyclic amplification (PMCA) platform was first described as a novel method for the in vitro replication of brain-derived prions [1], which are infectious PrP aggregates that self-propagate in the host at the expense of the endogenous and natively folded PrP. Recently, PMCA was standardized for the detection of human prions in urine and blood, with sensitivities and specificities near 100% [2, 3]. Moreover, blood detection has been shown to be possible during preclinical stages of the disease [4]. Since protein aggregates associated with other neurodegenerative diseases follow the same prion-like mechanism for misfolding and spreading [5–8], PMCA has been standardized for detecting these aggregates in the CSF: specifically, Aβ-PMCA for the detection of amyloid-beta (Aβ) aggregates in Alzheimer's disease (AD) patients [9], tau-PMCA for the detection of 3-repeat tau amyloid aggregates in patients with Pick's disease [10], and αS-PMCA for the detection of α-synuclein (αS) protein aggregates in various synucleinopathies (i.e., Parkinson's disease (PD), dementia

Tim Bartels (ed.), *Alpha-Synuclein: Methods and Protocols*, Methods in Molecular Biology, vol. 1948,
https://doi.org/10.1007/978-1-4939-9124-2_4,

with Lewy bodies (DLB), and multiple systemic atrophy (MSA)) [11, 12]. Remarkably, Shahnawaz and collaborators showed that αS-PMCA had 89% sensitivity and 97% specificity in samples from clinically diagnosed PD patients. Furthermore, αS oligomers were suggested as a biomarker for preclinical detection [11]. If the prion-like behavior of these misfolded aggregates extends to their bodily distribution, detection of these preclinical biomarkers might be possible by testing peripheral fluids, such as blood, urine, or saliva. Synucleinopathies are particularly interesting in this regard, since peripheral deposition of αS aggregates has been consistently described in the literature [13–15]. However, so far, the misfolded αS aggregates have been only detected in cerebrospinal fluid (CSF) and brain homogenates [11, 12].

Misfolded αS aggregates (also called αS-seeds) are incubated with an excess of monomeric recombinant human α-synuclein (rec-αS) and subjected to intermittent shaking/incubation cycles during αS-PMCA. The periodic shaking fragments misfolded αS aggregates into smaller species, increasing the number of αS seeds in the reaction. During the incubation, these seeds elongate into larger aggregates using rec-αS as substrate (Fig. 1a). Therefore, the amplification occurs by cyclical fragmentation and elongation of the aggregates, which is conceptually similar to PCR. Due to the amplification, in vitro-formed αS aggregates are easily detectable in time by fluorescence measurements with the amyloid-specific probe thioflavin T (ThT). Under the conditions used, in the absence of αS seeds, there is no self-aggregation of the rec-αS and only background levels of fluorescence are detected for more than 400 h. Given the high specificity of the misfolding process occurring in the αS-PMCA conditions, other amyloid protein aggregates such as Aβ and tau seeds do not generate a positive signal in the assay.

Seeds in the CSF of synucleinopathy patients induce aggregation of the monomeric αS added in excess to the αS-PMCA reaction following a classic aggregation curve, defined by a lag phase, exponential growth, and a final plateau (Fig. 1b). In the absence of αS seeds, there is no aggregation. Therefore, an easy readout of the assay when analyzing CSF samples is the maximum fluorescence detected at plateau (F_{max}). This is a measurement of the amount of amyloid generated in the αS-PMCA reaction and relative fluorescence units (RFU) values over a predefined threshold are considered positive. When amplifying known amounts of in vitro-generated seeds, the parameter that better correlates with the amount of αS-seeds is the time it reaches 50% of F_{max} (T_{50}) (Fig. 2). Interestingly, T_{50} values defining PD-CSF aggregation curves correlate with the Hoehn and Yahr scale, which is a clinical score that defines different stages of the PD progression [11]. This suggests that quantification of the misfolded αS aggregate levels

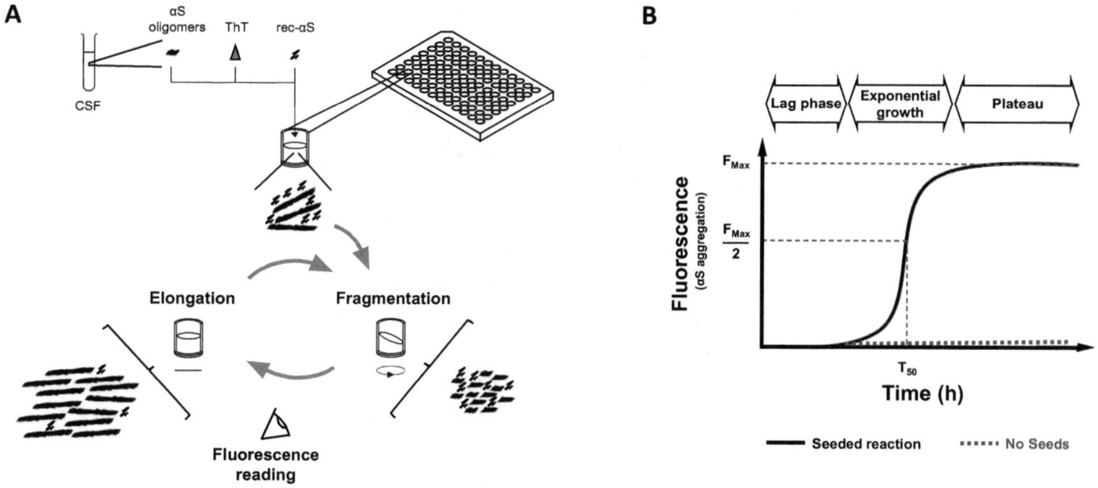

Fig. 1 Schematic explanation of the αS-PMCA technique for the detection of misfolded αS-soluble oligomers. (**a**) CSF from synucleinopathy patients is mixed with recombinant αS (rec-αS) and thioflavin T (ThT) in a 96-well ELISA plate. Then, the plate is intermittently shaken to fragment the αS aggregates in CSF and incubated to allow elongation of the smaller generated seeds. This process is repeated cyclically, which results in the amplification of the initial αS seed. (**b**) Typical aggregation curve observed in αS-PMCA conditions, where nucleation occurs during the lag phase. The lag phase is followed by an exponential increase in the amount of aggregates that eventually reaches plateau. Kinetic parameters, such as the maximum fluorescence (F_{max}) and the time to reach $F_{max}/2$ (T_{50}), are used to characterize the aggregation curve

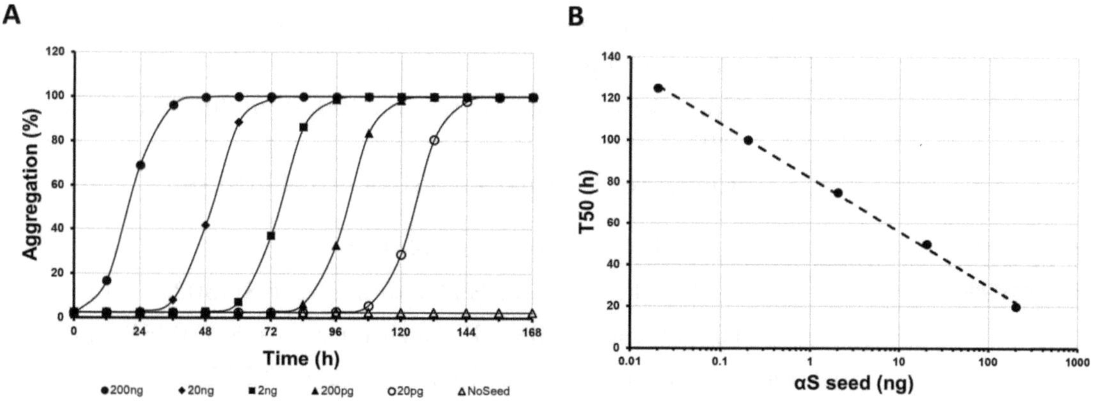

Fig. 2 Expected aggregation curves of known amounts of αS seeds. (**a**) 200 ng to 20 pg of synthetic αS seeds aggregate following classical aggregation curves characterized by a F_{max} and a T_{50}. The reaction without seeds should not aggregate. (**b**) T_{50} calculated for each curve correlates to the base-10 logarithm of the αS seed mass used in the reaction

could be an unbiased measurement to determine disease severity for synucleinopathies and perhaps to monitor the effect of treatments.

The purpose of this book chapter is to provide detailed instructions for the use of αS-PMCA in a basic research environment.

First, we describe a protocol for the generation of αS seeds to be used for standardizing the technique. Then, we describe in detail the procedure to establish the αS-PMCA technique using synthetic seeds in the absence of CSF. This is an important step to test the experimental setup and to check the quality of the αS seeds and rec-αS substrate. Lastly, we describe the procedure for the sensitive and specific amplification of endogenous αS seeds in CSF from patients affected by synucleinopathies. The protocols and instructions described are for the standard αS-PMCA reaction, which enables to distinguish positive from negative samples as well as to estimate the quantity of misfolded aggregates present in the sample. When the goal is only to separate positive from negative samples, the reaction can be done much faster (within 48 h) by modifying the conditions of the assay. However, this rapid αS-PMCA assay does not enable good separation between samples containing different quantities of seeds and thus cannot be used to correlate with disease progression or to estimate the effect of a treatment aimed to decrease the amount of these structures.

2 Materials

2.1 Equipment and Labware

1. Thermomixer R (Eppendorf).
2. Gemini SpectraMax EM Fluorometer (Molecular Devices).
3. Incubator (Thermo Fisher, MaxQ4000).
4. Microplate sonicator with Ti horn (QSonica, Q700).
5. Liquid nitrogen.
6. Pipetman micropipette set and 200 μL multichannel micropipette (Gilson).
7. Rubber roller (Amazon, ASIN# B01N9AL149).

2.2 Buffers and Reagents

1. 0.5 M PIPES pH 6.5 (*see* **Note 1**).
2. 1× HyClone PBS (GE Healthcare, cat# SH30256.01).
3. 10× HyClone PBS (GE Healthcare, cat# SH30258.01).
4. 5 M AccuGENE NaCl 1 L (Lonza, cat# 51202).
5. NERL Reagent-Grade Water (Thermo Fisher, cat# 9800-5).
6. Thioflavin T prepared in water (*see* **Note 2**).
7. Recombinant C-histag human α-synuclein in PBS (*see* **Note 3**).

2.3 Consumables

1. COSTAR 96-well ELISA plates (Corning, cat# 3916).
2. MicroAmp Film (Applied Biosystems, cat# 4311971).
3. 1 mL to 10 μL TipOne RPT tips (USA Scientific).

4. 200 μL DF200ST tips (Fisher Scientific, cat# F171503G) (*see* **Note 4**).
5. 1.5 mL LoBind tubes (Eppendorf cat# 022431081).
6. 0.5 mL LoBind tubes (Eppendorf cat# 022431064).
7. 5 mL LoBind tubes (Eppendorf, cat# 0030108302).
8. 0.2 mL PCR tubes (Eppendorf, cat# 2231000257).

3 Methods

3.1 Generation of Synthetic αS Misfolded Aggregates (αS Seeds)

1. Thaw rec-αS in water at room temperature (RT) and maintain on ice (*see* **Note 5**).
2. Prepare 500 μL of 1 mg/mL rec-αS in 1× PBS in a 1.5 mL LoBind tube. Incubate at 37 °C in a Thermomixer R with constant orbital shaking (600 rpm) for 5 days (*see* **Note 6**).
3. Measure and compare the ThT fluorescence of the generated αS seeds to freshly thawed rec-αS control and buffer control. In a COSTAR 96-well ELISA plate, prepare 200 μL of a mixture 153 μg/mL αS seed (10 μM considering molecular weight of the monomer), 1× PBS, and 5 μM ThT. Prepare a second mixture with freshly thawed monomeric rec-αS instead of the synthetic αS seeds and a third mixture with PBS 1× instead of protein. Measure fluorescence ($435_{EX}/485_{EM}$) using the Gemini Spectra fluorometer and compare the relative fluorescence units (RFU) of the three wells (*see* **Note 7**). αS seeds should produce a RFU at least 200 times higher than the monomeric rec-αS, which in turn should be similar to the buffer control (background).
4. If the ThT fluorescence assay is positive, then sonicate the newly generated αS seeds ten times with ON/OFF cycles of 30 s/60 s and an amplitude of 40 (*see* **Note 8**). Next, prepare 10 μL one-use aliquots in PCR tubes and snap-freeze them in liquid nitrogen. Store at −80 °C until use (*see* **Note 9**).

3.2 αS-PMCA for Amplification of Synthetic αS Seeds in Buffer

1. The 200 μL αS-PMCA reaction for amplification of synthetic αS seeds is 1× PBS, 5 μM ThT, and 0.1 mg/mL rec-αS, including 40 μL of αS seeds in 1× PBS. A master mix is prepared with 10× PBS, 5 mM ThT, and NERL water. This mix is maintained on ice and the rec-αS is added right before starting αS-PMCA.
2. Place a COSTAR 96-well ELISA plate on ice and add 200 μL of water to the outer wells (*see* **Note 10**).
3. Thaw synthetic αS seeds in water at RT and then prepare tenfold serial dilutions in 1× PBS. The dilutions are prepared in such a way that 40 μL of the serial dilutions contain a

20 ng–2 pg range of αS seeds. When preparing the serial dilutions, pipette up and down 20 times to remove as much αS seed from the inner wall of the tip and then vortex for 10 s to mix the suspension (*see* **Note 11**). Maintain the αS seeds on ice.

4. To add the 40 μL of seed into the plate wells, vortex the αS seeds for 10 s and then dispense the volume of αS seeds in the lower inner wall of each well. Use new tips for each well, even when the same αS seed dilution is added multiple times. When adding the same seed multiple times, it is not necessary to vortex the seeds in between wells.
5. As controls, the following sham reactions need to be analyzed alongside the seeded reactions: "NoSeed" (1× PBS instead of 40 μL αS-seeds), "NoSub" (20 ng of αS seeds and 1× PBS instead of rec-αS), and "Blank" (1× PBS instead of αS seeds and rec-αS).
6. Thaw rec-αS in RT water and keep on ice (*see* **Note 5**).
7. Prepare the αS-PMCA substrate by adding rec-αS to the chilled master mix and vortex at low speed. Using a multichannel pipette and a reservoir, add 160 μL of the substrate into each well (*see* **Note 12**).
8. Cover and seal the plate applying a MicroAmp film using a rubber roller. Measure fluorescence ($435_{EX}/485_{EM}$) at time zero in a fluorometer with the incubation chamber set at 37 °C (*see* **Note 7**).
9. Place the PMCA plate on a Thermomixer R maintained inside an incubator set at 37 °C and cover the plate using aluminum foil to protect ThT from light degradation. Set the Thermomixer R to 37 °C and a shaking cycle of 1 min ON at 500 rpm and 29 min OFF (incubation time).
10. Preheat the fluorometer chamber to 37 °C and measure fluorescence ($435_{EX}/485_{EM}$) every 12 h until completion of the experiment (*see* **Note 13**). Stop the Thermomixer R near the end of the incubation time and measure fluorescence. Place the plate back in the Thermomixer R and restart the shacking/incubation cycle (*see* **Note 14**).

3.3 αS-PMCA for Detection of Synthetic αS Seeds in CSF

1. The 200 μL αS-PMCA reaction for amplification of synthetic αS seeds is 100 mM PIPES pH 6.5, 0.5 M NaCl, 5 μM ThT, and 1 mg/mL rec-αS, including 40 μL of CSF. A master mix is prepared with 0.5 M PIPES pH 6.5, 5 M NaCl, 5 mM ThT, and NERL water. This mix is maintained on ice and the rec-αS is added right before starting αS-PMCA.
2. Place a COSTAR 96-well ELISA plate on ice and add 200 μL of water to the outer wells (*see* **Note 10**).

3. Thaw CSF samples in RT water and maintain them on ice (*see* **Note 5**). Add 40 μL of each CSF sample to plate wells using new low-binding tips for each well, even when the same CSF is added multiple times. Dispense the CSF in the lower sidewall of each well.
4. Thaw the rec-αS in RT water, and then keep on ice (*see* **Note 5**).
5. Prepare the αS-PMCA substrate by adding rec-αS to the chilled master mix and vortex at low speed. Using a multichannel pipette and a reservoir, add 160 μL of the substrate into each well (*see* **Note 12**).
6. Cover and seal the plate applying a MicroAmp film using a rubber roller. Measure fluorescence ($435_{EX}/485_{EM}$) at time zero in a fluorometer with the incubation chamber set at 37 °C (*see* **Note 7**).
7. Place the PMCA plate on a Thermomixer R maintained inside an incubator set at 37 °C and cover the plate using aluminum foil to protect ThT from light degradation. Set the Thermomixer R to 37 °C and a shaking cycle of 1 min ON at 500 rpm and 29 min OFF (incubation time).
8. Preheat the fluorometer chamber to 37 °C and measure fluorescence ($435_{EX}/485_{EM}$) every 12 h until completion of the experiment (*see* **Note 13**). Stop the Thermomixer R near the end of the incubation time and measure fluorescence. Place the plate back to the Thermomixer R and restart the shacking/ incubation cycle (*see* **Note 14**).

3.4 Data Analysis (Seeds in Buffer)

1. Plot the RFU measured at 485 nm versus the time in hours. The seeded reactions should present a classic amyloid aggregation curve with a lag phase, an exponential growth, and a final plateau. As controls, NoSeed (1× PBS instead of 40 μL αS seeds), NoSub (20 ng of αS seeds and 1× PBS instead of rec-αS), and Blank (1× PBS instead of αS seeds and rec-αS) reactions should be analyzed alongside (Fig. 2a).
2. Fit the data to the Boltzmann exponential equation $[F(t) = F_{min} + ((F_{max} - F_{min}))/(1 + \exp((T_{50}-t)/\text{slope}))]$ and determine the T_{50} values that define the aggregation curves for different amounts of αS seeds.
3. T_{50} values correlate with the logarithm of the mass of αS seeds used in each well (Fig. 2b). In our experience, the coefficient of determination (r^2) normally varies between 0.88 and 0.99, with a 2 pg limit of quantification (LOQ) and a 20 fg limit of detection (LOD). A high r^2 value and no self-aggregation of the rec-αS ("NoSeed" showing no aggregation) indicate that the seeds and substrate are optimal for αS-PMCA.

3.5 Data Analysis (CSF Samples)

1. Plot the RFU measured at 485 nm versus the time in hours. αS-PMCA-positive CSF samples should present a classic amyloid aggregation curve with a lag phase, an exponential growth, and a final plateau. αS-PMCA-negative CSF samples present background fluorescence values and therefore T_{50} values cannot be calculated for control reactions. Therefore, F_{max} measured at a time in which aggregation is completed in all samples that can aggregate (usually beyond 300 h) has been used as readout to determine positive and negative result of the assay (*see* **Note 15**).
2. Fit the data to the Boltzmann exponential equation $[F(t) = F_{min} + ((F_{max}-F_{min}))/(1 + \exp((T_{50}-t)/\text{slope}))]$ and determine the F_{max} that defines the aggregation curves. The fluorescence values are largely dependent on the equipment, and thus the thresholds to define positivity need to be determined empirically for each system. In our experiments, we considered F_{max} values of 50 and higher as positive, meaning the CSF from the patient contained αS seeds.
3. If there is clinical data from the PD-CSF donors, the T_{50} of the positive reactions could be calculated and correlated to the clinical severity (i.e., Hoehn and Yahr scale).

4 Notes

1. PIPES should be adjusted to pH 6.5 using NaOH only.
2. ThT stock solution is 5 mM in NERL water, filtered with 0.2 μm syringe filters.
3. Rec-αS should be stored at −80 °C and at concentrations between 4 and 10 mg/mL in 1× PBS.
4. Gilson tips are used only to add the PMCA master mix containing the substrate to the CSF or synthetic seeds.
5. αS seeds, rec-αS, and CSF samples are thawed in water at RT to avoid extended partial thawing, which can cause increase in local concentration of protein.
6. For the generation of αS seeds, the Thermomixer R can be directly placed on the workbench and is not required to be inside an incubator.
7. To measure fluorescence in the Gemini Spectra EM Fluorometer, select endpoint fluorescence readings using 96-well standard opaque plates after 5-s shaking. The excitation is set at 435 nm and emission is set at 485 nm with an automatic cutoff at 475 nm. The PMT is set to the low and the top reading optics to five flashes per read.

8. Given that different horns produce different amounts of heat, modify the ON/OFF times to maintain the water in the horn at RT.
9. Freeze/cycles have been shown not to affect the seeding capacity of endogenous misfolded αS aggregates in CSF, since αS-PMCA curves were comparable even after six freeze/thaw cycles. Therefore, larger aliquots of synthetic αS seeds could be prepared and frozen/thawed according to use.
10. Even though the Thermomixer R is inside an incubator set at 37 °C, the temperature and humidity might be different in the outer wells. This is a normal practice when using ELISA plates for a long period of time.
11. We recommend the use of low-binding tips and tubes specially when handling αS seeds, since they are highly hydrophobic and sticky.
12. Avoid the formation of bubbles when adding the substrate mix to the plate wells. Mix the final mixture with the same multichannel pipette (10 up/downs).
13. To maintain the temperature of the plates being read at each time point, maintain two sealed plates with water in the wells inside the incubator at 37 °C. Use these sham plates to cover the top and bottom of the PMCA plate when removing the plates from the Thermomixer R.
14. Because the reactions are normally kept for long periods of times (several days) at 37 °C, we recommend to work with sterile materials and be careful to prevent contamination with microorganisms in the solutions. We normally work with autoclaved materials and prepare the solutions inside a biosafety cabinet.
15. The protocols and instructions provided here are for the standard αS-PMCA reaction which can serve both to distinguish positive from negative samples and to provide a good separation in signal among samples containing different amounts of seeds. When the focus is only positive/negative outcome, the reaction can be done much faster (within 48 h) by modifying the conditions of the assay. Unfortunately, this rapid αS-PMCA assay does not permit good separation between samples containing different quantities of seeds and cannot be used to correlate with disease progression.

Acknowledgments

This work was funded in part by a grant from the Michael J. Fox Foundation for Parkinson's disease to CS.

References

1. Saborio GP, Permanne B, Soto C (2001) Sensitive detection of pathological prion protein by cyclic amplification of protein misfolding. Nature 411:810–813
2. Moda F, Gambetti P, Notari S et al (2014) Prions in the urine of patients with variant Creutzfeldt-Jakob disease. N Engl J Med 371:530–539
3. Concha-Marambio L, Pritzkow S, Moda F et al (2016) Detection of prions in blood from patients with variant Creutzfeldt-Jakob disease. Sci Transl Med 8:370ra183
4. Bougard D, Brandel J-P, Belondrade M et al (2016) Detection of prions in the plasma of presymptomatic and symptomatic patients with variant Creutzfeldt-Jakob disease. Sci Transl Med 8:370ra182
5. Jucker M, Walker LC (2013) Self-propagation of pathogenic protein aggregates in neurodegenerative diseases. Nature 501:45–51
6. Masuda-Suzukake M, Nonaka T, Hosokawa M et al (2013) Prion-like spreading of pathological α-synuclein in brain. Brain 136:1128–1138
7. Prusiner SB, Woerman AL, Mordes DA et al (2015) Evidence for α-synuclein prions causing multiple system atrophy in humans with parkinsonism. Proc Natl Acad Sci U S A 112: E5308–E5317
8. Cope TE, Rittman T, Borchert RJ et al (2018) Tau burden and the functional connectome in Alzheimer's disease and progressive supranuclear palsy. Brain 141(2):550–567
9. Salvadores N, Shahnawaz M, Scarpini E et al (2014) Detection of misfolded Aβ oligomers for sensitive biochemical diagnosis of Alzheimer's disease. Cell Rep 7:261–268
10. Saijo E, Ghetti B, Zanusso G et al (2017) Ultrasensitive and selective detection of 3-repeat tau seeding activity in pick disease brain and cerebrospinal fluid. Acta Neuropathol 133:751–765
11. Shahnawaz M, Tokuda T, Waragai M et al (2017) Development of a biochemical diagnosis of Parkinson disease by detection of α-synuclein misfolded aggregates in cerebrospinal fluid. JAMA Neurol 74:163
12. Fairfoul G, McGuire LI, Pal S et al (2016) Alpha-synuclein RT-QuIC in the CSF of patients with alpha-synucleinopathies. Ann Clin Transl Neurol 3:812–818
13. Stokholm MG, Danielsen EH, Hamilton-Dutoit SJ et al (2016) Pathological α-synuclein in gastrointestinal tissues from prodromal Parkinson disease patients. Ann Neurol 79:940–949
14. Hilton D, Stephens M, Kirk L et al (2014) Accumulation of α-synuclein in the bowel of patients in the pre-clinical phase of Parkinson's disease. Acta Neuropathol 127:235–241
15. Schneider SA, Boettner M, Alexoudi A et al (2016) Can we use peripheral tissue biopsies to diagnose Parkinson's disease? A review of the literature. Eur J Neurol 23:247–261

Chapter 5

Stereotaxic Targeting of Alpha-Synuclein Pathology in Mouse Brain Using Preformed Fibrils

Bin Zhang, Victoria Kehm, Ron Gathagan, Susan N. Leight, John Q. Trojanowski, Virginia M.-Y. Lee, and Kelvin C. Luk

Abstract

The accumulation of intraneuronal inclusions containing misfolded alpha-synuclein (aSyn) within the central nervous system (CNS) is a common feature found in several neurodegenerative disorders including Parkinson's disease (PD). Emerging evidence indicates that aSyn amyloid fibrils, a configuration that is present within these characteristic inclusions, are capable of self-replicating by templating the conversion of endogenously expressed aSyn in neurons. Stereotaxic administration of synthetic α-synuclein preformed fibrils (PFFs) into the mouse brain has been shown to seed the formation of intracellular aSyn pathology reminiscent of Lewy body (LB) inclusions present in human PD and related synucleinopathies. Moreover, pathology can be targeted to specific CNS regions. This experimental approach provides a versatile platform for investigating PD-like LB pathology in vivo. We focus here on procedures for initiating aSyn inclusion formation at various regions of the mouse brain using computer-assisted motorized stereotaxic microinjection of aSyn PFFs and discuss appropriate strategies for controls and analysis.

Key words α-Synuclein, Transmission study, Stereotaxic, Microinjection, in vivo model, α-Synuclein preformed fibrils (PFFs)

1 Introduction

Alpha-synuclein (aSyn) is a major component of Lewy bodies (LBs) and Lewy neurites (LNs), the intracellular inclusions characteristic of Parkinson's disease (PD) and dementia with LBs (DLB). LBs/LNs are also present in the brains of a sizeable portion of Alzheimer's disease patients in addition to nearly all sporadic and familial PD patients [1, 2]. Since point mutations or amplification of the aSyn gene (SNCA) causes autosomal dominant forms of familial PD [3–6], this implies a central pathogenetic role for this protein in sporadic as well as familial PD.

A number of important observations have emerged from extensive postmortem studies on the neuroanatomical distribution of LBs/LNs at various stages of PD. Firstly, LBs/LNs affect multiple

Tim Bartels (ed.), *Alpha-Synuclein: Methods and Protocols*, Methods in Molecular Biology, vol. 1948, https://doi.org/10.1007/978-1-4939-9124-2_5, © Springer Science+Business Media, LLC, part of Springer Nature 2019

CNS regions in PD, DLB, and AD patients and although the patterns of distribution may be heterogeneous, they overlap considerably between individuals [7, 8]. Secondly, motor and non-motor symptoms strongly correlate with the extent of aSyn pathology and the function associated with the affected areas [9, 10]. Thirdly, aSyn pathology accumulates progressively, affecting new CNS regions over time, while pathology in previously affected areas increases in severity [10]. Most commonly, LB/LN pathology first develops in lower brainstem nuclei, olfactory nuclei, and peripheral neurons, affecting the SNpc in midcourse of disease, while neocortical involvement typically occurs last. Although some patients deviate from this pattern [8], the majority of patients appear to exhibit this stereotypic progression of aSyn pathology described by Braak and colleagues [7, 8, 11].

The progressive and sequential spread of LBs/LNs from affected to unaffected CNS regions over time is consistent with the transmission of a pathogenic process from diseased to healthy neurons [10, 12]. In fact, LBs/LNs are frequently detected in gastrointestinal, cardiac, as well as olfactory neurons during early stages of PD [13, 14], suggesting that spread can occur over long distances [12]. One of the first clues that the transmissible agent in PD might be misfolded aSyn itself comes from postmortem studies showing the time-dependent formation of LBs in mesencephalic neurons grafted into PD patients [15, 16].

In contrast to its highly soluble state in healthy brains [17], aSyn in LBs and LNs exist as β-sheet-rich amyloid fibrils, an ultrastructural arrangement shared with other misfolded proteins that form the signature aggregates found in other neurodegenerative diseases [18]. Recombinant aSyn, which has no native secondary structure [19], also assembles into fibrils at micromolar concentrations [20, 21]. A common property of amyloids is their ability to catalyze and template the assembly-soluble monomers into fibrils [22]. For aSyn, this activity is apparent in both cell-free and cell-based systems and we and others have previously demonstrated that the internalization of purified recombinant aSyn PFFs to a variety of cells expressing aSyn, including mouse primary neurons, results in the seeded formation of insoluble inclusions that share many of the same morphological features and biochemical markers as human LBs, such as aSyn phosphorylation at Ser129 (pSyn) and recruitment of ubiquitin and heat-shock proteins [23, 24].

Administration of synthetic aSyn PFFs in vivo also induces the formation of LB- and LN-like pathology in aSyn-overexpressing transgenic mice (e.g., M83 line), as well as accelerates the neurological phenotype and dramatically reduces their survival [25]. The pathology burden in PFF-injected mice is also significantly increased compared to untreated aged M83 mice and more widely distributed, consistent with the amplification and spread of pathological aSyn. More recently, we have shown that intrastriatal injection of mouse aSyn PFFs into wild-type mice from a variety of genetic backgrounds leads to LB/LN formation in multiple connected regions, including the substantia nigra pars compacta which progressively degenerates, and results in loss of striatal DA and impaired motor function [26]. Biochemical analysis shows that aSyn PFFs trigger the pathological conversion of host-expressed aSyn, whereas PFFs are nontoxic and do not induce aSyn pathology in the absence of endogenous aSyn expression in aSyn knockout mice [24–28].

Subsequent studies indicate that seeding with aSyn PFFs or misfolded aSyn isolated from the brains of patients with a range of synucleinopathies also effectively seeds the formation of intracellular pathology with distinct characteristics [27, 29, 30]. Similarly, conformational strains of PFFs have been generated using recombinant aSyn [31, 32]. Furthermore, this phenomenon has been recapitulated in a variety of laboratory animals including mice, rats, and nonhuman primates (e.g., macaque, marmoset) [27, 33, 34]. These models thus provide a versatile platform for studying the basic mechanisms associated with transmission as well as for evaluating therapeutics directed against PD processes which have been reviewed elsewhere [35–37].

The delivery of PFFs or brain-derived aSyn is typically achieved through stereotaxic methods, although induction of pathology via intramuscular [38] and enteric [39, 40] routes has also been described. Methods for stereotaxic microinjection of suspensions into rodent brain are well established and PFFs have been successfully targeted to many different CNS regions [41–45]. Here, we describe a method for injection of aSyn PFFs into the mouse brain and validation of the resultant seeded pathology using immunohistochemistry, and discuss considerations that can influence seeding efficiency in vivo.

2 Materials

2.1 Sonication of aSyn PFFs

1. Purified monomeric recombinant aSyn: Please see Powers et al., this issue.
2. Recombinant αSyn fibrils: Please see Volpicelli-Daley, this issue.
3. Sonicator: Prior to injection, aSyn PFFs are sonicated using a high-power bath sonicator system such as the Bioruptor Plus (Diagenode; Denville, NJ). Aerosols generated during PFF sonication are contained using these closed systems. Alternatively, a handheld probe can also be used. However, this should be done within the confines of a BSL2 safety cabinet to reduce the risk of exposure to aerosolized PFFs.
4. Sterile PBS without Ca^{2+} and Mg^{2+} (Mediatech, MT21031CV).

2.2 Stereotaxic Injection into Mouse Brain

1. Mouse lines for microinjections: The aSyn PFF injection approach has been successfully employed in wild-type (i.e., non-transgenic) mice from a number of genetic backgrounds. The most commonly used are B6C3 (F1) mice generated from a cross between female C57BL/6 and male C3H lines (Charles River Laboratories). Procedures described here were developed based on this mouse strain (*see* **Note 1**). Wild-type animals are suitable for injection from 2 to 3 months of age onwards. Approval from relevant local authorities is necessary for all procedures involving live animals. The procedures described here have been approved by the Institutional Animal Care and Use Committee (IACUC) at the University of Pennsylvania.
2. Motorized stereotaxic apparatus: The stereotaxic apparatus used consists of a KOPF Model 902LS Dual small animal stereotaxic instrument equipped with Lazy Susan Base Plate (David Kopf Instruments, Tujunga, CA). Attached to this is a motorized stereotaxic microdrive system (StereoDrive, NeuroStar, Tübingen, Germany) and electronic microinjection system (InjectoMate, NeuroStar). The equipment is housed within a HEPA-grade safety cabinet (*see* Fig. 1).
3. Injection needles: Microinjections are performed using syringes with cemented needles. The Hamilton Company (Reno, NV) produces a 33-gauge, 10 μL cemented needle, 0.5–1.0 in. long, point style #4, with 45° bevel (Hamilton catalog #80308) suited for this application. Syringes are flushed a minimum of 20 times with 70% ethanol by alternatively drawing up and expelling into the syringe. Clean syringes should be stored in 70% ethanol for 2 h or overnight before using. In addition, the use of separate syringes for different types/batches of aSyn is highly recommended to avoid cross-contamination since even trace quantities of aSyn PFFs can potentially initiate pathology.
4. Scissors (Fine Science Tools #14088-10).
5. Fine forceps (#5 size).
6. Needle holder (Miltex #17-1020).
7. Surgical gowns.
8. Gauze and cotton applicators.
9. Veterinary surgical drapes.
10. Balance with cage for body weight measurement.
11. Surgical shaver.
12. Meloxicam (5 mg/mL).
13. Ketamine (100 mg/mL).
14. Xylazine (20 mg/mL).
15. Acepromazine (10 mg/mL).

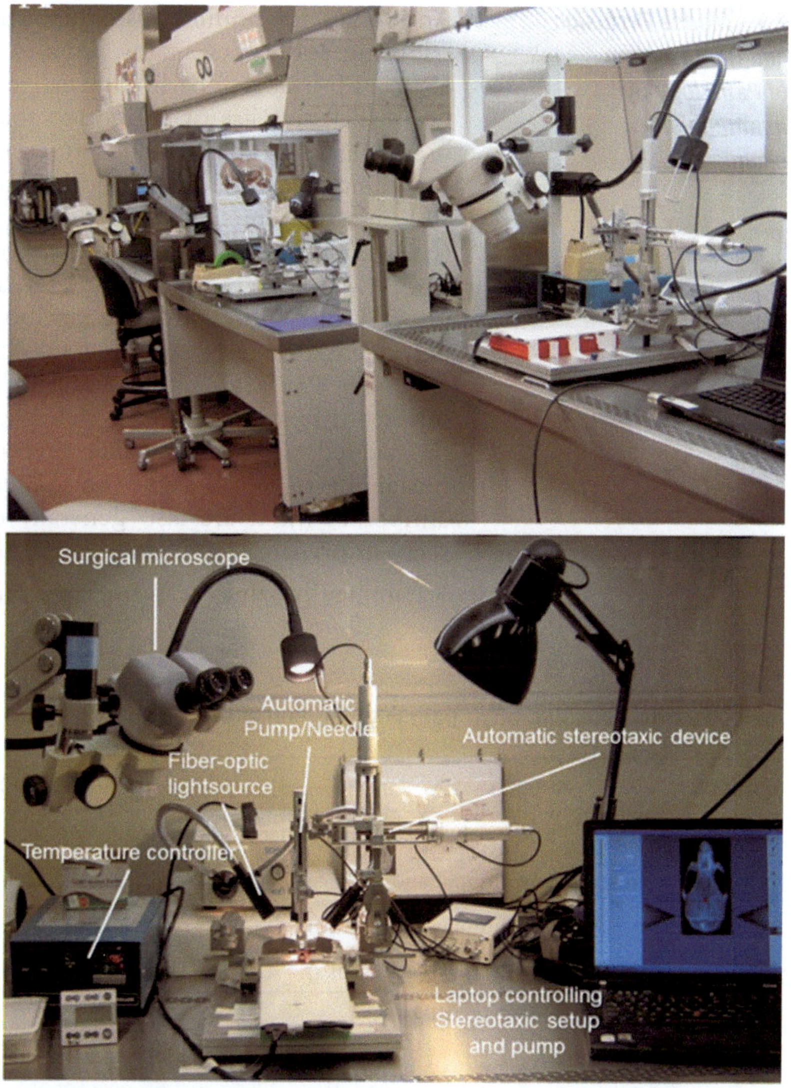

Fig. 1 Motorized stereotaxic system for PFF injection. Top: Apparatus housed with a HEPA-filtered biological cabinet. Bottom: The setup consists of a Kopf small-animal stereotaxic instrument with Lazy Susan Base Plate fitted with a motorized stereotaxic microdrive system and electronic microinjection system, and is controlled on a laptop computer

16. Bupivacaine (0.25%)/lidocaine (0.5%) mixture.
17. Povidone-iodine solution.
18. Heating pad.
19. Yohimbine solution (2 mg/mL).
20. Nylon sutures.
21. Warming lamp.
22. 70% Ethanol solution.
23. 1% SDS solution.

2.3 Mouse Perfusion and Confirmation of Pathology by Immunohistochemistry

1. Antibodies for immunohistochemistry: Immunohistochemistry using antibodies against aSyn is used to directly visualize aSyn PFF-induced pathology in injected brain samples. A non-exhaustive list of antibodies that have been successfully used is listed (Table 1). Phosphorylation of aSyn at Ser129 is a modification observed in both human synucleinopathy and PFF-injected animals. Antibodies recognizing this epitope (e.g., clone 81A) are sensitive for revealing induced pathology since injected PFFs do not undergo significant levels of phosphorylation. Other suitable antibodies include those that detect aSyn disease-associated conformations (e.g., Syn506) (Fig. 2). Pan-aSyn antibodies can also be used, although injected PFFs may also be detected and results should be interpreted with care. Information provided is based on immunostaining of paraffin-embedded brain sections, although the antibodies listed are also generally compatible with floating sections prepared from freshly fixed or cryopreserved tissue.
2. Ketamine/xylazine/acepromazine mixture.

Table 1
Antibodies for detecting aSyn pathology in mouse CNS tissue

Antibody	Antigen	Host species	Dilution	Source
$Syn^{pser129}$ (81A)	Phospho-Ser129 aSyn (human and mouse)	Mouse	1:10,000	Ref. 15
Syn303	aSyn (human and mouse)	Mouse	1:7,000	Ref. 16
Syn506	aSyn (human and mouse)	Mouse	1:20,000	Ref. 16
SNL-4	aSyn (human and mouse)	Rabbit	1:10,000	Ref. 16

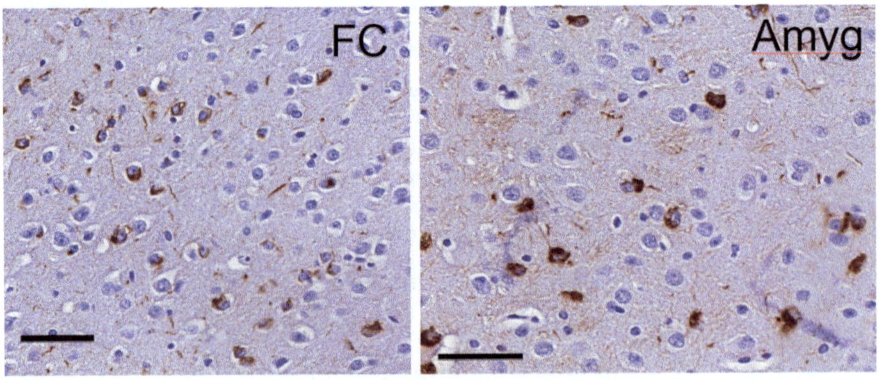

Fig. 2 PFF-induced pathology in mouse brain. Top: Mouse aSyn PFFs in the dorsal striatum 3 days after stereotaxic injection into a aSyn knockout mouse, revealed by immunohistochemistry using Syn506. Bottom: Example of aSyn inclusions in the frontal cortex (FC) and amygdala (Amyg) of a B6C3 (F1) mouse sacrificed 90 days after striatal PFF injection. An antibody against pSer129 was used. Scale bars: 60 μm (top), 10 μm

3. Blunt-tipped 18-gauge butterfly needle, connected to perfusion tubing to syringe and peristaltic pump.
4. Saline solution with heparin (10 U/mL).
5. Fixative solution such as 70% ethanol (in 150 mM NaCl) or 4% paraformaldehyde (in PBS).

3 Methods

3.1 Sonication of aSyn PFFs

If previously frozen, PFFs should be allowed to be thawed in a 37 °C water bath on the day of use. All other procedures can be performed at room temperature unless stated. The aSyn PFF stock is diluted from 5 mg/mL to 2 mg/mL in sterile PBS in a 1.5 mL Eppendorf tube. This preparation is sonicated using the Bioruptor Plus at high power for 10 cycles (30 s on, 30 s off) at a constant temperature of 10 °C. As PFFs can settle over time, gently flick the closed Eppendorf tube to mix contents with a vortex before drawing into the syringe for injection. The sonicated PFF solution should be used on the day of sonication [46]. Extended storage is not recommended as this increases the risk of microbial contamination.

3.2 Stereotaxic Injection into Mouse Brain

1. Prior to entering the surgery/procedure room, ensure that all persons have proper attire and personal protective equipment (PPE), including face mask, hair net, and shoe covers. The surgeon should wash hands extensively with sterile soap before putting on a sterile gown and sterile gloves (*see* **Note 2**).
2. Weigh the animal to be treated and inject mouse subcutaneously with meloxicam (5 mg/kg).
3. Place mouse under appropriate general anesthesia using a mixture of ketamine/xylazine/acepromazine (60–100 mg/kg; 8–12 mg/kg; 0.5–2 mg/kg) administered as an intraperitoneal (i.p.) injection.
4. After the mouse has reached a surgical plane (i.e., nonresponsive to toe pinch or ear flick), inject a mixture of bupivacaine: lidocaine (3.5 and 4 mg/kg, respectively) subcutaneously along the incision site. For intrastriatal injections, this is along the midline slightly between the eyes and the ears. After allowing this local anesthetic to take effect, remove hair on the middle of the head with shaver.
5. Disinfect the shaved skin area with povidone-iodine solution using a sterile cotton swab, and then wash the area with sterile saline. Repeat this step a total of three times.
6. Place the anesthetized mouse on a heating pad to maintain body temperature at 37 °C during surgery.

7. Cover the mouse with a sterile drape. Maintain a small opening for access to the scalp. All efforts should be made to maintain sterility of this surgical field.
8. Open the scalp using a scalpel blade (#11) to expose the skull.
9. Turn on the StereoDrive computer and open the StereoDrive interface software that controls the motorized stereotaxic system.
10. Calibrate the stereotaxic frame, making sure that the readings on the StereoDrive software match those on the actual stereotaxic device/frame.
11. Flush the Hamilton syringe 20 times with sterile PBS.
12. Initialize the InjectoMate on the motorized stereotaxic system. Load no more than 9.1 μL of PFF solution into the syringe. Unassembled (monomeric) aSyn has minimal seeding capacity and can also be used as a negative injection control.
13. Ensure that the top surface of the skull is level (i.e., horizontal) using the StereoDrive software. To do this, move the syringe needle to measure the heights at Bregma and Lambda for anterior-posterior levels, and at 2 mm on either side of the midline to ensure that the left and right sides are level. Adjust the heights of these landmarks to the same level as needed. Tolerance should be ≤0.1 mm between Bregma/Lambda and left/right axes (*see* **Note 3**).
14. Input the desired target coordinates into StereoDrive (*see* Table 2 for suggested coordinates of common CNS targets). The motorized stereotaxic arms holding the syringe needle will automatically move to the indicated location on the skull and place the needle tip on the surface to skull.
15. Move the stereotaxic arm holding the syringe needle away and make a small V-shaped notch on the surface of the skull with a disposable 30-gauge needle. Then use needle to manually bore

Table 2
Coordinates for different CNS targets

Injection site	Distance from Bregma (mm)	Distance from midline (mm)	Depth from brain surface (mm)
Dorsal striatum	+0.2	2.0	−2.6
Ventral striatum	+0.2	2.0	−3.6
Neocortex	+0.2	2.0	−0.8
Hippocampus (CA1)	−2.5	2.0	−1.8
Lateral accumbens (shell)	+1.45	1.75	−4.0

a small hole into the skull overlaying the injection site sufficient for the needle to enter (*see* **Note 4**).

16. Return the stereotaxic arm with the syringe needle above the borehole and push a small drop (0.1 μL) of injection material out of needle to prime. Remove drop with a cotton-tipped applicator (*see* **Note 5**). Insert the needle into brain to desired depth using the software interface. The system is now ready for PFF injection.
17. Using the InjectoMate interface on the motorized stereotaxic system, inject the desired volume of PFF solution. Typical injections contain a total of 5 μg aSyn PFFs (in 2.5 μL sterile PBS) or the same volume of monomeric aSyn or PBS alone as a negative control. Injections made into the striatum, hippocampus, and neocortex are done so at a rate of 0.4 μL/min. For smaller nuclei such as pons or locus coeruleus, a flow rate of 0.1 μL/min is more suitable. The injection needle should remain in place for at least an additional 1 min after injection is complete before it is gently withdrawn over a period of at least 30 s which minimizes leakage (*see* **Note 6**).
18. After injection, close the scalp using individual nylon stitches.
19. While the mouse is still anesthetized, provide an injection of yohimbine (xylazine antagonist; 2.0 mg/kg, i.p.) and a subcutaneous injection of buprenorphine SR (1 mg/kg) for analgesia. This should be followed by a subcutaneous injection of warm sterile 0.9% sodium chloride solution (1 mL total).
20. Place animal in a recovery cage under a warming lamp. Temperature and condition of the animal should be checked every 5 min until the mouse can self-ambulate.
21. Flush the used Hamilton syringe 20 times with sterile water followed by 20 times with 70% ethanol solution. Cleaned syringes can be stored loaded with ethanol.
22. Post-surgery, mice should be housed individually for 10 days or until sutures are eliminated from the skin, at which time normal housing arrangements can be resumed.

3.3 Coordinates for Different CNS Sites

LB/LN pathology formation and accumulation following the injection of misfolded aSyn is a highly dynamic process. This should be taken into consideration, together with the goals of the experiment, when selecting the anatomical location for stereotaxic injection. The accumulation of inclusions typically begins in axons in the vicinity of the injection site and propagates in a retrograde fashion over time but anterograde propagation is also observed. Cell-to-cell spread of pathology may also occur in a time-dependent manner, thus expanding the distribution of inclusions well beyond the initial injection site.

A list of injection sites that lead to productive aSyn pathology is given in Table 2. Note that these coordinates are based on established stereotaxic atlases but significant variation may be present between animals, even those that share similar genetic backgrounds. Particular attention should be paid to the age and weight of the mice used as this also affects brain dimensions. Where the surgeon is unfamiliar with a target, it is recommended that pilot injections using a dye (e.g., Brilliant Blue) be performed as a proxy and the target inspected immediately to confirm the accuracy of the coordinates. Pilot studies are also recommended to assess the distribution of PFFs at and around a particular target site. This is particularly important if using brain homogenates as these specimens are often more viscous than PFF suspensions.

3.4 Mouse Perfusion and Confirmation of Pathology by Immunohistochemistry

Intraneuronal inclusions are expected to develop in the CNS within 30 days after injection with active aSyn PFFs in wild-type mice. Examination of pathology should be confirmed by immunohistochemistry after sacrifice and transcardial perfusion/fixation of the tissue as previously described [25].

1. Anesthetize the mouse using a single combined i.p. injection of ketamine/xylazine/acepromazine (100 mg/kg/10 mg/kg/10 mg/kg). Once the animal is no longer responsive to toe pinch or ear flick, restrain the mouse in a supine position on a dissection block using tape to secure the fore- and hind limbs.
2. Using a pair of clean surgical scissors, cut open the thorax to expose the heart. Avoid damaging major blood vessels close to the heart. Gently introduce a small hole in the right atrium with scissors as an exit port for the blood. Insert a blunt-tipped 18-gauge butterfly needle connected to perfusion tubing, syringe, and pump into the left ventricle.
3. Pump heparinized saline through the mouse to rinse out the blood. This step helps reduce the background staining.
4. Remove brain and spinal cord from the mouse and fix in 70% ethanol (in 150 mM NaCl) or 4% paraformaldehyde (in PBS) overnight at 4 °C. Tissues should then be processed for paraffin embedding, sectioning, and collection on slides. Dispose of carcass according to local and institutional regulations.
5. Tissue sections on the slides are processed using routine immunohistochemical procedures as described in Sanderson et al., this issue.

3.5 Decontamination of Instruments and Work Surfaces

Given that small quantities of aSyn PFFs can trigger aggregation in permissive hosts, it is essential that residual PFFs be thoroughly removed from all instruments and work surfaces after procedures. Common detergents such as SDS effectively solubilize recombinant aSyn PFFs [47] and thus greatly reduce their activity. Any spills or aerosol contamination should be neutralized immediately.

1. Clean all exposed work surfaces with 1% SDS after use. Allow solution to sit for 1 min before wiping with paper towels. Dispose of towels and repeat procedure using distilled water to remove any residue (*see* **Note 7**).
2. Microinjection needles should be rinsed extensively before storage. To further avoid contamination, it is essential that each needle assembly be used for only one type of injection material (e.g., PFF strain, monomer).

4 Notes

1. To date, efficient seeding has also been reported in C57BL/6J [28, 43], C3H/Hej [48], and CD-1 [42] mice. aSyn knockout mice are nonpermissive to seeding with PFFs and can be used as negative controls in many cases. Additional considerations should be made when using transgenic lines (e.g., those overexpressing aSyn) as their neuroanatomy may deviate from wild-type animals and/or develop age-dependent phenotypes. In addition, mice may differ in their response to anesthesia depending on background strain.
2. Surgical tools and materials should be steam sterilized at 120 °C for 30 min prior to surgery.
3. Levelling the skull is critical for precise stereotaxic injections, especially when targeting relatively small regions or nuclei. A tolerance of no more than 0.01–0.03 mm for Bregma/Lambda and left/right should be maintained.
4. To prevent damage to the skull and brain surface during surgery, manually bore a small hole into the skull overlying the injection site with the smallest needle that can get least damage which is better when imaging the brain sections.
5. Priming of the needle by advancing a small drop of injection material from needle tip prior to inserting into brain helps avoid introducing air bubbles.
6. Using a faster infusion rate will lead to increased leakage of the PFF suspension beyond the target site. Slow removal of the needle also reduces leakage into areas surrounding the injection tract due to drawback and is especially critical when targeting smaller nuclei.
7. Dispose unused PFFs and PFF-contaminated waste (e.g., paper towels, scalpels) according to local regulations.

Acknowledgments

This work was supported in part by NIH grants NS088322 and NS053488.

References

1. Toledo JB, Gopal P, Raible K et al (2016) Pathological α-synuclein distribution in subjects with coincident Alzheimer's and Lewy body pathology. Acta Neuropathol 131:393–409
2. Jellinger KA (2004) Lewy body-related alphasynucleinopathy in the aged human brain. J Neural Transm (Vienna) 111:1219–1235
3. Polymeropoulos MH, Lavedan C, Leroy E et al (1997) Mutation in the alpha-synuclein gene identified in families with Parkinson's disease. Science 276:2045–2047
4. Singleton AB, Farrer M, Johnson J et al (2003) alpha-Synuclein locus triplication causes Parkinson's disease. Science 302:841
5. Zarranz JJ, Alegre J, Gómez-Esteban JC et al (2004) The new mutation, E46K, of alphasynuclein causes Parkinson and Lewy body dementia. Ann Neurol 55:164–173
6. Lesage S, Anheim M, Letournel F et al (2013) G51D α-synuclein mutation causes a novel parkinsonian-pyramidal syndrome. Ann Neurol 73:459–471
7. Halliday GM, Del Tredici K, and Braak H (2006) Critical appraisal of brain pathology staging related to presymptomatic and symptomatic cases of sporadic Parkinson's disease. J Neural Transm Suppl 70:99–103
8. Dickson DW, Uchikado H, Fujishiro H et al (2010) Evidence in favor of Braak staging of Parkinson's disease. Mov Disord 25(Suppl 1): S78–S82
9. Spillantini MG, Crowther RA, Jakes R et al (1998) Filamentous alpha-synuclein inclusions link multiple system atrophy with Parkinson's disease and dementia with Lewy bodies. Neurosci Lett 251:205–208
10. Braak H, Del Tredici K, Rüb U et al (2003) Staging of brain pathology related to sporadic Parkinson's disease. Neurobiol Aging 24:197–211
11. Burke RE, Dauer WT, Vonsattel JPG (2008) A critical evaluation of the Braak staging scheme for Parkinson's disease. Ann Neurol 64:485–491
12. Braak H, Rüb U, Gai WP et al (2003) Idiopathic Parkinson's disease: possible routes by which vulnerable neuronal types may be subject to neuroinvasion by an unknown pathogen. J Neural Transm (Vienna) 110:517–536
13. Wakabayashi K, Takahashi H, Takeda S et al (1988) Parkinson's disease: the presence of Lewy bodies in Auerbach's and Meissner's plexuses. Acta Neuropathol 76:217–221
14. Duda JE, Shah U, Arnold SE et al (1999) The expression of alpha-, beta-, and gamma-synucleins in olfactory mucosa from patients with and without neurodegenerative diseases. Exp Neurol 160:515–522
15. Kordower JH, Chu Y, Hauser RA et al (2008) Lewy body-like pathology in long-term embryonic nigral transplants in Parkinson's disease. Nat Med 14:504–506
16. Li J-Y, Englund E, Holton JL et al (2008) Lewy bodies in grafted neurons in subjects with Parkinson's disease suggest host-to-graft disease propagation. Nat Med 14:501–503
17. Baba M, Nakajo S, Tu PH et al (1998) Aggregation of alpha-synuclein in Lewy bodies of sporadic Parkinson's disease and dementia with Lewy bodies. Am J Pathol 152:879–884
18. Spillantini MG, Crowther RA, Jakes R et al (1998) alpha-Synuclein in filamentous inclusions of Lewy bodies from Parkinson's disease and dementia with Lewy bodies. Proc Natl Acad Sci U S A 95:6469–6473
19. Fauvet B, Mbefo MK, Fares M-B et al (2012) α-Synuclein in central nervous system and from erythrocytes, mammalian cells, and Escherichia coli exists predominantly as disordered monomer. J Biol Chem 287:15345–15364
20. Crowther RA, Jakes R, Spillantini MG et al (1998) Synthetic filaments assembled from C-terminally truncated alpha-synuclein. FEBS Lett 436:309–312
21. Conway KA, Harper JD, Lansbury PT (1998) Accelerated in vitro fibril formation by a mutant alpha-synuclein linked to early-onset Parkinson disease. Nat Med 4:1318–1320
22. Come JH, Fraser PE, Lansbury PT (1993) A kinetic model for amyloid formation in the prion diseases: importance of seeding. Proc Natl Acad Sci U S A 90:5959–5963
23. Luk KC, Song C, O'Brien P et al (2009) Exogenous alpha-synuclein fibrils seed the formation of Lewy body-like intracellular inclusions

in cultured cells. Proc Natl Acad Sci U S A 106:20051–20056
24. Volpicelli-Daley LA, Luk KC, Patel TP et al (2011) Exogenous α-synuclein fibrils induce Lewy body pathology leading to synaptic dysfunction and neuron death. Neuron 72:57–71
25. Luk KC, Kehm VM, Zhang B et al (2012) Intracerebral inoculation of pathological α-synuclein initiates a rapidly progressive neurodegenerative α-synucleinopathy in mice. J Exp Med 209:975–986
26. Luk KC, Kehm V, Carroll J et al (2012) Pathological -synuclein transmission initiates Parkinson-like neurodegeneration in nontransgenic mice. Science 338:949–953
27. Recasens A, Dehay B, Bové J et al (2014) Lewy body extracts from Parkinson disease brains trigger α-synuclein pathology and neurodegeneration in mice and monkeys: LB-induced pathology. Ann Neurol 75:351–362
28. Masuda-Suzukake M, Nonaka T, Hosokawa M et al (2013) Prion-like spreading of pathological α-synuclein in brain. Brain J Neurol 136:1128–1138
29. Bernis ME, Babila JT, Breid S et al (2015) Prion-like propagation of human brain-derived alpha-synuclein in transgenic mice expressing human wild-type alpha-synuclein. Acta Neuropathol Commun 3:75
30. Watts JC, Giles K, Oehler A et al (2013) Transmission of multiple system atrophy prions to transgenic mice. Proc Natl Acad Sci U S A 110:19555–19560
31. Bousset L, Pieri L, Ruiz-Arlandis G et al (2013) Structural and functional characterization of two alpha-synuclein strains. Nat Commun 4:2575
32. Guo JL, Covell DJ, Daniels JP et al (2013) Distinct α-synuclein strains differentially promote tau inclusions in neurons. Cell 154:103–117
33. Paumier KL, Luk KC, Manfredsson FP et al (2015) Intrastriatal injection of pre-formed mouse α-synuclein fibrils into rats triggers α-synuclein pathology and bilateral nigrostriatal degeneration. Neurobiol Dis 82:185–199
34. Shimozawa A, Ono M, Takahara D et al (2017) Propagation of pathological α-synuclein in marmoset brain. Acta Neuropathol Commun 5:12
35. Koprich JB, Kalia LV, Brotchie JM (2017) Animal models of α-synucleinopathy for Parkinson disease drug development. Nat Rev Neurosci 18:515–529
36. Luna E, Luk KC (2015) Bent out of shape: α-Synuclein misfolding and the convergence of pathogenic pathways in Parkinson's disease. FEBS Lett 589:3749–3759
37. Dehay B, Vila M, Bezard E et al (2016) Alpha-synuclein propagation: new insights from animal models. Mov Disord 31:161–168
38. Sacino AN, Brooks M, Thomas MA et al (2014) Intramuscular injection of -synuclein induces CNS -synuclein pathology and a rapid-onset motor phenotype in transgenic mice. Proc Natl Acad Sci 111:10732–10737
39. Holmqvist S, Chutna O, Bousset L et al (2014) Direct evidence of Parkinson pathology spread from the gastrointestinal tract to the brain in rats. Acta Neuropathol 128:805–820
40. Manfredsson FP, Luk KC, Benskey MJ et al (2018) Induction of alpha-synuclein pathology in the enteric nervous system of the rat and non-human primate results in gastrointestinal dysmotility and transient CNS pathology. Neurobiol Dis 112:106–118
41. Mason DM, Nouraei N, Pant DB et al (2016) Transmission of α-synucleinopathy from olfactory structures deep into the temporal lobe. Mol Neurodegener 11:49
42. Nouraei N, Mason DM, Miner KM et al (2018) Critical appraisal of pathology transmission in the α-synuclein fibril model of Lewy body disorders. Exp Neurol 299:172–196
43. Rey NL, Steiner JA, Maroof N et al (2016) Widespread transneuronal propagation of α-synucleinopathy triggered in olfactory bulb mimics prodromal Parkinson's disease. J Exp Med 213:1759–1778
44. Osterberg VR, Spinelli KJ, Weston LJ et al (2015) Progressive aggregation of alpha-synuclein and selective degeneration of Lewy inclusion-bearing neurons in a mouse model of Parkinsonism. Cell Rep 10:1252–1260
45. Masuda-Suzukake M, Nonaka T, Hosokawa M et al (2014) Pathological alpha-synuclein propagates through neural networks. Acta Neuropathol Commun 2:88
46. Polinski NK, Volpicelli-Daley LA, Sortwell CE et al (2018) Best practices for generating and using alpha-synuclein pre-formed fibrils to model Parkinson's disease in rodents. J Parkinsons Dis 8(2):303–322
47. Bousset L, Brundin P, Böckmann A et al (2016) An efficient procedure for removal and inactivation of alpha-synuclein assemblies from laboratory materials. J Parkinsons Dis 6:143–151
48. Abdelmotilib H, Maltbie T, Delic V et al (2017) α-Synuclein fibril-induced inclusion spread in rats and mice correlates with dopaminergic Neurodegeneration. Neurobiol Dis 105:84–98

Chapter 6

Quantification of Alpha-Synuclein in Biological Fluids by Electrochemiluminescence-Based Detection

Niels Kruse and Brit Mollenhauer

Abstract

Several potential marker candidates for Parkinson's disease (PD) in cerebrospinal fluid (CSF) have been identified. These include α-synuclein, a major constituent of the intracellular aggregates Lewy bodies, a neuropathological hallmark of PD, and others. The extracellular presence of α-synuclein in blood, CSF, saliva, and conditioned media motivated the hypothesis that the quantification of CSF α-synuclein is a biomarker for α-synuclein-related disorders. We here describe the development of an electrochemiluminescence-based assay by conversion of an established ELISA for quantification of α-synuclein. The assay not only works with CSF but can also be used to quantify α-synuclein in different biological fluids, i.e., whole blood and blood products, saliva, as well as brain homogenates and cell culture material.

Key words Parkinson's disease, α-Synuclein, ELISA, Immunodetection, Biomarker, CSF

1 Introduction

In the field of neurodegenerative diseases objective laboratory risk and progression markers are lacking [1]. Movement disorders and dementias often feature clinical phenotypes with overlapping clinical symptoms making their accurate differentiation challenging, especially in the early stages. With the emergence of specific therapeutic interventions, it will be necessary to measure the response and progression of diseases more objectively. In the early 1990s the extracellular presence of pathologically relevant amyloid-β (Aβ) and tau proteins was first described in cerebrospinal fluid (CSF) [2, 3]. These findings pioneered the search for biomarkers in the field of Alzheimer's disease (AD) as a neurodegenerative disease. Extensive validation of biochemical markers within the last decade (together with the development of new imaging techniques) has led to the revision of AD research criteria incorporating these markers [4, 5]. The identification of new marker candidates is the first step in biomarker research and can either be driven by a specific

Tim Bartels (ed.), *Alpha-Synuclein: Methods and Protocols*, Methods in Molecular Biology, vol. 1948,
https://doi.org/10.1007/978-1-4939-9124-2_6,

hypothesis [such as for CSF tau and Aβ in AD or for CSF α-synuclein (aSyn) in Parkinson's disease] or be hypothesis free and unbiased [6]. An important issue in biomarker assay development is the identification of marker candidates and validation of the generated assays. Recent investigations indicate that the use of multiple parameters seems to be more informative for the diagnosis of many medical conditions and also necessary due to clinically heterogeneous diseases [7, 8]. Several platforms have been developed for multiplexing different classes of biomolecules ([9–15]; for review *see* [6]). With these platforms, nucleic acids as well as proteins, lipids, and additional molecules can be measured or characterized. Here, we describe our general approach toward developing singleplex assays in a first step for the quantification of aSyn in CSF and other biological fluids by transforming established ELISA assays into an electrochemiluminescence (ECL)-based detection system. Details about the CSF α-synuclein-specific ECL-based assay based on the published ELISA [16, 17] are given. The underlying technology presented here was developed by Meso Scale Discovery™ (MSD). Compared to conventional ELISA formats for protein quantification the MSD platform makes use of carbon surfaces for coating of capture antibodies to the solid phase of special microtiter plates. Due to the high-binding capacity the assays display high sensitivity combined with a large dynamic range, usually comprising several orders of magnitude. The high sensitivity reduces the amount of biological fluid required to perform the assay (25–50 μL, sometimes less). It therefore is suitable to work with precious material derived from patients. The whole assay is completed in 4–5 h and is therefore much faster than previously established assays [17].

2 Materials

1. Labeling reaction buffer 1: Dissolve 0.96 g PBS in distilled water, adjust pH to 7.9, and fill up to 0.1 L.
2. Zeba Spin Desalting Columns, 40 K MWCO (Cat. No. 87766, Thermo Scientific GmbH, Schwerte, Germany).
3. Antibody labeling solution: Dissolve 150 nmol MSD SULFO-TAG NHS Ester (Cat. No. R91AN-1, MSD) in 50 μL ice-cold distilled water and use immediately.
4. Labeling reaction buffer 2: Dissolve 0.96 g PBS in distilled water, adjust pH to 7.4, add 5 mL 1% NaN_3, and fill up to 0.1 L.
5. Capture antibody solution: Dilute MJF-1 antibody (Cat. No. ab138501, Abcam, Cambridge, UK) to 3 μg/mL in PBS.
6. MULTI-ARRAY® 96-well Plates (Cat. No. L15XA-3, Meso Scale Discovery, Gaithersburg, MD, USA) (*see* **Note 1**).

7. MSD Tris wash buffer (10×), Cat. No. R61TX-2, MSD): Dilute tenfold with distilled water.
8. PBS-T wash buffer: Dissolve 9.6 g PBS in 1 L distilled water and add 0.5 mL Tween-20 (*see* **Note 2**).
9. Sample dilution buffer: Dissolve 1% BSA (Cat. No. AP-4510-80, SeraCare Life Sciences, Gaithersburg, MD, USA) in 100 mL PBS-T wash buffer.
10. Recombinant α-synuclein monomer standards (available from several vendors).
11. Detection antibody solution: Dilute SULFO-TAG-labeled Syn1 clone 42/α-synuclein antibody (Cat. No. 610787, BD Biosciences, Heidelberg, Germany) to 1 μg/mL in sample dilution buffer supplemented with 0.1% mouse IgG (Cat. No. D609-0100, Rockland, Limerick, PA, USA) and 0.1% goat IgG (SLG66, Equitech-Bio, Kerrville, TX, USA).
12. 2× MSD read buffer T: Dilute 4× MSD read buffer T (Cat. No. R92TC-3, MSD) with an equal volume of distilled water (*see* **Note 3**).

3 Methods

3.1 Preparation of Biospecimen

CSF: Obtain CSF by lumbar puncture in the morning in polypropylene tubes.

To compare with other studies standard operating procedures should be followed, as suggested [18]. Briefly: Centrifuge CSF samples at 3500 × *g* for 10 min at 4 °C. Make 0.5 mL aliquots and store at −80 °C within 30 min after lumbar puncture.

Whole blood: Obtain blood samples by venipuncture using EDTA tubes. Make 0.5 mL aliquots of whole blood in 1.5 or 2.0 mL reaction tubes and store them at −80 °C.

Plasma: Obtain plasma samples by centrifugation of EDTA tubes at 3500 × *g* for 10 min at 4 °C. Make 0.5 mL aliquots from the supernatant and store them at −80 °C.

Serum: Leave serum tubes for 20 min at room temperature to allow clotting of blood cells. Centrifuge serum tubes for 10 min at 3500 × *g* at 4 °C. Make 0.5 mL aliquots and store them at −80 °C.

3.2 SULFO-TAG Labeling of Detection Antibodies

Detection antibodies are labeled with ruthenium (II) tris-bipyridine-(4-methylsulfonate) NHS ester. The ester is an amine-reactive N-hydroxysuccinimide molecule which couples to primary amine groups of proteins comparable to biotin.

For tagging antibodies with MSD SULFO-TAG NHS ester antibodies have to be buffer-exchanged against labeling reaction buffer 1. This can be done using Zeba Spin Desalting Columns (*see* **Note 4**).

Storage buffer is removed from spin columns by centrifugation at 1500 × *g* for 1 min at 2-8 °C in a swinging bucket rotor after placing the column in a 1.5 mL collection tube. Wash/equilibrate the column matrix by adding 300 μL labeling reaction buffer 1 in a dropwise manner and centrifuge as above. Repeat this step for two more times with 3-min centrifugation for the third wash. Transfer column to a new collection tube. Apply up to 100 μL antibody solution and centrifuge at 1500 × *g* for 4 min.

Antibody concentrations should be in the range of 1-2 mg/mL (*see* **Note 5**). We usually incubate antibodies with a 20-fold molar excess of the ester (=challenge ratio), but the optimal challenge ratio might have to be optimized for each antibody. Add the required volume of antibody labeling solution according to the manufacturer's instruction. Incubate for 2 h at 23 °C. After completion of the labeling reaction another buffer exchange is performed against labeling reaction buffer 2. The procedure is the same as described above with labeling reaction buffer 2 replacing labeling reaction buffer 1. By this procedure the labeled antibodies are separated from unincorporated SULFO-TAG ester by size exclusion. Antibodies can now be stored at 4 °C. Once optimized, we found that the number of incorporated ester molecules is very reproducible.

3.3 Coating of Assay Plates

Apply 30 μL of the capture antibody solution to each well of the assay plate. Verify that the solution is spread over the entire carbon surface at the bottom of the wells. If this is not the case, tap the assay plate with each side on the bench without splashing the antibody solution and verify homogeneous distribution by visual inspection of all wells. Store the plates overnight at 4 °C to allow for binding of the antibodies to the plates.

3.4 Washing

After coating empty the plate over a sink and pat dry on paper towels. Apply 150 μL PBS-T wash buffer to each well using a multichannel pipet (*see* **Notes 6** and 7). Empty over a sink and pat dry on paper towels. Repeat two times for a total of three washes.

3.5 Blocking

Block unspecific binding of biomolecules by adding 150 μL sample dilution buffer to each well and incubate for 1 h with shaking at 300 rpm at room temperature (*see* **Note 8**). Wash as described above.

3.6 Preparation of Standard Curves

We recommend the preparation of 25 μL single-use aliquots at 1 μg/mL aSyn (*see* **Note 9**). Take a 20 μL single-use aliquot of 1 μg/mL aSyn, dilute tenfold with 180 μL sample dilution buffer, and mix by pipetting up and down using a pipet. Prepare eight 1.5 mL Eppendorf tubes marked #1 to #8 containing 300 μL sample dilution buffer. Take 100 μL from the diluted single-use

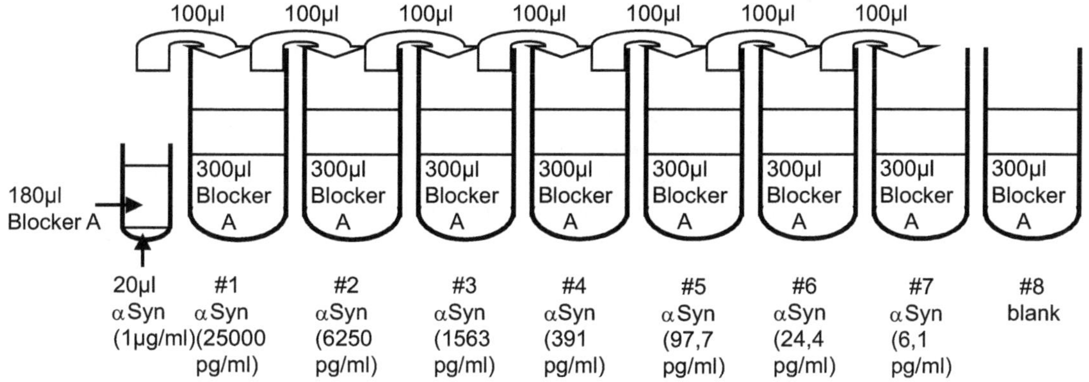

Fig. 1 Preparation of standard curves

aliquot, transfer to Eppendorf tube #1, and mix by vortexing. Repeat the step six times with the other tubes and leave tube #8 with pure sample dilution buffer. This creates the aSyn standard curve starting with a concentration of 25,000 pg/mL in serial fourfold dilutions (Fig. 1). Place standards on ice.

3.7 Sample Dilution

Depending on the specimen biosamples may have to be diluted appropriately to match the concentration range of the standard curve or the range where dilution linearity is met (*see* **Note 10**).

CSF, serum, and plasma have to be diluted at least 1 in 8. Whole-blood samples have to be diluted 20,000-fold. For dilution sample dilution buffer is used.

3.8 Application of Standards and Samples to the Assay Plate

Apply 25 µL of both standard dilutions and samples or sample dilutions to the wells of the assay plate. Incubate for 1 h with shaking at 700 rpm at room temperature. Wash as described above.

3.9 Application of Detection Antibody

Apply 25 µL detection antibody solution to each well and incubate for 1 h with shaking at 700 rpm at room temperature (*see* **Notes 11** and **12**). Wash as described above.

3.10 Measurement

Apply 150 µL 2× MSD read buffer to the wells of the assay plate and measure the plate within a few minutes in a Meso Scale Discovery Sector Imager (MSD). Within the Sector Imager an electrochemical stimulation initiates a redox reaction mediated by tripropylamine at the electrode surface of the microtiter plates. The SULFO-TAG changes between different oxidation states and emits a light signal upon reduction. Multiple excitation cycles of each label enhance the light levels and improve sensitivity. The light signal is measured at 620 nm.

3.11 Data Analysis

Results are analyzed using MSD's Discovery Workbench. The software allows for assignment of sample types (standards, blanks, unknowns, etc.) to individual wells. A standard curve is generated from the signal intensities and the indicated standard concentrations. Concentrations of unknown samples are determined from the standard curves by interpolating the respective concentration from the measured signal intensity.

4 Notes

General note: The procedure described here is easily adaptable to other target molecules. We have recently described our general procedure to establish electrochemiluminescence-based assays for the quantification of target molecules in different biological specimens [19]. The reader is referred to this manuscript as a general guideline.

1. MULTI-ARRAY® 96-well plates contain carbon surfaces and allow for denser coating with biomolecules, e.g., antibodies, as compared to conventional polystyrene ELISA plates. This feature accounts for an increased sensitivity and a broad dynamic range of this assay.
2. For pipetting viscous fluids, e.g., Tween-20, we recommend to cut the tip of a micropipette tip slightly to widen the diameter of the tip.
3. 2× MSD read buffer should be gently inverted several times after dilution, not vortexed. Vortexing generates air bubbles that interfere with measurements.
4. Before removing the storage solution from ZEBA Spin columns we recommend to invert the column several times in order to dissolve the column matrix homogeneously. Doing this will create an even surface of the matrix during the first centrifugation enabling good performance of the size-exclusion chromatography.
5. High antibody concentrations are required for efficient SULFO-TAG labeling. If concentrations exceeding 1 mg/mL are not available, custom preparations of antibodies may be required. Some antibodies may be purchased lyophilized. In these cases the antibodies can be dissolved at high concentrations. Antibodies should be carefully dissolved in labeling reaction buffer 1. Free amines contained in stabilizing proteins or contaminating cell culture media, Tris, glycine, and NaN_3 as preservative compete with the antibodies for incorporation of the SULFO-TAG label. In addition glycerol must be avoided.

6. For the assay described here we compared both MSD Tris wash buffer and PBS-T wash buffer as wash buffers. Both gave equivalent results.
7. For washing of the assay plates we use an Eppendorf eight-channel pipet (Cat. No. 3122 000.051, Eppendorf, Hamburg, Germany).
8. For repeated pipetting of same volumes, e.g., 30 μL capture antibody solution or 150 μL 2× MSD read buffer, we use an Eppendorf Multipette Plus (Cat. No. 4982 000.012, Eppendorf) with Eppendorf combitips. We verified that it works with high reproducibility.
9. We strongly recommend preparing single-use standards. We usually prepare 25 μL aliquots of defined concentration (e.g., 1 μg/mL) in 0.2 mL PCR tubes and store them at −80 °C. On the day of an experiment we prepare a standard curve, place it on ice, and use this for multiple assays on the same day. The diluted standards are discarded at the end of the day. This procedure reduced the number of freeze/thaw cycles applied to the standards and yields highly reproducible standard curves (Fig. 2).

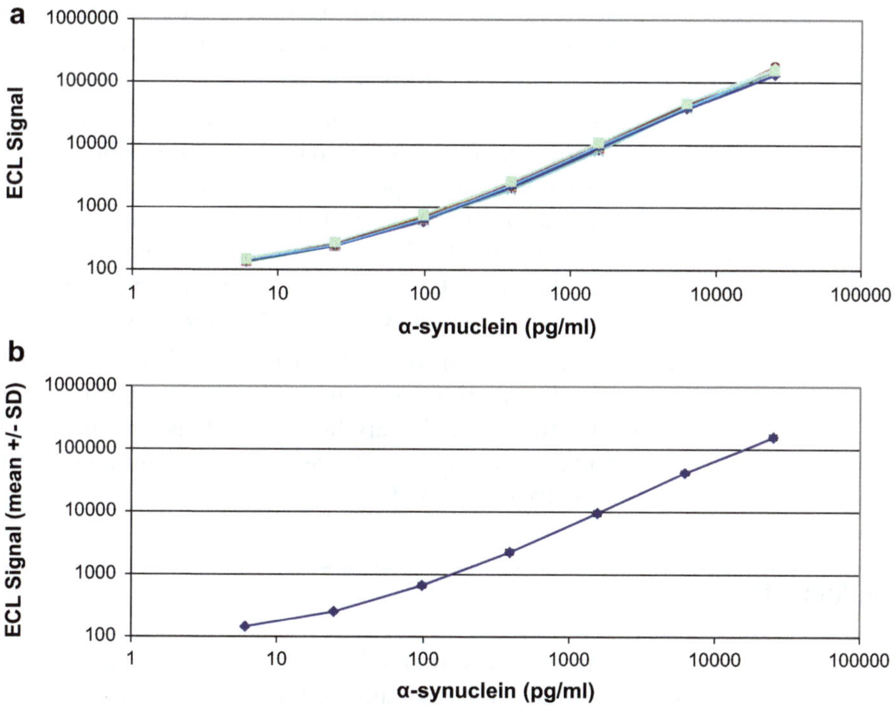

Fig. 2 Reproducibility of α-synuclein standard curves. A. Individual standard curves from 11 independent experiments. B. Mean values of ECL signals ±standard deviation for defined α-synuclein concentrations from the 17 experiments. Note that CVs were less than 20% over the entire range

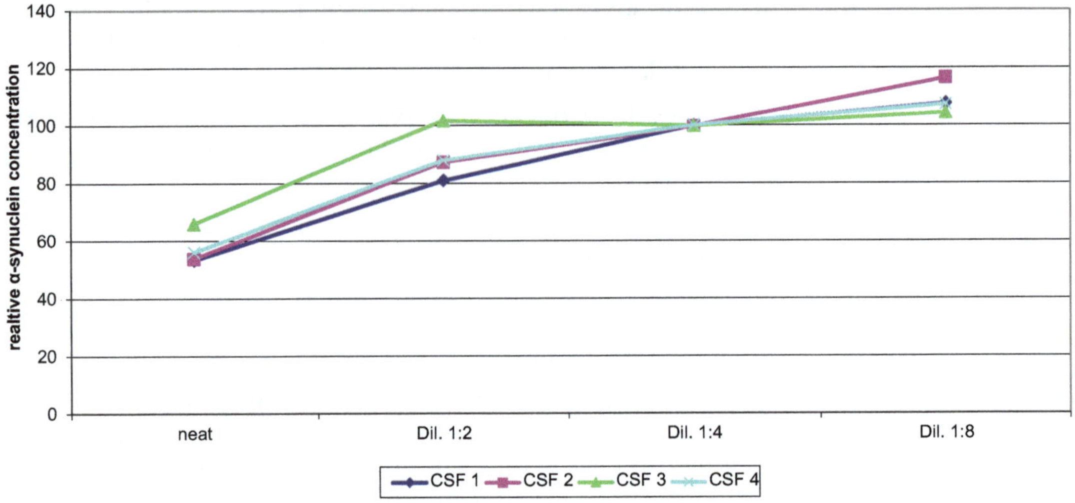

Fig. 3 Determination of CSF dilution linearity. Four CSF samples were analyzed neat and in increasing dilutions. Protein concentrations were normalized to data from samples diluted 1:4

10. Dilution linearity of biospecimen needs to be verified before quantification of analytes in biological material. The biological material should be quantified at increasing dilutions and results plotted against the dilution factor. Reduced analyte concentrations at high concentration indicate inhibition of detectability by matrix compounds (Fig. 3).
11. If capture and detection antibodies are derived from different host species detection of analytes can be done using SULFO-TAG-labeled anti-species antibodies (e.g., SULFO-TAG-labeled anti-rat antibody if the detection antibody was generated from rat hybridomas). Alternatively biotinylated detection antibodies may be used in combination with SULFO-TAG-labeled streptavidin (Cat. No. R32AD-5, MSD).
12. In some cases it is possible to apply both sample and detection antibodies at the same time. When doing this the detection antibody should be applied first. This will reduce the time required to complete the assay. It is recommended to increase the incubation time from 1 to 2 h.

Acknowledgments

The authors want to thank Sabine Lange for helpful discussions and advice. Omar el-Agnaf and Liyu Wu provided us kindly with α-synuclein standards and antibody MJFR1. We thank Omar El-Agnaf and Michael Schlossmacher who developed the first prototype ELISA assay and Walter Schulz-Schaeffer for his fruitful input.

References

1. Lee JW, Weiner RS, Sailstad JM, Bowsher RR, Knuth DW, O'Brien PJ, Fourcroy JL, Dixit R, Pandite L, Pietrusko RG, Soares HD, Quarmby V, Vesterqvist OL, Potter DM, Witliff JL, Fritche HA, O'Leary T, Perlee L, Kadam S, Wagner JA (2005) Method validation and measurement of biomarkers in nonclinical and clinical samples in drug development: a conference report. Pharm Res 22:499–511
2. Motter R, Vigo-Pelfrey C, Kholodenko D, Barbour R, Johnson-Wood K, Galasko D, Chang L, Miller B, Clark C, Green R et al (1995) Reduction of beta-amyloid peptide42 in the cerebrospinal fluid of patients with Alzheimer's disease. Ann Neurol 38:643–648
3. Seubert P, Vigo-Pelfrey C, Esch F, Lee M, Dovey H, Davis D, Sinha S, Schlossmacher M, Whaley J, Swindlehurst C et al (1992) Isolation and quantification of soluble Alzheimer's beta-peptide from biological fluids. Nature 359:325–327
4. Dubois B, Feldman HH, Jacova C, Cummings JL, Dekosky ST, Barberger-Gateau P, Delacourte A, Frisoni G, Fox NC, Galasko D, Gauthier S, Hampel H, Jicha GA, Meguro K, O'Brien J, Pasquier F, Robert P, Rossor M, Salloway S, Sarazin M, de Souza LC, Stern Y, Visser PJ, Schelters P (2010) Revising the definition of Alzheimer's disease: a new lexicon. Lancet Neurol 9:1118–1127
5. McKhann GM, Knopman DS, Chertkow H, Hyman BT, Jack CR Jr, Kawas CH, Klunk WE, Koroshetz WJ, Manly JJ, Mayeux R, Mohs RC, Morris JC, Rossor MN, Scheltens P, Carrillo MC, Thies B, Weintraub S, Phelps CH (2011) The diagnosis of dementia due to Alzheimer's disease: recommendations from the National Institute on Aging-Alzheimer's Association workgroups on diagnostic guidelines for Alzheimer's disease. Alzheimers Dement 7:263–269
6. Dieterle F, Marrer E (2008) New technologies around biomarkers and their interplay with drug development. Anal Bioanal Chem 390:141–154
7. Lewczuk P, Wiltfang J (2008) Neurochemical dementia diagnostics: State of the art and research perspectives. Proteomics 8:1292–1301
8. Gerlach M, Maetzler W, Broich K, Hampel H, Rems L, Reum T, Riederer P, Stoffler A, Streffer J, Berg D (2012) Biomarker candidates of neurodegeneration in Parkinson's disease for the evaluation of disease-modifying therapeutics. J Neural Transm 119:39–52
9. Steemers FJ, Gunderson KL (2007) Whole genome genotyping technologies on the BeadArray platform. Biotechnol J 2:41–49
10. Abramovitz M, Barwick BG, Willis S, Young B, Catzavelos C, Li Z, Kodani M, Tang W, Bouzyk M, Moreno CS, Leyland-Jones B (2011) Molecular characterisation of formalin-fixed paraffin-embedded (FFPE) breast tumour specimens using a custom 512-gene breast cancer bead array-based platform. Br J Cancer 105:1574–1581
11. Shi M, Bradner J, Hancock AM, Chung KA, Quinn JF, Peskind ER, Galasko D, Jankovic J, Zabetian CP, Kim HM, Leverenz JB, Montine TJ, Ginghina C, Kang UJ, Cain KC, Wang Y, Aasly J, Goldstein D, Zhang J (2011) Cerebrospinal fluid biomarkers for Parkinson disease diagnosis and progression. Ann Neurol 69:570–580
12. Cotte S, von Ahsen N, Kruse N, Huber B, Winkelmann A, Zettl UK, Starck M, Konig N, Tellez N, Dorr J, Paul F, Zipp F, Luhder F, Koepsell H, Pannek H, Montalban X, Gold R, Chan A (2009) ABC-transporter gene-polymorphisms are potential pharmacogenetic markers for mitoxantrone response in multiple sclerosis. Brain 132:2517–2530
13. Kruse N, Pette M, Toyka K, Rieckmann P (1997) Quantification of cytokine mRNA expression by RT PCR in samples of previously frozen blood. J Immunol Methods 210:195–203
14. Trune DR, Larrain BE, Hausman FA, Kempton JB, MacArthur CJ (2011) Simultaneous measurement of multiple ear proteins with multiplex ELISA assays. Hear Res 275:1–7
15. Roberts R, Stewart AFR, Wells GA, Williams KA, Kavaslar N, McPherson R (2007) Can J Cardiol (Suppl. A);7A–15A
16. Mollenhauer B, Cullen V, Kahn I, Krastins B, Outeiro TF, Pepivani I, Ng J, Schulz-Schaeffer W, Kretzschmar HA, McLean PJ, Trenkwalder C, Sarracino DA, Vonsattel JP, Locascio JJ, El-Agnaf OM, Schlossmacher MG (2008) Direct quantification of CSF alpha-synuclein by ELISA and first cross-sectional study in patients with neurodegeneration. Exp Neurol 213:315–325
17. Mollenhauer B, Locascio JJ, Schulz-Schaeffer-W, Sixel-Doring F, Trenkwalder C, Schlossmacher MG (2011) alpha-Synuclein and tau concentrations in cerebrospinal fluid of patients presenting with parkinsonism: a cohort study. Lancet Neurol 10:230–240

18. Mollenhauer B, El-Agnaf OM, Marcus K, Trenkwalder C, Schlossmacher MG (2010) Quantification of alpha-synuclein in cerebrospinal fluid as a biomarker candidate: review of the literature and considerations for future studies. BiomarkMed 4:683–699

19. Kruse N, Schulz-Schaeffer WJ, Schlossmacher MG, Mollenhauer B (2012) Development of electrochemiluminescence-based singleplex and multiplex assays for the quantification of alpha-synuclein and other proteins in cerebrospinal fluid. Methods 56:514–518

Chapter 7

Alpha-Synuclein Proximity Ligation Assay (AS-PLA) in Brain Sections to Probe for Alpha-Synuclein Oligomers

Rosalind F. Roberts, Nora Bengoa-Vergniory, and Javier Alegre-Abarrategui

Abstract

Alpha-synuclein oligomers are thought to be toxic mediators of Parkinson's disease and other alpha-synucleinopathies, but their histological detection *in situ* in diseased brain has been a challenge in the field for some time. Here we describe a method, the alpha-synuclein proximity ligation assay (AS-PLA), to detect alpha-synuclein oligomers in paraffin-embedded brain sections. Using AS-PLA previously unobserved alpha-synuclein oligomeric pathology is revealed.

Key words Alpha-synuclein, Oligomers, Aggregates, Proximity ligation assay, Pathology, Parkinson's disease, Synucleinopathy

1 Introduction

Lewy bodies, proteinaceous inclusions that contain fibrillized alpha-synuclein [1], are the pathological hallmark of Parkinson's disease (PD). Studies over the past two decades have provided evidence that Lewy bodies are markers of a late stage of the disease and may form as a protective mechanism, placing smaller aggregates of alpha-synuclein, or oligomers, as the toxic species mediating disease [2]. However, while Lewy bodies can be easily detected using alpha-synuclein immunohistochemistry, alpha-synuclein oligomers have proved more challenging to detect in patient brain.

To address this, we developed the alpha-synuclein proximity ligation assay (AS-PLA) [3]. Proximity ligation assays (PLAs) were originally developed for the sensitive detection of proteins, but soon adapted to detect protein-protein interactions [4, 5]. PLA probes are generated from antibodies raised against the protein (s) of interest, one for each of the proteins involved in the putative interaction, which are conjugated to short oligonucleotides. If the probes bind interacting proteins, the oligonucleotides are

Tim Bartels (ed.), *Alpha-Synuclein: Methods and Protocols*, Methods in Molecular Biology, vol. 1948,
https://doi.org/10.1007/978-1-4939-9124-2_7,

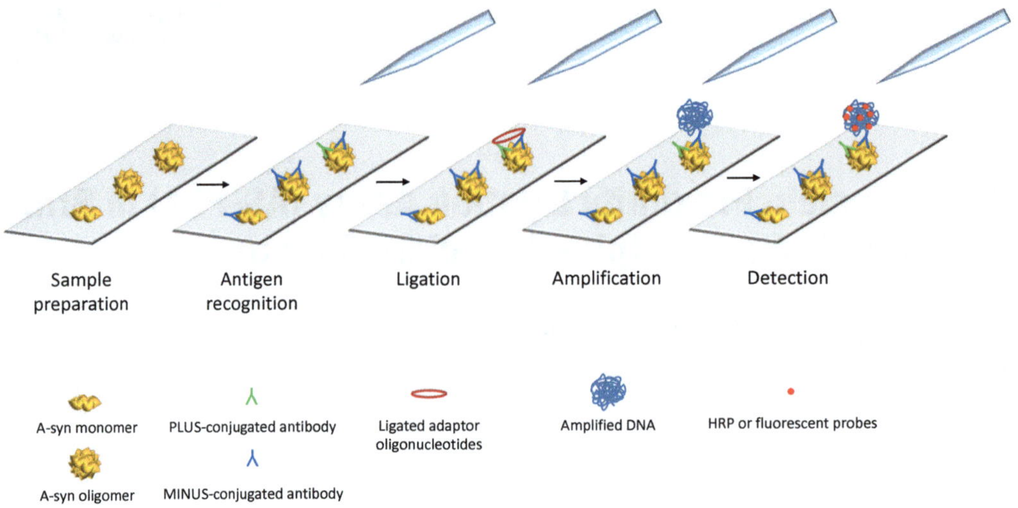

Fig. 1 Schematic representation of the alpha-synuclein proximity ligation assay procedure (AS-PLA). After the sample has been adequately prepared (sample preparation), AS-PLA consists of four main steps, namely antigen recognition, ligation, amplification, and detection. Antigen recognition: During this step (normally performed overnight) antibodies conjugated to probes are allowed to hybridize with their antigens. As epitope-blocking antibodies are used, further antibody molecules will be prevented from binding, ensuring that monomers are not detected. For oligomers, the reaction will be able to proceed, provided that the epitopes are appropriately exposed and no more than 40 nm apart from each other. Note that the stoichiometry of the reaction only allows the detection of 50% of binding events, as PLUS-PLUS events and MINUS-MINUS events that form will not render productive ligation. Ligation: When an PLUS-conjugated antibody binds its target less than 40 nm away from a MINUS-conjugated antibody, ligation with connector oligonucleotides can occur, which will produce circular DNA that acts as a template for the amplification step. Amplification: Productive ligations will be able to serve as primers for a rolling circle amplification reaction. Detection: Oligonucleotides complementary to the amplified DNA and tagged with either HRP or fluorescent tags bind the product of the amplification step and can be visualized as puncta using light or fluorescent microscopy, respectively

sufficiently close to prime an amplification reaction, which can be detected by tagged oligonucleotides and observed as punctate signal, with each punctum representing an interaction.

We adapted PLA in order to detect alpha-synuclein oligomers by using the same epitope-blocking anti alpha-synuclein antibody to generate both probes (Fig. 1) and performed extensive *in vitro* characterization to demonstrate the specific detection of alpha-synuclein oligomers. Using AS-PLA we were able to describe for the first time the presence of abundant diffusely deposited oligomeric pathology in the medulla and cingulate cortex of Parkinson's disease postmortem brain tissue [3] (Fig. 2). AS-PLA also labeled very early perikaryal aggregates in morphologically intact neurons that may precede the development of classical Parkinson's disease, whereas brain stem LBs, considered heavily compacted late lesions, were only stained on their periphery, if at all, indicating that oligomers exist in the cytoplasm surrounding the more compact

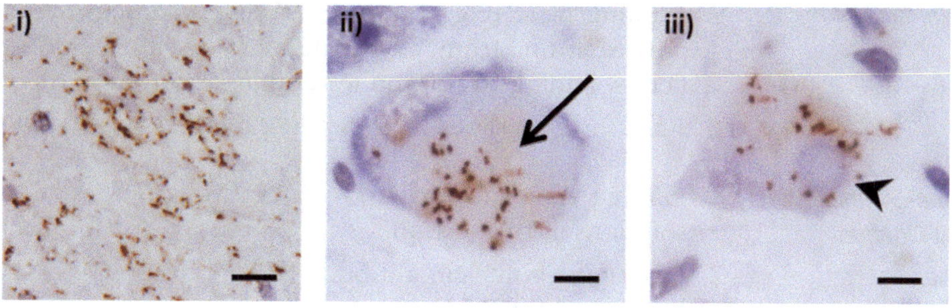

Fig. 2 Expected results of AS-PLA staining in Parkinson's disease brain tissue. Diffusely distributed signal is present in the neuropil (**a**, scale bar 20 μm), particularly abundantly in areas that do not have a heavy Lewy body load. Staining of pale bodies (**b**) is also frequently observed. Lewy bodies are not stained using AS-PLA, although occasional staining may be observed at their periphery (**c**). Scale bars for **b** + **c** are 10 μm

structures of LBs (Fig. 2). AS-PLA is a robust technique for the detection of alpha-synuclein oligomers and can be performed in a similar time frame to and using equipment for standard immunohistochemistry, making it widely applicable.

2 Materials

2.1 Preparation of Tissue for Staining

1. Sections of formalin-fixed paraffin-embedded tissue should be cut using a microtome at 4-6 μm (*see* **Note 1**) and deposited on positively charged slides (e.g., Fisher SuperFrost slides).
2. Hot plate to warm slides (e.g., VWR catalogue #: 720-2422).
3. Solutions for dehydration and rehydration of tissue: Xylenes, 100% (Sigma); Histo-Clear clearing agent, 100% (Diamed); ethanol, 100% (Sigma); reverse osmosis (RO)-purified water or filtered deionized Milli-Q water (Millipore). Make up solutions for rehydration and dehydration of tissue in Coplin staining jars (Sigma). Make ethanol- and water-based solutions fresh each time and reuse xylenes and Histo-Clear up to a maximum of five times.
4. For mounting: DPX mounting medium (Sigma) and rectangular coverslips (Fisher).
5. 30% Hydrogen peroxide (Sigma): Store at 4 °C in the dark.
6. PBS: Phosphate-buffered saline (137 mM NaCl, 2.7 mM KCl, 10 mM Na_2HPO_4, 1.8 mM KH_2PO_4; PBS).
7. Antigen retrieval buffer: Citrate buffer pH 6.0 (e.g., Abcam catalogue #: ab93678).
8. Hydrophobic barrier pen (e.g., Vector Laboratories catalogue #: H-4000).

9. Staining plate (e.g., StainTray™ Slide Staining System, Simport Scientific from VWR catalogue #: 75838-966), humidified by adding RO or tap water in the bottom of the plate.
10. Rack for glass slides (e.g., Sigma catalogue #: Z710989).

2.2 Alpha-Synuclein Proximity Ligation Assay

1. Filtered deionized Milli-Q water (Millipore) is used on all solutions in this section.
2. Anti-alpha-synuclein antibody 211 (1 mg/mL): Abcam catalogue #: ab80627 (*see* **Note 2**).
3. Duolink Probemaker kit (Sigma catalogue #: DUO92009 and DUO92010) (*see* **Note 3**).

   Components: lyophilized probe oligonucleotides, conjugation buffer, stop reagent, storage solution, blocking solution, and probe diluent.
4. Duolink bright-field detection reagents (Sigma catalogue #: DUO92012) (*see* **Note 3**): Note that some reagents are stored at −20 °C and some at 4 °C. Components: 5× ligation stock, ligase (1 U/μL), 5× amplification stock, polymerase (10 U/μL), 5× detection stock, substrates A (66.6×), B (100×), C (100×), and D (50×).
5. TBS: Tris-buffered saline (10 mM Tris base, 137 mM NaCl, pH 7.4) + 0.05% tween (TBS-T).
6. Hematoxylin (e.g., Vector Labs catalogue #: H-3404, Hematoxylin QS; we use this modified Mayer's hematoxylin as it does not require a bluing step).

3 Methods

3.1 Preparation of AS-PLA Probes

AS-PLA probes should start to be prepared at least the day before experiment, as an overnight incubation is required. Testing each new batch of probes using, for example, Parkinson's brain tissue, before running experiments is advised. The protocol described here utilizes the Duolink Probemaker kit.

1. Add 20 μL of the antibody syn211 to 2 μL of conjugation buffer, mix by gently pipetting, then add the solution to the freshly opened PLUS or MINUS lyophilized probe oligonucleotides, and mix by gentle pipetting.
2. Incubate at room temperature overnight (16 h minimum).
3. Add 2 μL of stop reagent to the reaction and incubate at room temperature for 30 min.
4. Add 24 μL of storage solution and store at 4 °C.

3.2 Preparation of Tissue

1. Warm slides for 30 min on the hot plate at 55 °C.
2. Dewax by briefly dipping slides in 100% xylenes five times, then leaving in the xylene for 2 min, and then dipping five times again.
3. Repeat the procedure in point 2 with Histo-Clear.
4. Rehydrate by briefly dipping slides in 100% ethanol five times, then leaving in the ethanol for 2 min, and then dipping five times again. Repeat with 95% ethanol and then 70% ethanol. Then repeat with RO or Milli-Q water, in this case leaving three times, 1 min each time.
5. Quench endogenous peroxidases with 10% H_2O_2 diluted in PBS for 15 min at room temperature in the dark.
6. Wash in running RO (or Milli-Q) water for 5 min (we place slides in the slide rack and immerse in a 20 cm long Tupperware on one side, immersing the end of the flexible tube of the water dispenser on the opposite side to avoid direct exposure of the tissue to the running water).
7. ***CRITICAL STEP*** (*see* **Note 4**): Antigen retrieval: Put slides in microwaveable box filled with citrate buffer pH 6 and microwave on a standard cycle for 4 min followed by 5-min rest. We use a Sharp 900 W microwave. Microwave again for 1.5 min, then rest for 5 min, and top up the buffer. Repeat microwaving for 1.5 min followed by a 5-min rest three times. Ensure that the tissue is covered by buffer throughout the retrieval.
8. Cool the box on ice until cool (approximately 45 min), and then wash the slides under running RO (or Milli-Q) water, as above.
9. Put slides onto humidified staining plate and circle the sample with a hydrophobic barrier pen.
10. Wash slides briefly a couple of times with TBS-T on the staining plate.

3.3 AS-PLA Staining

Do not allow the tissue to dry at any point during the staining protocol as this will increase background.

Defrost stock solutions fully at room temperature and mix before use. Always prepare immediately before use. Do not use diluted solutions unless fresh.

For steps with enzymes, add them to the mix immediately before addition to the sample. Once out of the freezer, ensure that they are placed on an ice block.

1. Cover the tissue with an appropriate amount of blocking solution (*see* **Note 5**) and incubate for 1 h at 37 °C. Blocking at lower temperatures or for shorter times will increase background. If 1 h at 37 °C results in diffuse background that is

not punctate (especially noticeable in the nucleus) block for longer.

2. Dilute AS-PLA probes appropriately in PLA probe diluent (1:100 for bright field). Add mix to slides and incubate overnight at 4 °C (*see* **Note 6**).
3. Wash slides by incubating 4 × 5 min with TBS-T on the staining plate.
4. Prepare the ligation solution consisting of 5× ligation stock diluted to 1× in Milli-Q water and ligase (1 U per 40 μL reaction). Add the solution to the sections and incubate for 1 h at 37 °C.
5. Wash 4 × 5 min with TBS-T, as above.
6. Prepare the amplification solution consisting of 5× amplification stock diluted to 1× in Milli-Q water and polymerase (5 U per 40 μL reaction) and incubate for 2.5 h at 37 °C.
7. Wash 4 × 5 min with TBS-T, as above.
8. Prepare the detection solution consisting of 5× detection stock diluted to 1× in Milli-Q water and incubate for 1 h at room temperature.
9. Wash 4 × 5 min with TBS-T, as above.
10. Prepare the substrate solution consisting of 66.6× substrate A stock, 100× substrate B stock, 100× substrate C stock, and 50× substrate D stock all diluted to 1× in Milli-Q water and incubate for 20 min at room temperature.
11. Wash slides by incubating 3 × 5 min with Milli-Q water.
12. Add hematoxylin for 5 min (*see* **Note 7**).
13. Wash under running tap water for 5 min, as above.
14. Dehydrate (RO water, 95% EtOH, 100% EtOH, xylenes, as rehydration procedure explained above but in reverse order) and mount using DPX.

4 Notes

1. We have tested using 20 μm thick frozen tissue and found that there was an order of magnitude less signal. Therefore, for optimal results we recommend using thinly cut paraffin-embedded sections.
2. The antibody must be in an amine-free buffer, ideally PBS. The manufacturer suggests that the buffer should be carrier and preservative free, but may contain up to 0.1% bovine serum albumin (BSA), 5% trehalose, and 0.02% sodium azide. It is best if the antibody is bought at a high concentration (ideally 1 mg/mL) as the yield of antibody following concentration by

filter column is poor. For murine applications alpha-synuclein 4D6 (Abcam catalogue #: ab1903) enables the detection of alpha-synuclein oligomers detecting mouse, rat and human alpha-synuclein [6]. Please note that the detection of different alpha-synuclein species requires careful validation of the conjugated antibodies, as the epitope of each antibody will determine which alpha-synuclein species is/are detected by the assay. It is also essential to perform *in vitro* validation of novel antibodies to ensure that they block the epitope following binding, and spurious signal does not arise from multiple antibody molecules binding a monomer. This can be achieved using the antibody for both capture and detection in an ELISA with monomeric protein (for ELISA setup, *see* chapter by Kruse and Möllenhauer in this issue).

3. It is possible to make the PLA reagents in-house, as previously described [5], but for most laboratories using kits is more convenient. As the sequence of the oligonucleotides for the PLA probes provided in the Probemaker kit is proprietary, it is not known if this is the same as in previous publications.
4. Antigen retrieval via microwave is a critical step in the success of AS-PLA. We have found that antigen retrieval needs to be more extensive than what is required for immunohistochemistry and is essential for the observation of AS-PLA signal. The antigen retrieval method described here has given us consistent results but may require optimization in each laboratory according to the microwave. Other tools, such as an autoclave, may be used as an alternative, but the specific conditions must be adjusted according to the tools available.
5. Ensure that the tissue is fully covered by the reagents. We have found that 120 μL is enough volume to cover tissue sections that cover most of a standard slide, and 60 μL is used for half a slide. If a tissue section covers the full area of the slide, 150 μL may be required. Volumes should be adjusted according to the size of tissue for each assay. Other trays or staining systems may be used. For example, we have achieved similar results using the Shandon Sequenza Coverplates. The technique is also amenable to automation.
6. Note that while we have described bright-field AS-PLA here, it is also possible to utilize fluorescent AS-PLA and complement with immunofluorescence to undertake co-localization studies. For this, a standard immunofluorescence protocol is undertaken first, followed by washing with TBS-T. Then the steps for AS-PLA described in this protocol are followed, until the amplification step and the amplification and detection are done together (use Duolink fluorescent reagents appropriate to your imaging solutions). Following amplification, wash with TBS

(0.2 M Tris and 0.1 M NaCl) including DAPI and then a further three times with TBS and mount using an aqueous solution.

7. It is important not to overdevelop the hematoxylin so its signal does not obscure the PLA signal.

References

1. Spillantini MG et al (1997) Alpha-synuclein in Lewy bodies. Nature 388(6645):839–840
2. Caughey B, Lansbury PT (2003) Protofibrils, pores, fibrils, and neurodegeneration: separating the responsible protein aggregates from the innocent bystanders. Annu Rev Neurosci 26:267–298
3. Roberts RF, Wade-Martins R, Alegre-Abarrategui J (2015) Direct visualization of alpha-synuclein oligomers reveals previously undetected pathology in Parkinson's disease brain. Brain 138(Pt 6):1642–1657
4. Gullberg M et al (2004) Cytokine detection by antibody-based proximity ligation. Proc Natl Acad Sci U S A 101(22):8420–8424
5. Soderberg O et al (2006) Direct observation of individual endogenous protein complexes in situ by proximity ligation. Nat Methods 3 (12):995–1000
6. McMillan KJ et al (2017) Loss of microRNA-7 regulation leads to alpha-synuclein accumulation and dopaminergic neuronal loss in vivo. Mol Ther 25(10):2404–2414

Chapter 8

Studying α-Synuclein Conformation by Intact-Cell Cross-Linking

Thibaut Imberdis, Saranna Fanning, Andrew Newman, Nagendran Ramalingam, and Ulf Dettmer

Abstract

β-Sheet-rich aggregates of α-synuclein (αS) are the hallmark neuropathology of Parkinson's disease (PD) and related synucleinopathies, whereas the native conformations of αS in healthy cells are under debate. Cross-linking analyses in intact cells detect a large portion of endogenous αS in apparent multimeric states, most notably as putative tetramers (αS60) that run around 60 kDa on SDS-PAGE, but also point at the dynamic nature of cellular αS states. Standardization of αS cross-linking methods will facilitate efforts to study the effects of genetic, pharmacological, and environmental factors on αS conformation. Here, we present detailed protocols for cross-linking cellular αS multimers in cultured cells and brain tissues. These protocols will benefit future studies aimed at characterizing αS conformation in its cellular environment, both at steady state and upon perturbation, be it chronic or acute.

Key words α-Synuclein, Multimer, Tetramer, Chemical cross-linking, Protein homeostasis, Parkinson's disease

1 Introduction

Lewy bodies and Lewy neurites, the hallmarks of familial and sporadic PD and several other fatal brain diseases [1], are characterized by accumulation of the 14 kDa protein αS. Missense mutations, copy number variants, and otherwise upregulated expression of αS have each been associated with familial PD (fPD) [2–6]. While αS assembly has traditionally only been studied from the angle of pathology, unexpected findings from several laboratories [7–14] have suggested that αS can occur physiologically in multimeric states that presumably have α-helical conformation and resist aggregation. We previously reported that treating fresh human and rodent brain, primary rodent neurons, and a variety of immortalized cell types with cell-penetrant chemical cross-linkers such as disuccinimidyl glutarate (DSG) markedly shifts the SDS-PAGE running behavior of αS from 14 kDa (the expected

Tim Bartels (ed.), *Alpha-Synuclein: Methods and Protocols*, Methods in Molecular Biology, vol. 1948,
https://doi.org/10.1007/978-1-4939-9124-2_8,

position of the 14.5 kDa αS monomer) to higher molecular weight positions [7, 9, 15–17]. In particular, αS has been routinely trapped at 60, 80, and 100 kDa positions (αS60, αS80, and αS100), with 60 kDa typically being the most prominent species. The principle of parsimony suggests that αS60 represents the tetrameric ($4 \times 14.5 = 58$ kDa), helical αS structure that has been purified by several labs [7, 8, 18]. Accordingly, αS80 and αS100 are either hexamers ($6 \times 14.5 = 87$ kDa) or octamers ($8 \times 14.5 = 116$ kDa), respectively, or tetramers that are cross-linked in a different mode, possibly due to a different conformation.

The assumed native, not aggregated, nature of αS60 (as well as 80 and 100) is supported by several observations: (1) αS in fresh non-PD human brain tissue, rodent brain, and primary neurons is trapped as αS60 [9, 16]; (2) the non-disease-causing homologs β- and γ-synuclein are also trapped at prominent ~60 kDa positions [9, 17]; (3) αS60 also forms in the absence of the aggregation-mediating NAC domain [9]; (4) all known fPD-linked αS missense mutations shift the αS60:14 ratio toward αS14 [16]; (5) strategic αS variants that lack αS60 entirely are neurotoxic and form inclusions [17, 19]; and (6) PD-relevant GBA1 deficiency negatively affects αS60 [20].

In our cross-linking studies, known monomeric proteins (e.g., Parkin, Ran, casein-kinase 1) were not altered by the cross-linking, while known multimeric proteins (e.g., DJ-1; voltage-dependent anion channel; dynamin-related protein 1) were trapped as the expected multimers. Interestingly, αS60 as well as αS80 and αS100 were depolymerized to monomers when cross-linkers were applied after cell lysis, unless the lysis was done at high protein concentrations ("molecular crowding") [9]. Only αS80 displayed relevant lysis stability. Other endogenous multimeric proteins that we tested did not show this striking cell lysis sensitivity, except for the homolog β-synuclein [9]. These findings led us to postulate that dynamic intracellular populations of αS monomers and metastable multimers coexist normally [7, 9]. Such a dynamic equilibrium has been proposed for well-known tetrameric proteins such as p53 [21, 22] and hemoglobin [23]. Consequently, αS cross-linking studies are ideally carried out in intact, live cells and fresh tissues.

Here, we present detailed protocols for trapping αS *in situ* in its intracellular conformation. The methods differ significantly between tissues and cultured cells, but both approaches have in common that post-lysis cross-linking of αS is avoided because of the dynamic nature of cellular αS homeostasis which is disturbed by cell lysis.

There are several applications of these protocols that will provide valuable insight into αS conformation in health and disease. For example, efforts to develop drugs that prevent the aggregation of αS would benefit from detailed characterization of how the drugs affect cellular αS homeostasis.

The basic principles of studying proteins by intact-cell cross-linking and the DSG cross-linking reaction itself are shown in Fig. 1.

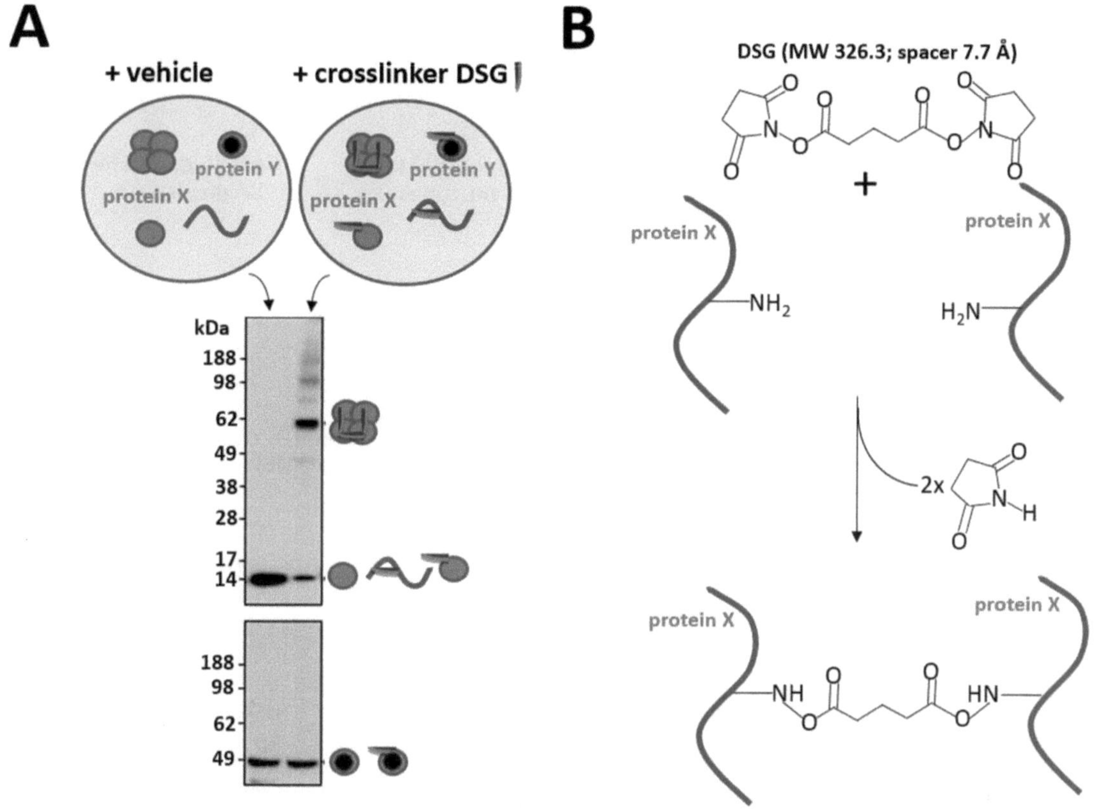

Fig. 1 Principles of intact-cell cross-linking. (**a**) Monomeric proteins (black) are modified by cell-penetrant cross-linkers such as DSG, but their running behavior on Western blot is unchanged. Multimeric proteins (grey; shown as a mixture of monomers and multimers), both hetero- and homomultimers, are covalently connected by the cross-linker, leading to a shift of SDS-PAGE running behavior toward the higher molecular weight range. (**b**) Simplified chemical reaction of the cross-linker DSG with primary amines (lysine residues) of proteins

Figures 2 and 3 illustrate the procedures for cross-linking of cultured cells (Subheading 3.1) and brain material (Subheading 3.2), respectively. Figure 4 shows an example of a cross-linking experiment: the determination of the optimal cross-linking and Western blot (WB) conditions for studying αS in human erythroleukemia (HEL) cells. Figure 5 exemplifies the densitometrical and statistical analysis of αS cross-linking experiments.

2 Materials

Unless stated otherwise, solutions are prepared using analytical grade reagents and ultrapure water (sensitivity of 18 MΩ cm at 25 °C). Unless stated otherwise, all reagents are prepared and stored at room temperature, and waste materials are discarded following disposal regulations.

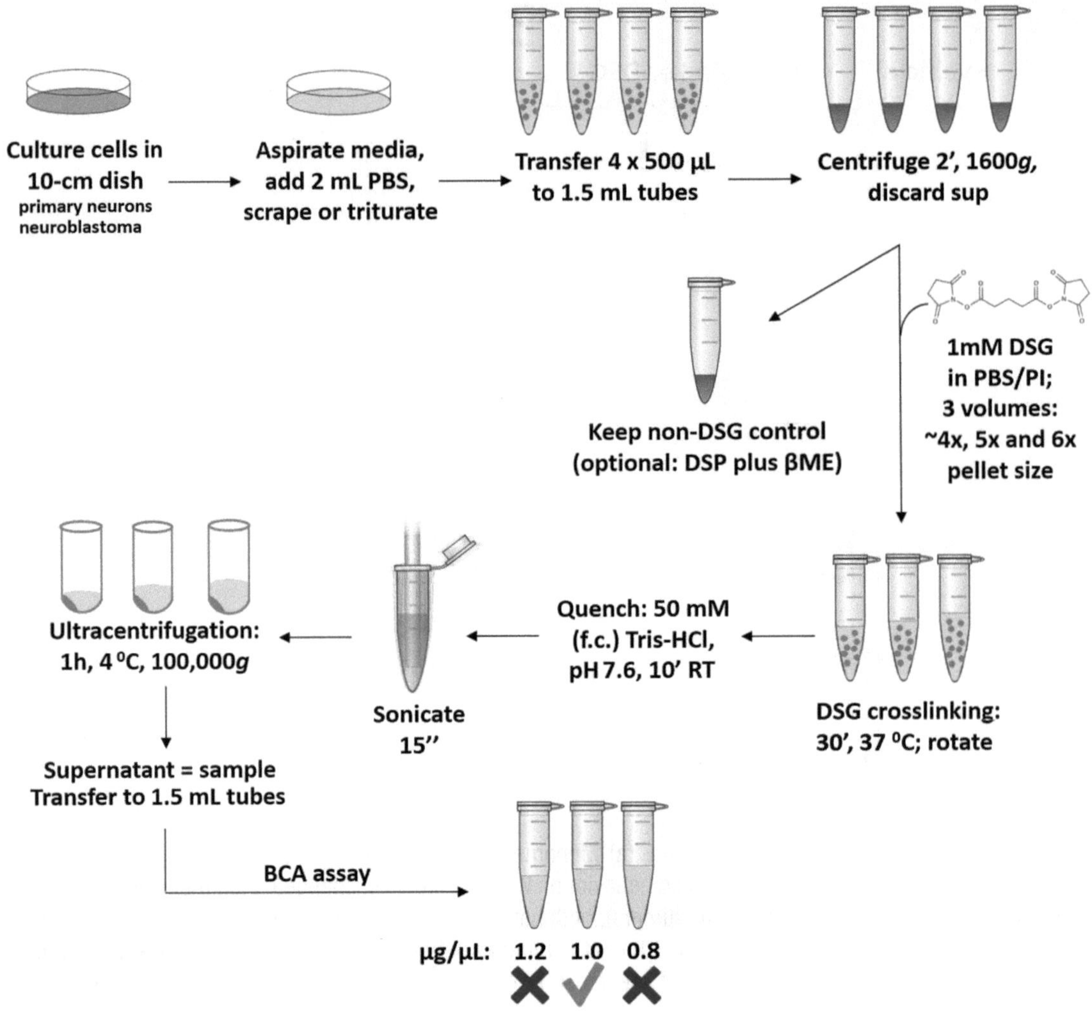

Fig. 2 Schematic of αS cross-linking in cultured cells. Aspirate media carefully, and triturate or scrape to collect cells grown in 10 cm dishes after adding respective volumes of PBS. Distribute triturated cells equally into four 1.5 mL tubes, followed by centrifugation. Subsequently, add different volumes of 1 mM DSG solution to three of the tubes (the fourth tube will serve as a negative control for cross-linking). Mix the contents well, and incubate at 37 °C for 30 min on a nutator. Afterwards, quench the cross-linker at room temperature for 10 min with 50 mM Tris pH 7.6. Lyse cells by sonication, and transfer the 100,000 × *g* supernatant (lysate) into new 1.5 mL tubes. Measure protein concentration by BCA assay. Use the lysates with a concentration at 1 μg/μL (or close) for WB analysis along with the negative control

2.1 Intact-Cell Cross-Linking of Cultured Cells

1. DSG: Disuccinimidyl glutarate (e.g., Thermo Fisher, ref. 20593); store at 4 °C under desiccation.
2. DSP (optional): Dithiobis (succinimidyl propionate) (e.g., Thermo Fisher, ref. 22585); store at 4 °C under desiccation.
3. Dimethyl sulfoxide (DMSO)

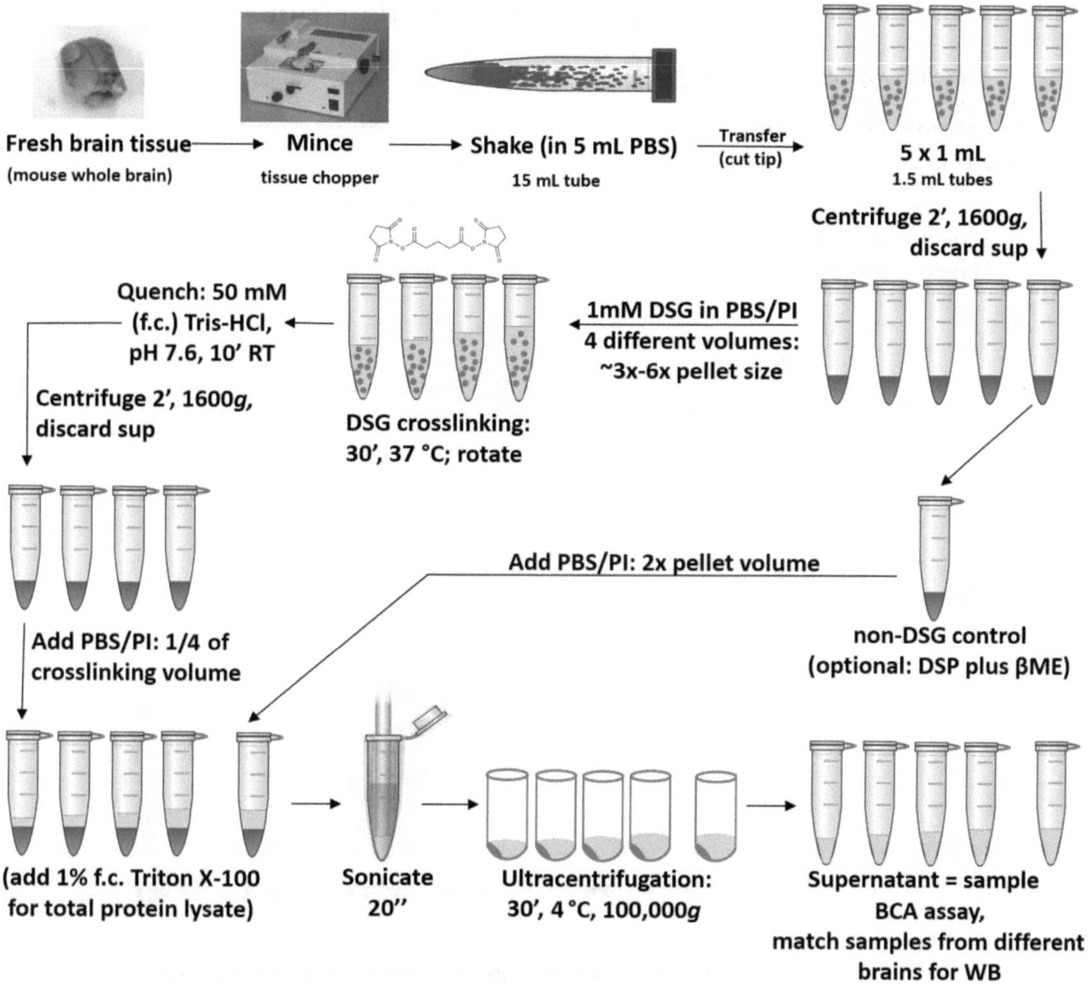

Fig. 3 Schematic of αS cross-linking in fresh brain tissue. Mince the freshly dissected brain using a tissue chopper. Transfer chopped material into a 15 mL tube containing 5 mL PBS, shake gently to mix, and use a cut tip distribute the contents equally into five 1.5 mL tubes. Centrifuge to collect minced tissue and, and add different volumes of 1 mM DSG solution to four of the tubes (the fifth tube will serve as a negative control for cross-linking). Mix the contents well, and cross-link αS by incubating at 37 °C for 30 min on a nutator. Afterwards, quench the cross-linker at room temperature for 10 min with 50 mM Tris pH 7.6, followed by centrifugation and disposal of the supernatant. Add reduced volumes of PBS (plus protease inhibitors; optional: plus 1% Triton X-100 for total protein lysates) and sonicate to lyse tissues. Transfer the 100,000 × *g* supernatant (lysate) into four new 1.5 mL tubes and measure protein concentration by BCA assay. Use lysates with a concentration close to 1 μg/μL for WB analysis along with the negative control

4. Desiccant for storage of DSG (and DSP).
5. 1.5 mL tubes (e.g., Eppendorf).
6. PBS: 10 mM Phosphate, 2.7 mM KCl, 137 mM NaCl, pH 7.4.
7. PI: Protease inhibitor mix, e.g., Complete Protease Inhibitor, EDTA-free (Roche, ref. 11873580001).

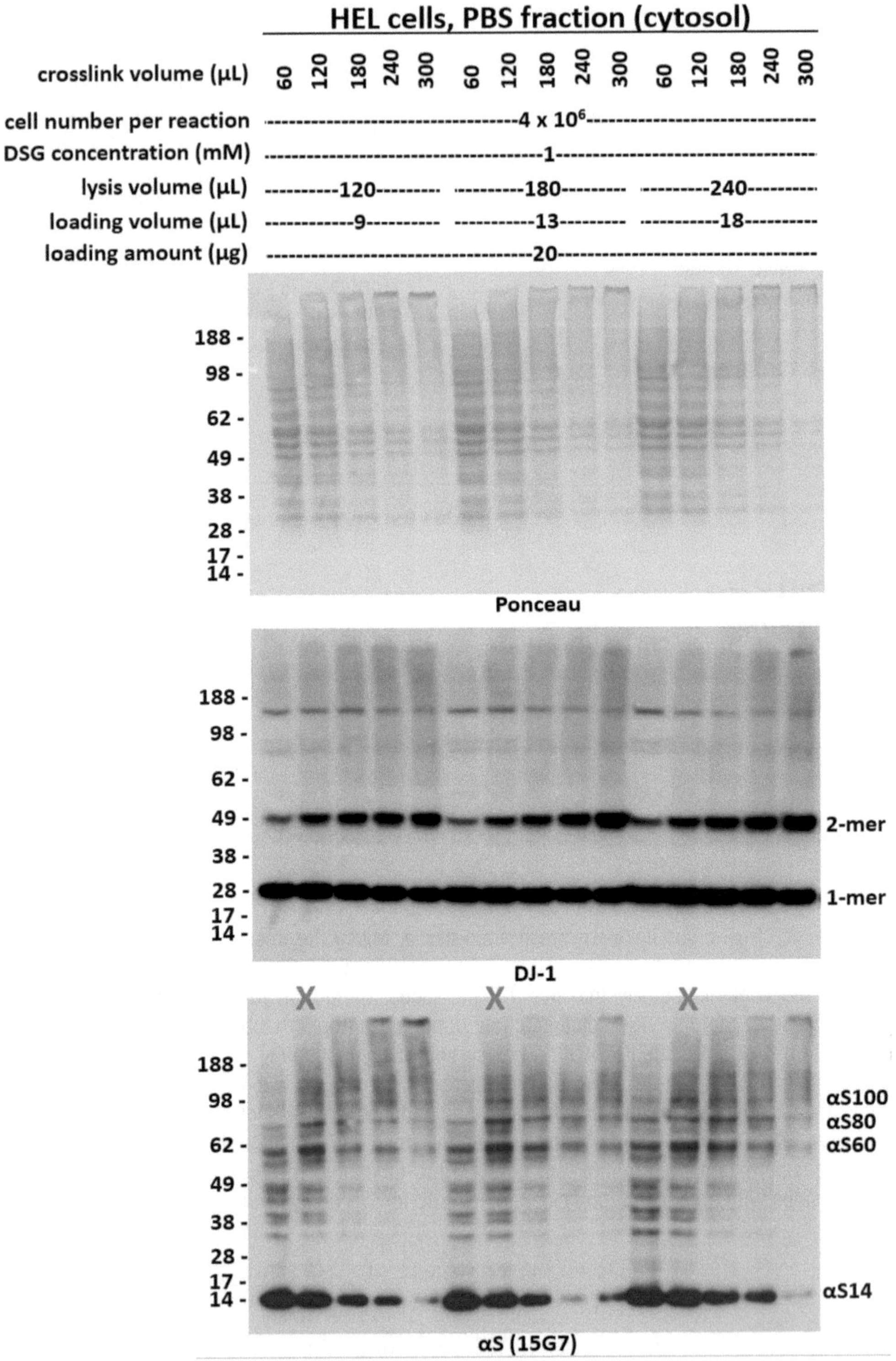

Fig. 4 Testing parameters of αS DSG cross-linking in HEL cells. Volume of the reaction during cross-linking, volume of the solution during lysis, and volume loaded were varied as indicated. Ponceau staining, DJ-1 WB, and αS WB (mAb 15G7) are shown. Optimal conditions are marked with an X. It is obvious that keeping the

8. 37 °C Incubator.
9. Nutating mixer.
10. Tris–HCl, pH 7.6.
11. Probe-type ultrasonicator (Sonic Dismembrator model 300 or similar) with microtip.

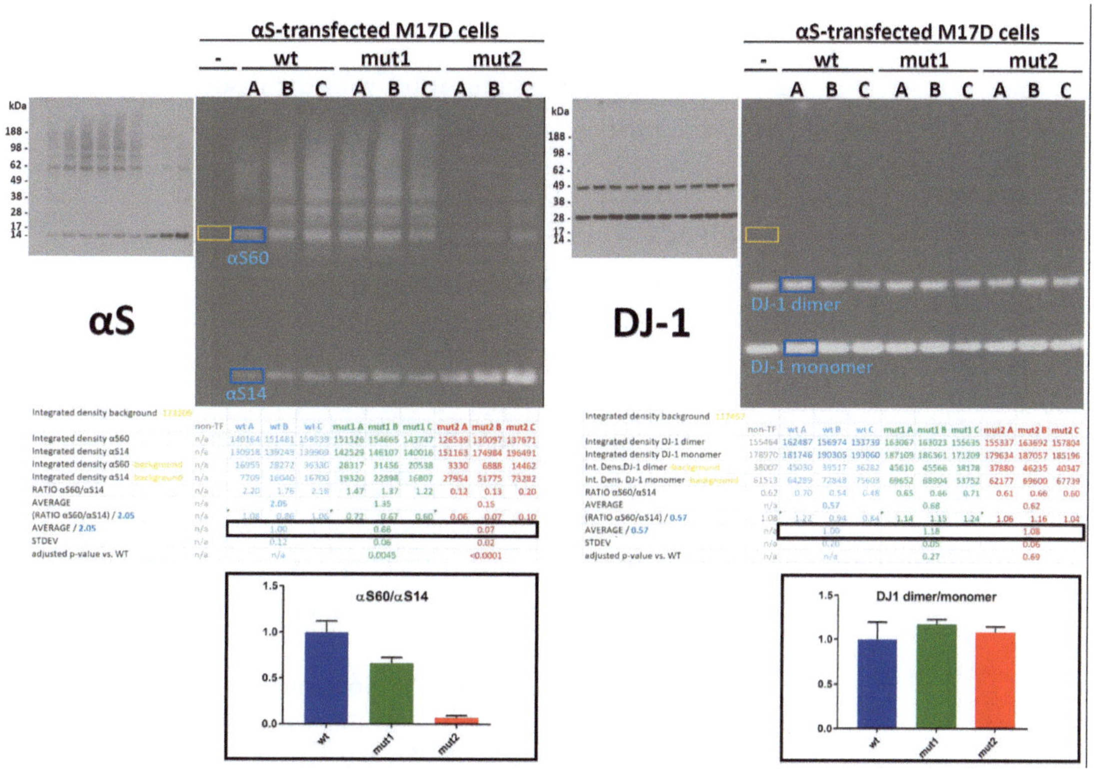

Fig. 5 Densitometrical and statistical analysis of αS cross-linking. Example: wt αS and mutants 1 and 2 (mut1, mut2) were transfected in triplicates into M17D neuroblastoma cells and intact cells underwent DSG cross-linking 48 h post-transfection ("-": vector-only control). Scanned WBs were "color inverted" and analyzed by ImageJ or a similar program. A background value (orange box) was subtracted from all values obtained for αS60 and αS14, and then αS60/αS14 ratios were calculated in Excel and normalized to 1 for the control samples. Similarly, DJ-1 dimer/monomer ratios were determined. While the mutants had significant αS60:αS14-lowering effects, no dimer:monomer changes for DJ-1 were observed. One-way ANOVA followed by Dunnett's post hoc test ($n = 3$); adjusted p-values are listed in the table

Fig. 4 (continued) DSG concentration at 1 mM alone is not sufficient for consistent results. Instead, the ratio of cell number and cross-linker amount needs to be tightly controlled. "Under-cross-linking" leads to false-positive monomers, while "over-cross-linking" traps nonspecific interactions and causes high molecular weight "smears" and gel exclusion of αS and many other proteins (*see* Ponceau staining; DJ-1 is unusually resistant to "over-cross-linking")

12. Ultracentrifuge (e.g., TLA-100 rotor) and 1.5 mL polypropylene tubes.
13. BCA assay: Bicinchoninic acid kit to determine protein concentration and 2.0 mg/mL bovine serum albumin (BSA) as a standard solution (*see* **Note 1**).

2.2 Intact-Cell Cross-Linking of Fresh Brain Tissue

In addition to the materials listed in Subheading 2.1:

1. McIlwain Tissue Chopper (model MTC/2E, Mickle Laboratory Engineering Co., Gomshall, UK; blade interval: 100 μm) or similar model.
2. Razor blade that is compatible with the tissue chopper.
3. 15 mL tubes (e.g., Falcon).

2.3 SDS-PAGE and Western Blot

1. 4× NuPAGE LDS sample buffer or similar.
2. NuPAGE 4–12% Bis-Tris gels.
3. NuPAGE MES-SDS running buffer.
4. SeeBlue Plus2 MW marker.
5. Immobilon-Psq 0.2 μm PVDF membrane (Millipore) or iBlot PVDF gel transfer stacks (Invitrogen).
6. Transfer buffer: 25 mM Tris, 192 mM glycine, 20% methanol.
7. Fixation solution: 0.4% Paraformaldehyde in PBS.
8. Ponceau solution: 0.1% Ponceau S in 5% acetic acid.
9. PBS-T: PBS containing 0.1% (v/v) Tween 20 (PBS-T).
10. 0.2% I-Block solution: PBS-T containing 0.2% (w/v) I-Block (Tropix).
11. Suitable antibodies to αS: Monoclonal Syn1 (Clone 42, Becton-Dickinson; 1:3000 in WB for transfectants, 1:400 in WB for endogenous αS); 2F12 (MABN1817, Millipore/Sigma; 0.09 μg per mL for transfectants and 0.9 μg per mL for endogenous); 15G7 [24] (hybridoma supernatants were 1:2000 on WB for transfectants, 1:200 for endogenous αS); polyclonal antibody C20 (Santa Cruz; 1:1000 for endogenous αS and 1:5000–1:10,000 for transfectants).
12. Suitable positive control antibodies: Anti DJ-1 antibody [25] (1:3000)—standard positive control (dimer); PA1-954A to VDAC (Affinity Bioreagents; 1:1000 on WB); VDAC is a dimeric transmembrane protein in mitochondria; DLP1 to DRP-1 (Becton-Dickinson; 1:1000 on WB), sc-126 to p53 (Santa Cruz, 1:1000 in WB), EP1537Y to β-synuclein (Novus Biologicals, 1:3000 on WB).
13. Suitable negative control antibodies: Anti-Ran (4462, Cell Signaling; 1:1000 on WB), anti-casein kinase 1α (Sc-6477, Santa Cruz; 1:2000 on WB), anti-Parkin (PRK8, Santa Cruz, 1:1000 on WB).

14. Horseradish peroxidase-conjugated secondary antibody (GE Healthcare) diluted 1:10,000 in 0.2% I-Block.
15. ECL Prime (GE Healthcare-Amersham Biosciences) or Super-Signal West Dura (Thermo Scientific).

2.4 Densitometry and Statistical Analysis

1. ImageJ or similar image analysis program.
2. GraphPad Prism or similar statistics program.

3 Methods

3.1 Intact-Cell Cross-Linking of Cultured Cells (Exemplified for 10 cm Dishes, Fig. 2)

1. Plate cells (neuroblastoma, primary neurons) in 10 cm dishes, and incubate overnight or longer. Suitable cell densities: neuroblastoma ~5 × 10^6 or primary neurons ~8 × 10^6 per 10 cm dish. If necessary, treat cells (transfection, drugs, stressors, other treatments), and incubate for an appropriate amount of time.
2. Estimate the volume of 1 mM DSG cross-linking solution that will be needed. Rule of thumb: 2 mL per confluent 10 cm dish of cultured cells.
3. Weigh out 1.6 mg DSG per 5 mL final cross-linking solution into 1.5 mL tube, and store until needed (*see* **Note 2**).
4. Aspirate media from the cell cultures, and add 2 mL PBS.
5. Lift off cells by trituration (or scraping, but only if necessary because less gentle), and transfer to four 1.5 mL tubes (*see* **Note 3**).
6. Centrifuge cells at 1600 × *g* for 2 min at RT.
7. While cells are centrifuging: Add 50 μL DMSO per 1.6 mg DSG followed by vortexing to obtain a 100× DSG solution. Important: Always add DMSO *immediately* before using: DSG solutions are unstable and cannot be stored! (*see* **Note 2**).
8. Dilute the 100× DSG solution 1:100 in PBS/PI (protease inhibitor mix, Roche) to obtain the 1× DSG solution (e.g., 50 μL in 5 mL). Mix well.
9. After centrifuging: Estimate cell pellet volume. Discard supernatant.
10. Save one of the four replicates as non-DSG control; optional: subject this sample to 2 mM DSP cross-linking followed by reductive cleavage using βME (*see* **Note 4**).
11. Add 1 mM DSG solution to the three remaining replicates. Use three different volumes: ~4×, 5×, and 6× the volume of the pellet.
12. Mix by vigorous shaking (manually, not by vortexing; *see* **Note 5**).

13. Incubate for 30 min at 37 °C while rotating.
14. Quench reaction by adding 50 mM (final concentration) Tris–HCl, pH 7.6, and incubate for 10 min at RT.
15. Lyse by sonication (Sonic Dismembrator 300; setting = 40; 15 s; *see* **Note 6**); alternative: add 1% (final concentration) Triton-X 100 (TX-100) for total protein lysates.
16. (Optional: Nuclear spin at 800 × *g* for 5 min, discard pellet).
17. Transfer to 1.5 mL polypropylene (ultracentrifuge) tube.
18. Ultracentrifuge at 100,000 × *g* for 30 min at 4 °C, and keep supernatants. Pellets of cells lysed in the absence of TX-100 may be lysed in 1% TX-100 to study membrane proteins (*see* **Note** 7).
19. Determine protein concentrations, e.g., BCA assay (*see* **Note 1**).
20. For each sample keep the replicate (1, 2, or 3) that is the closest to 1 μg/μL (0.9–1.2 is acceptable; ~1.0-1.6 for total protein lysates that were made in the presence of TX-100). Discard others or freeze for future analysis (*see* **Note 8**).
21. Add 1/3 volume of 4× NuPAGE LDS sample buffer. Boil samples (5–10 min). Proceed to SDS-PAGE (*see* Subheading 3.3).

3.2 Intact-Cell Cross-Linking of Fresh Brain Tissue (Fig. 3)

Exemplify for one mouse brain, 200–300 mg mass; modify for other samples.

1. Estimate the volume of 1 mM DSG cross-linking solution that will be needed. Rule of thumb: 1.5 mL per 100 mg of brain tissue.
2. Weigh out 1.6 mg DSG per 5 mL final cross-linking solution into 1.5 mL tube, and store until needed (*see* **Note 2**).
3. Mince fresh brain sample by two rounds on a McIlwain Tissue Chopper (model MTC/2E, Mickle Laboratory Engineering Co., Gomshall, UK; blade interval: 100 μm); turn sample 90° after the first round (two brains can be chopped in parallel).
4. Transfer minced samples to 15 mL tubes containing 5 mL PBS. Mix by shaking until the solution is homogeneous.
5. Transfer aliquots to five (or six) 1.5 mL tubes. Make sure (by repeated shaking) that tissue in the 15 mL tube does not settle while aliquoting. Spin at 1600 × *g* for 2 min at RT. Each tube contains 40–50 mg brain material.
6. While cells are centrifuging: Add 50 μL DMSO per 1.6 mg DSG followed by vortexing to obtain a 100× DSG solution. Important: Always add DMSO *immediately* before using: DSG solutions are unstable and cannot be stored! (*see* **Note 2**).

7. Dilute the 100× DSG solution 1:100 in PBS/PI (protease inhibitor mix, Roche) to obtain the 1× DSG solution (e.g., 50 μL in 5 mL). Mix well.
8. After centrifuging: Discard supernatants, and subject the pellets (intact tissue bits) to cross-linking at ratios of, e.g., 400, 600, 800, and 1000 μL 1 mM DSG cross-linking solution (PBS containing 1 mM DSG) per 50 mg tissue. Keep one sample as non-DSG control. Optional: Subject this sample to 2 mM DSP cross-linking followed by reductive cleavage using βME (*see* **Note 4**).
9. Incubate at 37 °C for 30 min with shaking.
10. Quench reactions by adding 50 mM (final concentration) Tris–HCl, pH 7.6, and incubate for 10 min at RT.
11. Centrifuge at 1600 × *g* for 2 min at RT. Discard supernatants, and resuspend pellets in PBS/PI, 25% of cross-linking solution, e.g., 100, 150, 200, and 250 μL per 50 mg tissue.
12. Lyse intact tissue by sonication (20″, amplitude 40) (*see* **Note 6**); alternative: add 1% (final concentration) TX 100 for total protein lysates, followed by sonication.
13. (Optional: Nuclear spin at 1000 × *g* for 5 min, and discard pellet). Transfer to 1.5 mL polypropylene (ultracentrifuge) tube.
14. Ultracentrifuge at 100,000 × *g* for 30 min at 4 °C, and keep supernatants (transfer to 1.5 mL tubes) (*see* **Note 7**).
15. Determine protein concentrations (e.g., BCA assay). (Optional: Pellets may be lysed in 1% TX-100 to study membrane proteins) (*see* **Note 1**).
16. To compare different brains, run replicates (1, 2, 3, or 4) with similar concentration (0.7–1.2 μg/μL is generally acceptable; other concentrations can be tested).
17. Boil samples (5–10 min) after adding 1/3 volume of 4× NuPAGE LDS sample buffer. Proceed to SDS-PAGE.

3.3 SDS-PAGE and Western Blot (Fig. 4)

1. Prepare an appropriate electrophoresis system, e.g., NuPAGE 4–12% Bis-Tris gels with NuPAGE MES-SDS running buffer and the SeeBlue Plus2 MW marker.
2. Load appropriate amounts of total protein per lane, e.g.: 20 μg per lane of NuPAGE 26-well gel or 30 μg per lane of 20-well gel.
3. Run gels at constant 100 volts (*see* **Note 9**).
4. After electrophoresis, transfer (electroblot) proteins onto Immobilon-Psq 0.2 μm PVDF membranes for 90 min at 400 mA constant current at 4 °C in transfer buffer. Alternatively, use a semidry fast transfer system such as the iBlot system

(Invitrogen) using PVDF gel transfer stacks and the P0 program for 7 min total (20 V for 1 min, 23 V for 4 min, and 25 V for 2 min).

5. After transfer, incubate membranes in 0.4% paraformaldehyde in PBS for 30 min at RT, rinse twice with PBS, stain with 0.1% Ponceau S in 5% acetic acid, rinse with water, and block in 0.2% I-Block solution (PBS containing 0.1% (v/v) Tween 20 (PBS-T) and 0.2% (w/v) I-Block) for either 30 min at RT or overnight at 4 °C (*see* **Note 10**).
6. After blocking, incubate membranes in primary antibody in 0.2% I-Block with 0.02% sodium azide in PBS-T for either 1 h at RT or overnight at 4 °C. Wash membranes 3 × 10 min in PBS-T at RT, and incubate (45 min at RT) in horseradish peroxidase-conjugated secondary antibody (GE Healthcare) diluted 1:10,000 in 0.2% I-Block in PBS-T. Then wash membranes 3 × 10 min in PBS-T, and develop with ECL Prime (GE Healthcare-Amersham Biosciences) or SuperSignal West Dura (Thermo Scientific) according to the manufacturer's instructions.

3.4 Densitometry and Statistical Analysis (Fig. 5)

Scan WB with a positive film setting on the scanner, "invert" the image, and analyze by ImageJ or similar program. Quantify αS60 and αS14 bands separately (subtract background), calculate the αS60/αS14 ratios, and normalize to 1 for the control samples (optional: do the same for αS80/αS14 or (αS60 + αS80 + αS100)/αS14). Similarly, quantify dimeric and monomeric bands of DJ-1 (or other control proteins) separately, calculate the dimer/monomer ratio, and normalize to 1 for the control samples. As stated previously, no difference for DJ-1 dimer/monomer ratio (or any other multimeric control protein) should be observed; otherwise the observed effects are not αS specific, and the data are probably inconclusive. The normalization to 1 serves the purpose of facilitating pooling data obtained on different days. Perform T-test or one-way ANOVA to compare the ratios, e.g., using GraphPad Prism 7 (ANOVA if >2 comparisons). Replicates analyzed on the same day are typically considered "n," while experiments done on different days are considered "N".

4 Notes

1. Alternative approaches to determine protein concentrations can be used, but linear range and potential interference with lysis buffer or cross-linking must be examined.
2. Store DSG at 4 °C, desiccated. Aliquoting can be done in advance, but DSG has to be dissolved in DMSO for each experiment right before application.

3. In-dish cross-linking (without detaching the cells) is generally possible if cells are grown as a monolayer. However, initial optimization of the method suggested that cross-linking of detached cells is more efficient.
4. For the non-cross-linked control, DSP cross-linking followed by its reduction with 4% βME helps monomers attach to membranes by neutralizing the positive charges of lysine [15].
5. An efficient strategy is to cover the rack containing the tubes with another rack, and move ten times up and down fast.
6. We used a Sonic Dismembrator Model 300 (Fisher Scientific, USA) at 40% power. Begin the sonication with the microtip submerged at the bottom of the tube. Each stroke consists of a downward movement of the tube to bring the sonicator microtip almost to the surface of the liquid followed by an upward movement to return to the original tube position and should last approximately 1 s. Removing and reinserting the sonicator microtip from the solution will create unwanted bubbles. The microtip should be wiped between samples.
7. To save time, if an ultracentrifuge is unavailable, or if sample numbers are very high, centrifugation using a standard benchtop centrifuge for 30 min at 20,000 × *g* is an alternative.
8. 1 μg/μL protein concentration when lysing directly in the 1 mM DSG cross-linking solution followed by ultracentrifugation has consistently been correlating with good cross-linking results (neither "under-" nor "over-cross-linking"). For tabletop centrifugation, this value is closer to 1.3 μg/μL. When lysis is in the presence of a detergent, the values will increase by 20–30%. While we assume that these values might be similar for a large variety of cell types, it may be necessary to determine the correct parameters for each system analogous to Fig. 4.
9. Running the gel as far as possible results in a better separation of the αS60, αS80 and αS100, making the bands easier to quantify by densitometry.
10. Ponceau staining serves the purpose of checking for equal cross-linking and equal loading of the samples. Stained membranes can be scanned and presented as a control.

References

1. Spillantini MG, Schmidt ML, Lee VM, Trojanowski JQ, Jakes R, Goedert M (1997) Alpha-synuclein in Lewy bodies. Nature 388:839–840. https://doi.org/10.1038/42166
2. Polymeropoulos MH, Lavedan C, Leroy E, Ide SE, Dehejia A, Dutra A, Pike B, Root H, Rubenstein J, Boyer R, Stenroos ES, Chandrasekharappa S, Athanassiadou A, Papapetropoulos T, Johnson WG, Lazzarini AM, Duvoisin RC, Di Iorio G, Golbe LI, Nussbaum RL (1997) Mutation in the alpha-synuclein gene identified in families with Parkinson's disease. Science 276:2045–2047
3. Krüger R, Kuhn W, Müller T, Woitalla D, Graeber M, Kösel S, Przuntek H, Epplen JT,

Schöls L, Riess O (1998) Ala30Pro mutation in the gene encoding alpha-synuclein in Parkinson's disease. Nat Genet 18:106–108. https://doi.org/10.1038/ng0298-106

4. Zarranz JJ, Alegre J, Gómez-Esteban JC, Lezcano E, Ros R, Ampuero I, Vidal L, Hoenicka J, Rodriguez O, Atarés B, Llorens V, Gomez Tortosa E, del Ser T, Muñoz DG, de Yebenes JG (2004) The new mutation, E46K, of alpha-synuclein causes Parkinson and Lewy body dementia. Ann Neurol 55:164–173. https://doi.org/10.1002/ana.10795
5. Singleton AB, Farrer M, Johnson J, Singleton A, Hague S, Kachergus J, Hulihan M, Peuralinna T, Dutra A, Nussbaum R, Lincoln S, Crawley A, Hanson M, Maraganore D, Adler C, Cookson MR, Muenter M, Baptista M, Miller D, Blancato J, Hardy J, Gwinn-Hardy K (2003) Alpha-Synuclein locus triplication causes Parkinson's disease. Science 302:841. https://doi.org/10.1126/science.1090278
6. Fuchs J, Tichopad A, Golub Y, Munz M, Schweitzer KJ, Wolf B, Berg D, Mueller JC, Gasser T (2008) Genetic variability in the SNCA gene influences alpha-synuclein levels in the blood and brain. FASEB J Off Publ Fed Am Soc Exp Biol 22:1327–1334. https://doi.org/10.1096/fj.07-9348com
7. Bartels T, Choi JG, Selkoe DJ (2011) α-Synuclein occurs physiologically as a helically folded tetramer that resists aggregation. Nature 477:107–110. https://doi.org/10.1038/nature10324
8. Wang W, Perovic I, Chittuluru J, Kaganovich A, Nguyen LTT, Liao J, Auclair JR, Johnson D, Landeru A, Simorellis AK, Ju S, Cookson MR, Asturias FJ, Agar JN, Webb BN, Kang C, Ringe D, Petsko GA, Pochapsky TC, Hoang QQ (2011) A soluble α-synuclein construct forms a dynamic tetramer. Proc Natl Acad Sci U S A 108:17797–17802. https://doi.org/10.1073/pnas.1113260108
9. Dettmer U, Newman AJ, Luth ES, Bartels T, Selkoe D (2013) In vivo cross-linking reveals principally oligomeric forms of α-synuclein and β-synuclein in neurons and non-neural cells. J Biol Chem 288:6371–6385. https://doi.org/10.1074/jbc.M112.403311
10. Gurry T, Ullman O, Fisher CK, Perovic I, Pochapsky T, Stultz CM (2013) The dynamic structure of α-synuclein multimers. J Am Chem Soc 135:3865–3872. https://doi.org/10.1021/ja310518p
11. Wang L, Das U, Scott DA, Tang Y, McLean PJ, Roy S (2014) α-synuclein multimers cluster synaptic vesicles and attenuate recycling. Curr Biol CB 24:2319–2326. https://doi.org/10.1016/j.cub.2014.08.027
12. Burre J, Sharma M, Sudhof TC (2014) Alpha-Synuclein assembles into higher-order multimers upon membrane binding to promote SNARE complex formation. Proc Natl Acad Sci U S A 111:E4274–E4283. https://doi.org/10.1073/pnas.1416598111
13. Gould N, Mor DE, Lightfoot R, Malkus K, Giasson B, Ischiropoulos H (2014) Evidence of native α-synuclein conformers in the human brain. J Biol Chem 289:7929–7934. https://doi.org/10.1074/jbc.C113.538249
14. Iljina M, Tosatto L, Choi ML, Sang JC, Ye Y, Hughes CD, Bryant CE, Gandhi S, Klenerman D (2016) Arachidonic acid mediates the formation of abundant alpha-helical multimers of alpha-synuclein. Sci Rep 6:33928. https://doi.org/10.1038/srep33928
15. Newman AJ, Selkoe D, Dettmer U (2013) A new method for quantitative Immunoblotting of endogenous α-synuclein. PLoS One 8:e81314. https://doi.org/10.1371/journal.pone.0081314
16. Dettmer U, Newman AJ, Soldner F, Luth ES, Kim NC, von Saucken VE, Sanderson JB, Jaenisch R, Bartels T, Selkoe D (2015) Parkinson-causing α-synuclein missense mutations shift native tetramers to monomers as a mechanism for disease initiation. Nat Commun 6:7314. https://doi.org/10.1038/ncomms8314
17. Dettmer U, Newman AJ, von Saucken VE, Bartels T, Selkoe D (2015) KTKEGV repeat motifs are key mediators of normal α-synuclein tetramerization: their mutation causes excess monomers and neurotoxicity. Proc Natl Acad Sci U S A 112:9596–9601. https://doi.org/10.1073/pnas.1505953112
18. Westphal CH, Chandra SS (2013) Monomeric synucleins generate membrane curvature. J Biol Chem 288:1829–1840. https://doi.org/10.1074/jbc.M112.418871
19. Dettmer U, Ramalingam N, von Saucken VE, Kim T-E, Newman AJ, Terry-Kantor E, Nuber S, Ericsson M, Fanning S, Bartels T, Lindquist S, Levy OA, Selkoe D (2017) Loss of native α-synuclein multimerization by strategically mutating its amphipathic helix causes abnormal vesicle interactions in neuronal cells. Hum Mol Genet 26:3466–3481. https://doi.org/10.1093/hmg/ddx227
20. Kim S, Yun SP, Lee S, Umanah GE, Bandaru VVR, Yin X, Rhee P, Karuppagounder SS, Kwon S-H, Lee H, Mao X, Kim D, Pandey A, Lee G, Dawson VL, Dawson TM, Ko HS (2018) GBA1 deficiency negatively affects

physiological α-synuclein tetramers and related multimers. Proc Natl Acad Sci U S A 115:798–803. https://doi.org/10.1073/pnas.1700465115

21. Kamada R, Nomura T, Anderson CW, Sakaguchi K (2011) Cancer-associated p53 tetramerization domain mutants: quantitative analysis reveals a low threshold for tumor suppressor inactivation. J Biol Chem 286:252–258. https://doi.org/10.1074/jbc.M110.174698
22. Gaglia G, Guan Y, Shah JV, Lahav G (2013) Activation and control of p53 tetramerization in individual living cells. Proc Natl Acad Sci U S A 110:15497–15501. https://doi.org/10.1073/pnas.1311126110
23. Manning LR, Russell JE, Padovan JC, Chait BT, Popowicz A, Manning RS, Manning JM (2007) Human embryonic, fetal, and adult hemoglobins have different subunit interface strengths. Correlation with lifespan in the red cell. Protein Sci Publ Protein Soc 16:1641–1658. https://doi.org/10.1110/ps.072891007
24. Kahle PJ, Neumann M, Ozmen L, Muller V, Jacobsen H, Schindzielorz A, Okochi M, Leimer U, van Der Putten H, Probst A, Kremmer E, Kretzschmar HA, Haass C (2000) Subcellular localization of wild-type and Parkinson's disease-associated mutant alpha -synuclein in human and transgenic mouse brain. J Neurosci 20:6365–6373
25. Baulac S, LaVoie MJ, Strahle J, Schlossmacher MG, Xia W (2004) Dimerization of Parkinson's disease-causing DJ-1 and formation of high molecular weight complexes in human brain. Mol Cell Neurosci 27:236–246. https://doi.org/10.1016/j.mcn.2004.06.014

Chapter 9

C. elegans as a Model for Synucleinopathies and Other Neurodegenerative Diseases: Tools and Techniques

Mandy Koopman, Renée I. Seinstra, and Ellen A. A. Nollen

Abstract

Caenorhabditis elegans is widely used to investigate biological processes related to health and disease. Multiple *C. elegans* models for human neurodegenerative diseases do exist, including those expressing human α-synuclein. Even though these models do not feature all pathological and molecular hallmarks of the disease they mimic, they allow for the identification and dissection of molecular pathways that are involved. In line with this, genetic screens have yielded multiple modifiers of proteotoxicity in *C. elegans* models for neurodegenerative diseases. Here, we describe a set of common screening approaches and tools that can be used to study synucleinopathies and other neurodegenerative diseases in *C. elegans*. RNA interference and mutagenesis screens can be used to find genes that affect proteotoxicity, while relatively simple molecular, cellular (fractionation studies), metabolic (respiration studies), and behavioral (thrashing and crawling) readouts can be used to study the effects of disease proteins and modifiers more closely.

Key words *C. elegans*, Misfolding diseases, Neurodegenerative diseases, Parkinson disease, α-Synuclein, Forward genetic screens, RNA interference, Fractionation, Thrashing, Phenotyping

1 Introduction

The nematode *Caenorhabditis elegans* has become an important tool in biomedical research [1–7]. Longevity and phenotypic studies have led to major discoveries in neurodegenerative diseases, but also revealed regulators in various processes, e.g., programmed cell death (as reviewed in [8]). The genetics of *C. elegans* are well established; it was the first multicellular organism to have its complete genome sequenced. This genetic information plus the ease of manipulation of gene expression makes *C. elegans* a powerful organism for genome-wide genetic screens. Indeed, genetic studies are widely used in *C. elegans* to discover gene function [9–12]. At least 83% of the *C. elegans* proteome is predicted, by pairwise comparisons, to have human homolog genes [13]. A more recent study estimated that 38% of the *C. elegans* protein-coding genes have a unique functional ortholog in humans [14]. Thus, biological

Tim Bartels (ed.), *Alpha-Synuclein: Methods and Protocols*, Methods in Molecular Biology, vol. 1948,
https://doi.org/10.1007/978-1-4939-9124-2_9, © Springer Science+Business Media, LLC, part of Springer Nature 2019

processes unraveled in the small nematode can actually provide insight into human biology. Another important advantage of *C. elegans* is that it is amenable to generate "humanized" models of diseases. Worms that mimic for example Parkinson's disease (PD)—expressing human α-synuclein—do already exist [4, 15–20]. Although these models normally do not feature all the pathological and clinical aspects of the disease, they do provide a means to gain insight into molecular processes contributing to the pathology. Forward genetic screens in these models have yielded multiple modifiers of proteotoxicity, which were also found to be reproducible in higher model organisms [6, 12].

Here, we describe a set of screening approaches and tools (Fig. 1) that can be used to study synucleinopathies in our *C. elegans* model for PD. In this model, human α-synuclein fused to yellow fluorescent protein (YFP) is expressed under the control of the *unc-54* promotor (strain: OW40 or NL5901) [6]. The expression of α-synuclein in muscle tissue provides us with multiple advantages; muscle cells are large, which allows us to visually (due to YFP) detect α-synuclein expression and localization. Moreover, RNA interference (RNAi) by means of feeding works more efficiently in muscles than in neurons, making such a muscular model more amenable for RNAi screens. Importantly, even though expression of disease proteins may not mimic the human situation (in which mainly neurons are affected), previous studies have shown that transgene expression in the muscle can be successfully used to identity evolutionarily conserved modifier genes and to model misfolding diseases [1, 4, 21].

In the first method sections, we describe two methods (RNAi and EMS) to perform genome-wide screens. Within the third method we focus on behavior that can be used as a screening phenotype, namely crawling and thrashing (the last one is also known as "swimming" or "motility") behavior. Importantly, this behavior correlates very well with toxicity [11, 22] and therefore also provides a mean of characterizing and exploring toxicity in novel models for (neurodegenerative) diseases. Finally, we describe a fractionation method to assess nuclear versus cytosolic localization of specific proteins in *C. elegans.*

Importantly, multiple other techniques do exist to study aggregation-prone proteins like α-synuclein in *C. elegans.* Even though we do not describe these methods specifically here, they are worthwhile to mention. One of these techniques is fluorescence recovery after photobleaching (FRAP). For example, FRAP can be used to address whether α-synuclein is aggregated within inclusions (immobile) or not (mobile). In fact, FRAP has already been successfully used in *C. elegans* as described in [6]. In addition, recently a technique has been developed to measure mitochondrial respiration in *C. elegans* with Seahorse XF Analyzers. It is becoming more and more apparent that a lot of disease proteins associated with

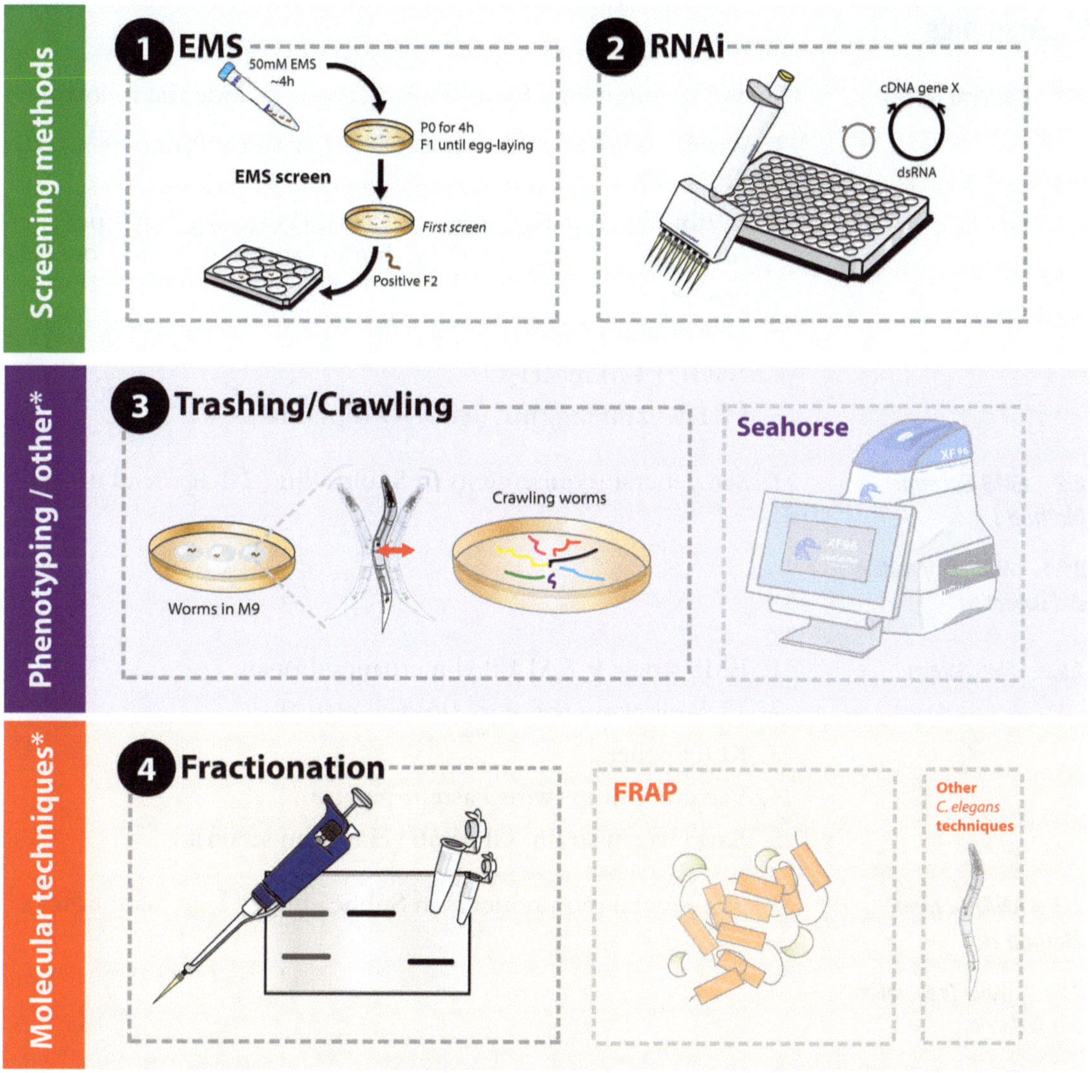

Fig. 1 An overview of the described methods. Multiple methods are described in this chapter. Methods I and II are screening approaches (EMS and RNAi), method III is a phenotypic method (thrashing and crawling), and method IV is a molecular technique (cellular fractionation). The overview shows also two other techniques, namely measuring mitochondrial respiration with Seahorse XF Analyzers and FRAP, two techniques that are only shortly mentioned. *Some illustrations are transparent: these methods are not described in detail in this chapter

neurodegenerative diseases are interacting with mitochondria [23–25]. In fact, mitochondrial dysfunction is associated with a broad spectrum of diseases [26–29]. Therefore, screening and testing worms for mitochondrial respiration (a functional parameter for mitochondrial health) may be a novel way to assess (neurodegenerative) diseases. A detailed protocol is described by Koopman et al. [30].

2 Materials

2.1 General Notes

General requirements for all four methods include the following:

1. Worm strains: For example OW40 (zgIs15[P(unc-54)::asyn::YFP] IV).
2. M9 buffer: 3 g KH_2PO_4, 6 g Na_2HPO_4, 5 g NaCl per liter. After autoclaving, cool the buffer and add 1 mL of 1 M $MgSO_4$.
3. 13% Sodium hypochloride.
4. NaOH: 4 M in dH_2O.
5. FUDR: 200 mg/mL in dH_2O (optional).

2.2 EMS Screen: Method I

2.2.1 Worm Preparation and Collection

1. See general requirements in Subheading 2.1 (general notes).

2.2.2 EMS Screen

1. EMS stock: 9.7 M Ethyl methanesulfonate.
2. 12-Well plates (or, e.g., 96-well plates).
3. KOH pellets.
4. Vacuum pump (with Pasteur pipette).
5. Extra worm strain: CB4856 (Hawaiian strain).

2.3 RNAi Screen: Method II

2.3.1 Worm Preparation and Collection

1. See general requirements in Subheading 2.1 (general notes).

2.3.2 RNAi Screen

1. RNAi library: For example Ahringer [31] or Vidal [32].
2. Control RNAi: *Ht115* with an empty vector: L4440.
3. 96-deep-well plates.
4. 12-Well plates (or, e.g., 6-well, 24-well).
5. Isopropyl β-D-1-thiogalactopyranoside (IPTG): 12.5 mg/mL in dH_2O.
6. Ampicillin: 50 mg/mL in dH_2O.

2.4 Trashing (and Crawling): Method III

1. See general requirements in Subheading 2.1 (general notes).

2.4.1 Worm Preparation and Collection

2.4.2 Thrashing and Crawling

1. Microscope (*manual).*
2. Mechanical palm click counter *(manual).*
3. Timer *(manual).*
4. GS3-U3-41C6C 1 Grasshopper USB 3.0 monochrome camera (Point Grey, Richmond, CA; 24 bits; 2048 × 2048 pixels) with 16 mm focal length high-resolution lens f/1.8–f/16 under bright-field illumination (8″ × 8″ white AI side-fired backlight) (Edmund Optics Ltd.) [33] *(automated).*
5. Tracking software: Wide Field-of-View Nematode Tracking Platform (WF-NTP). Software is available as download: see Perni et al. [33] *(automated).*
6. Recording software: Fly Capture Software Development Kit (SDK) USB 3.0. (Point Grey Research Inc., Richmond, Canada) *(automated).*

2.5 Nuclear Versus Cytosolic Fractionation: Method IV

1. See general requirements in Subheading 2.1 (general notes).
2. PBS.

2.5.1 Worm Preparation and Collection

2.5.2 Fractionation

1. Hypotonic buffer: 15 mM Hepes KOH pH 7.6, 10 mM KCl, 5 mM $MgCl_2$, 0.1 mM EDTA, 350 mM sucrose.
2. Hypertonic buffer: 15 mM Hepes KOH pH 7.6, 400 mM KCl, 5 mM $MgCl_2$, 0.1 mM EDTA, 0.1% Tween 20, 10% glycerol.
3. DTT: 1 M in dH_2O.
4. Protease inhibitors: 25× in dH_2O.
5. Pellet pestle.
6. Benzonase nuclease (optional).

2.5.3 12% SDS Polyacrylamide Gel

1. Ammonium persulfate (APS): 10% Solution in dH_2O.
2. *N,N,N,N′*-Tetramethyl-ethylenediamine (TEMED).
3. SDS: 10% in dH_2O.
4. 40% Acrylamide/Bis solution (29:1).
5. Resolving gel buffer: 1.5 M Tris–HCl, pH 8.8.
6. Stacking gel buffer: 0.5 M Tris–HCl, pH 6.8.
7. Mini PROTEAN® 3 System glass plates.

2.5.4 Immunoblotting

1. Running buffer: 0.025 M Tris–HCl, 0.192 M glycine, 0.1% SDS.
2. Transfer buffer: 0.025 M Tris–HCl, 0.192 M glycine, 10% MeOH.
3. SDS lysis buffer (5×): 250 mM Tris–HCl (1 M, pH 6.8), 8% SDS, 0.1% bromophenol blue, 40% glycerol (80%, v/v), 7 μL/mL β-mercaptoethanol.
4. Nitrocellulose membranes.
5. PBS-T: PBS containing 0.01% Tween-20.
6. Blocking solution: 5% Milk in PBS-T.

2.5.5 Antibodies

1. Nuclear marker: LMN-1 antibody, Novus Biomedicals (Cat. 38530002).
2. Cytosolic marker: Monoclonal anti-α-tubulin, Sigma (Cat. T6074).
3. YFP antibody: Anti-GFP, Abcam (Cat. ab6556)/living colors® A.v. monoclonal antibody (JL-8), ClonTech (632381).
4. Anti-α-synuclein: syn204, Cell Signaling (Cat. #2647S)/α-synuclein mouse mAb (clone LB509), concentrate, Invitrogen (180215).
5. HRP-conjugated goat-anti-rabbit.
6. HRP-conjugated goat-anti-mouse.

3 Methods

3.1 General Notes

For equipment and protocols related to maintenance and growth of *C. elegans* (i.e., nematode growth medium (NGM) plates, FUDR plates) we kindly refer to Stiernagle [34]. The same accounts for basics related to worm development and larval stages [35].

3.2 EMS Screen: Method I

In *C. elegans*, genetic screens are widely used to either discover gene function or find genes that modify or suppress a specific phenotype. Screens have been performed in our *C. elegans* model for PD, using inclusion number and motility (*see* also method III, Subsection 3.4 in this chapter) as readouts [6, 11]. These screens were, however, RNAi screens. One of the first methods to mutate the genome of *C. elegans* was based on a treatment with the mutagen ethyl methanesulfonate (EMS) [36]. The mutagen induces mutations in the sperm and oocytes of hermaphrodites [9]. In a typical EMS screen, thousands of copies of any particular gene are assessed. With a frequency of one null mutation at any particular locus of 1 in 2000 copies, EMS screens are relatively efficient. It is very important to have a consistent, strong, and clearly visible phenotype to screen for. Screening can be very labor intensive, and having a weak phenotype to screen for will definitely add to the complexity.

3.2.1 Worm Preparation

1. Grow worms on multiple 9 cm NGM plates (*see* **Note 1**).
2. When plates are relatively full with adult worms (not starved) collect worms in a Falcon tube by washing them off the plates with M9 buffer. Wash (i.e. spin the worms down in a centrifuge: 20 °C, ~500 × *g* for 2 min, remove the supernatant and add fresh M9 buffer) several times until the supernatant is clear.
3. Bleach worms (bleaching time should be adjusted to individual worm strains (*see* **Note 2**)) by adding 1 mL 13% hypochlorite:4 M NaOH (750:500) solution to a 2 mL suspension of worms in M9.
4. Hatch worms overnight in 12-well plates in M9 for approximately 24 h.
5. Day 1: Pipette ~800 L1 worms per 9 cm NGM plate: two plates in total (*see* **Note 3**). It is recommended to plate the worms late in the afternoon (*see* **Note 4**).
6. Day 4: Check whether worms are in the L4 stage (*see* **Note 5**). If they are, continue with the next steps.

3.2.2 EMS Screen

1. Collect L4-staged worms with M9 in a Falcon tube and wash (i.e. spin the worms down in a centrifuge: 20 °C, ~500 × *g* for 2 min, remove the supernatant and add fresh M9 buffer) them until the supernatant is clear.
2. Resuspend the worms in 4 mL M9. Add 21 μL EMS (9.7 M, final concentration: 50 mM) to the worm suspension and incubate the solution for 4 h at room temperature (use a tube roller to ensure constant mixing). IMPORTANT: Parafilm the tubes. EMS is highly toxic! Make sure to wear protective gear (*see* **Note 6**).
3. After 4 h, wash the worms twice with M9. EMS waste should be collected in separate Falcon tubes. EMS waste is incubated overnight with pellets of KOH (in dH_2O; add multiple pellets per Falcon tube) before it can be discarded.
4. Transfer EMS-treated worms to 9 cm NGM plates (10–80×) and keep them at 20 °C overnight (*see* **Note 7**).
5. Day 5: Transfer 100–800 worms (P_0) to 9 cm NGM plates (10 worms/per plate). Worms are allowed to lay eggs for 4–16 h (*see* **Note 8**).
6. Day 5/6: Check whether there are eggs and L1 worms (F_1) on the plates. If there are eggs and/or L1 worms remove the P_0s from the plates with a vacuum pump (*see* **Note 9**) and allow the larvae/eggs to develop for 4 days at 20 °C.
7. Day 9/10: Check the worms daily; after one day of egg laying the F_1 parents and the hatched F_2 progeny are washed off with M9. Eggs (F_2) will stick to the plate (*see* **Note 10**).

8. Day X: Depending on the age of interest, F_2 worms will be tested at a specific age and thus day for a phenotype (*see* **Note 11**). As mentioned before, this phenotype should be strong, consistent, easily screenable, and visible before or within the reproductive phase of the worm strain (e.g., thrashing ability/ crawling ability in models that are paralyzed, aggregate counts).
9. Worms that show a reversal of the screenable phenotype are transferred to individual wells of a 12-well plate. Annotate each mutant in such a way that it is clear which individuals did arise from the same 9 cm NGM plate (and thus might have the same mother and mutation).
10. Let the F2 worms lay eggs for 6 h (F_3). Remove the F_2 worms with a vacuum pump.
11. Rescreen the F_3 worms, those that still show a reversal of the phenotype of interest are kept (character breeds through). Only one mutant per plate is used for further studies (*see* **step 9**, **Note 12**). Freeze the mutants as soon as possible.
12. From here on, the researcher can start experiments with the generated mutants. However, deep sequencing is required to find the mutated gene. To make the mapping of point mutations induced by EMS easier, one can use the SNP diversity between the initial strain (OW40, which has a N2 background) and the Hawaiian *C. elegans* strain CB4856, as described in [37, 38]. One also needs to cross the transgene-containing strain with CB4856 (outcrossed for at least ten generations).

3.3 RNAi Screen: Method II

Very similar to EMS screens, RNAi screens can be used to identify suppressors or enhancers of specific phenotypes. In contrast to EMS treatment, RNAi can be directed towards known and specific genes. It is therefore not necessary to perform a genome-wide screen if a researcher has a group of genes of interest to investigate. By feeding *C. elegans* with an *E. coli* strain expressing target gene dsRNA, it is possible to knock down genes in a systematic way. Clear advantages and disadvantages of EMS and RNAi screens in *C. elegans*, and comparisons between the two, are described in Sin et al. [39]. In short, because RNAi does not alter the genome it is possible to identify roles of essential genes in a post-developmental process. In addition, because the candidate gene is known, identification of the gene—such as deep sequencing in EMS—is not required. Finally, as stated before, dsRNA has limited penetrance in neurons and RNAi has limited efficiency if the protein that is encoded by the candidate gene is very stable over time [39].

3.3.1 Plate Preparation

1. When performing a genome-wide RNAi screen, one has to prepare enormous amounts of 12-well plates (around

500–600 per chromosome). It is highly recommended to split the experiment in smaller parts (*see* **Note 13**). Make sure that there is ~2 mL of NGM medium (with 50 μg/mL ampicillin and 15 μg/mL IPTG) per well (the use of a pump is advised) (*see* **Note 14**).

2. Grow *E. coli* strains from the RNAi library. Suggestion: When using the Ahringer library, clones are stored in 384-well format, by using a 96-well pin replicator one can easily transfer bacteria to 96-deep-well plates. Add 900 μL LB medium with ampicillin (100 μg/mL) per well (*see* **Note 15**) and then use a pin replicator to transfer bacteria from the library to the 96-well plates.
3. Grow the bacteria overnight at 37 °C on a shaker (200 rpm). Also grow a control *E. coli* strain: *Ht115* with an empty vector (L4440).
4. Concentrate the bacteria by spinning the plates at 3500 × *g* for 20 min. Remove the supernatant partially (8–10× concentrated stocks are recommended). Resuspend each pelleted *E. coli* clone.
5. Add 40 μL of each bacterial clone to an individual well from a NGM-filled 12-well plate. Suggestion: We tend to include one control per 12-well plate. L4440 is plated in A1.
6. Let the plates dry at room temperature for 2 days. Keep the plates in the dark; IPTG and ampicillin are light sensitive.

3.3.2 Worm Preparation and the RNAi Screen

1. Grow worms on multiple 9 cm NGM plates (*see* **Note 1**).
2. When plates are relatively full with adult worms (not starved), collect worms in a Falcon tube by washing them off the plates with M9 buffer. Wash (i.e. spin the worms down in a centrifuge: 20 °C, ~500 × *g* for 2 min, remove the supernatant and add fresh M9 buffer) several times until the supernatant is clear.
3. Bleach worms (bleaching time should be adjusted to individual worm strains (*see* **Note 2**)) by adding 1 mL 13% hypochlorite:4 M NaOH (750:500) solution to a 2 mL suspension of worms in M9.
4. Hatch worms overnight in 12-well plates in M9 for approximately 24 h.
5. Pipette ~20–30 L1 worms per well in a 12-well plate (*see* **Note 16**). The number of plates depends on the number of RNAi clones to be tested at the same time.
6. Grow and age the worms until they have the appropriate age (*see* **Note 11**).
7. Test the worms for a specific phenotype: for example flood the wells with M9 and look at the thrashing behavior of the worms

(*see* method III: Subsecction 3.4), or put the worms under a fluorescence microscope and check whether the number of α-synuclein aggregates (YFP signal) is reduced compared to the control.

8. Write down the RNAi clones that show the effect you are looking for. Always rescreen worms on these RNAi clones (use for example 9 cm plates) to ensure that the effect is real and reproducible (*see* **Note 17**).

3.4 Thrashing (and Crawling): Method III

A decline in muscle function is a common characteristic observed in aging *C. elegans* [40, 41]. Interestingly, a decreased protein quality control system or the expression of aggregation-prone proteins appears to accelerate this age-related decline in muscle function [11, 22]. Thrashing assays are a common way to examine muscle function in *C. elegans* [6, 11]. When transferring worms to a drop of liquid, they directly start swimming, making characteristic body bends (Fig. 2a) that can be counted. Next to thrashing behavior, we also describe a way to assess crawling behavior. It has been shown that the thrashing behavior is distinct from crawling behavior in both kinematics and pattern of muscle activity [42]. In other words, assessing crawling behavior may add to the characterization of worm models and/or may provide a novel phenotype to use as screening readout.

3.4.1 Worm Preparation and Collection

1. Grow worms on multiple 9 cm NGM plates (*see* **Note 1**).
2. When plates are relatively full with adult worms (not starved), collect worms in a Falcon tube by washing them off the plates with M9 buffer. Wash (i.e. spin the worms down in a centrifuge: 20 °C, ~500 × *g* for 2 min, remove the supernatant and add fresh M9 buffer) several times until the supernatant is clear.

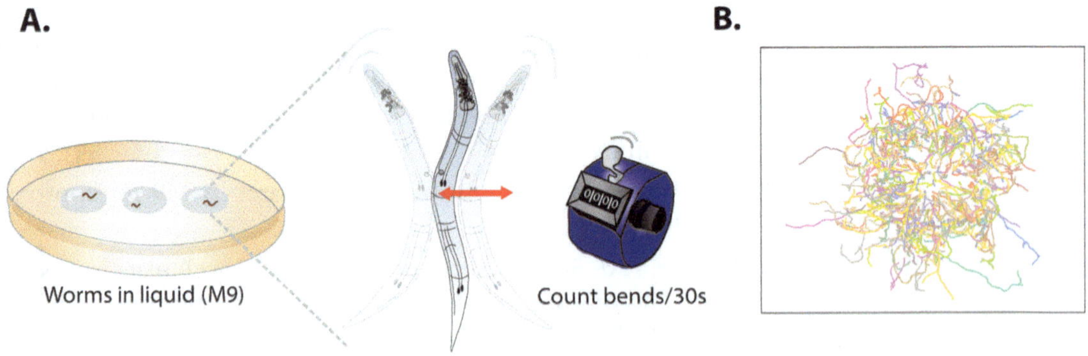

Fig. 2 Thrashing and crawling visualized. (**a**) Thrashing (also swimming or motility) is calculated by counting the C-shapes. Every time a worm bends towards one side (with both its head and tail) a C-shape is formed which is counted as a "bend." Automated software counts the same shapes. (**b**) A map (track.p file) which summarized 10 min of crawling. Every line represents the track of an individual worm. Worms were plated in the middle of a 9 cm NGM plate and allowed to crawl for 10 min (3 fps, 1800 frames)

3. Bleach worms (bleaching time should be adjusted to individual worm strains (*see* **Note 2**)) by adding 1 mL 13% hypochlorite:4 M NaOH (750:500) solution to a 2 mL suspension of worms in M9.
4. Hatch worms overnight in 12-well plates in M9 for approximately 24 h.
5. Pipette or manually transfer approximately ~800 L1 worms per 9 cm NGM plate or 10–150 per 6 cm plate (*see* **Note 18**).
6. Age the worms at 20 °C until they have the appropriate age (depending on the hypothesis and/or research question) (*see* **Note 19**). Make sure to transfer worms to FUDR plates at adulthood D1 when studying older worms (*see* **Note 11**).

3.4.2 Thrashing

1. For the **manual thrashing assay** procedure continue from **steps 2** to **4**; for the **automated thrashing assay** procedure go to **step 5** (*see* **Note 20**).
2. *Start manual:* Blind the different plates, to avoid an observer bias.
3. Take a test plate (6 or 9 cm NGM plate without *E. coli*) and add a drop of M9 buffer (~3–10 μL). Transfer one worm with a platinum wire into the drop of M9.
4. Allow the worm to acclimatize for 30 s. Then count the number of bends (Fig. 2a) for 30 s with a mechanical palm click counter. Repeat **steps 3–4** for 15–20 worms per condition. Present the results as bends/30s or bends per minute (BPM). *Stop manual.*
5. *Start automated:* take a test plate and add either ~2 mL M9 to the test plate (6 cm) or ~4 mL (9 cm).
6. Wash worms off a 9 cm NGM plate and pipette a fraction of those worms on the test plate that is already flooded with M9. Let the worms settle for 30 s.
7. Record the worms for 30 s at a fps of 20 fps (600 frames, video recording type: M-JPEG, JPEG Compression quality: 95).
8. Analyze the movies with the tracking software [33] and present the results as bends/30s or bends per minute (BPM).

3.4.3 Crawling

1. Take a growth plate (with approximately 600–800 worms; less is sometimes better).
2. Record the worms for 10 min at a fps of 3 fps (1800 frames, video recording type: M-JPEG, JPEG Compression quality: 95).
3. ADDITIONAL OPTION (*see* **Note 21**): Wash worms off a 9 cm NGM plate and pipette a small fraction of the worm pellet in the middle of a 9 cm test plate. Wait until the drop is

evaporated and worms start to crawl. Record the worms for 10 min at a fps of 3 fps (1800 frames, video recording type: M-JPEG, JPEG Compression quality: 95).

4. Analyze the movies with the tracking software [33].

   Tracks of the crawling worms can be visualized by loading the track.p file in the tracking software (utilities > plot path); *see* Fig. 2b.

3.5 Nuclear Versus Cytosolic Fractionation: Method IV

Nuclear extracts have been extensively used in different model organisms to identify and study protein function in nuclei. Importantly, specific aggregation-prone proteins may be more toxic when they are localized in the nucleus or the cytosol [43, 44]. Here, we provide a protocol (adapted from [45, 46]) to localize specific proteins in nuclear and cytoplasmic fractions.

3.5.1 Preparation and Collection of Worms

1. Grow worms on multiple 9 cm NGM plates (*see* **Note 1**).
2. When plates are relatively full with adult worms (not starved), collect worms in a Falcon tube by washing them off the plates with M9 buffer. Wash (i.e. spin the worms down in a centrifuge: 20 °C, ~500 × *g* for 2 min, remove the supernatant and add fresh M9 buffer) several times until the supernatant is clear.
3. Bleach worms (bleaching time should be adjusted to individual worm strains (*see* **Note 2**)) by adding 1 mL 13% hypochlorite:4 M NaOH (750:500) solution to a 2 mL suspension of worms in M9.
4. Hatch worms overnight in 12-well plates in M9 for approximately 24 h.
5. Pipette approximately 800–1000 L1 worms per 9 cm NGM plate. In order to get relatively large nuclear fractions, aim for 10–15 9 cm plates per strain (*see* **Note 22**).
6. Incubate worms at 20 °C for 4 days (*see* **Note 23**). At the day of collection, worms will be at the first day (D1) of adulthood.
7. Collect D1-staged worms with M9 in a Falcon tube and wash them until the supernatant is clear.
8. Wash the worm pellet two times with 1 mL cold hypotonic buffer.
9. Prepare "complete hypotonic buffer" by adding 1 M DTT (final concentration: 1 mM DTT) and 25× protease inhibitors (final concentration: 2× protease inhibitors) to a few mL of standard hypotonic buffer (*see* **Note 24**).
10. Remove all hypotonic buffer and add "complete hypotonic buffer," as prepared in the previous step (estimate the volume of the worm pellet and add the same volume of "complete hypotonic buffer"). Pipette the worm suspension in an Eppendorf tube.

11. Avoid freezing the worm pellet at this point, as this may damage the nuclei and cause leakage of nuclear proteins in the cytosol.

3.5.2 Fractionation

1. Homogenize worms with a pellet pestle for 1 min (repeat 5–10 × in total, chill the worms on ice after each 1-min cycle of strokes) (*see* **Note 25**). Pipette a few μL of the worm suspension on a glass slide after 2–3 stroke cycles to investigate the degree of worm destruction. Only some worm debris should be visible. Unless 80–90% of the worms are degraded and only a few worm remnants are still visible, continue with the pellet pestle.
2. Pellet the worm bodies and debris for 5 min at 500 g (4 °C).
3. Transfer the supernatant to a new Eppendorf and repeat **step 2**.
4. Transfer the supernatant to a new Eppendorf. Save 25–30 μL of this fraction in a new tube ("input fraction") and store on ice. Write down the total volume of supernatant.
5. Pellet the nuclei for 5 min at 4000 g (4 °C).
6. Transfer the supernatant to a new Eppendorf and spin it for 30 min at maximal speed (13 kprm, 4 °C). The supernatant will contain the "cytoplasmic fraction."
7. Wash the pellet (**step 5**) with 500 μL "complete hypotonic buffer" (as prepared in Subsection 3.5.1; **step 9**) and spin the samples for 5 min at 4000 g (4 °C).
8. Remove the supernatant and add 500 μL of fresh "complete hypotonic buffer". Dissolve the nuclear pellet and pipette the suspension in a new Eppendorf tube (*see* **Note 26**). Spin the samples for 5 min at 4000 g (4 °C).
9. Prepare "complete hypertonic buffer" and add 1 M DTT (final concentration: 1 mM DTT) and 25× protease inhibitors (final concentration: 2× protease inhibitors) to the hypertonic buffer.
10. Remove the supernatant and resuspend the pellet in a small volume of "complete hypertonic buffer" (half of—or less—the volume that was noted in **step 4**). Again, write down the volume that was used to resuspend the nuclear fraction. Transfer the suspension to a new Eppendorf: "nuclear fraction."
11. OPTIONAL: For a better protein extraction add 25 U/μL of benzonase nuclease to the "nuclear fraction" and incubate the sample at 4 °C for 45 min on a shaker.
12. OPTIONAL: Depending on the research question (*see* **Note 27**) one can determine the protein concentration of the "input," "cytosolic fraction," and "nuclear fraction" with a BCA kit.
13. Snap-freeze the samples and store them at −20 °C until immunoblotting will be performed.

3.5.3 Western Blotting

1. Make a 12% SDS polyacrylamide gel (*see* **Note 28**) by mixing 2.2 mL dH_2O, 1.25 mL Tris–HCl (1.5 M, pH 8.8), 1.5 mL 40% acrylamide (29:1), 50 μL 10% SDS, 25 μL 10% APS, and 2.5 μL TEMED. Cast the gel within a gel cassette (the volumes are based on a 1.0 mm cassette) and let it polymerize.
2. Prepare a stacking gel solution by mixing 1.74 mL dH_2O, 0.625 mL Tris–HCl (0.5 M, pH 6.8), 0.25 mL 40% acrylamide (29:1), 25 μL 10% SDS, 25 μL 10% APS, and 2.5 μL TEMED. Add the solution on top of the polymerized gel and immediately insert a 10- or 15-well gel comb.
3. Thaw the samples containing the "input," "cytosolic fraction," and "nuclear fraction." Pipette a volume of the samples (either the same protein concentration for all fractions or the same volume, *see* **Note 27**) in new tubes and add SDS lysis buffer (5×) at a dilution of 4:1. Optional: Boil the samples at 95 °C for 10 min.
4. Load the samples on a 12% SDS polyacrylamide gel. Fill the plastic tank with running buffer. Electrophorese at 100 V until the samples have entered the gel and then continue at 120–140 V until the dye front reaches the bottom of the gel (*see* **Note 29**).
5. Blot the gel to a nitrocellulose membrane at 95 V for 55–60 min at 4 °C.
6. Block the nitrocellulose membrane for 1 h in blocking solution. OPTIONAL: Cut the nitrocellulose membrane just below the 70 kDa marker (LMN-1 normally runs around 70 kDa and tubulin around 55 kDa).
7. Incubate the membrane with the appropriate antibodies for 1 h at room temperature or overnight at 4 °C. LMN-1 is used as a nuclear marker (~70 kDa) and tubulin (~55 kDa) as a cytosolic marker. When interested in the localization of α-synuclein (~48 kDa), one can use an antibody either for α-synuclein itself or for YFP (*see* **Note 30**).
8. Wash the incubated membranes 3 × 10 min with PBS-T.
9. Incubate the membranes with appropriate HRP-conjugated secondary antibodies (α-synuclein and tubulin: anti-mouse, GFP, and LMN-1: anti-rabbit) for 1 h at room temperature.
10. Wash the incubated membranes 3 × 10 min with PBS-T.
11. Develop the blots. OPTIONAL: When not cutting the membranes, one can strip the membranes and reprobe with a different antibody (suggested order: YFP/α-syn > tubulin).

4 Notes

1. The higher the number of 9 cm NGM plates with worms, the higher the yield of L1 worms after bleaching.
2. Strains like N2 (commonly used "wild-type" worms) are normally exposed to the alkaline hypochlorite solution for approximately 4 min. Bleaching for a shorter period will decrease the yield of L1 worms, as a substantial part of the adult worms will not be dissolved yet. At the other end of the spectrum, bleaching too long may damage the eggs, which will also result in a lower yield of L1 worms. For the OW40 strain, 4 min is also an appropriate bleaching time. Note: Always follow the bleaching process with a microscope, as soon as worms are dissolving and eggs are becoming visible: stop the reaction immediately.
3. As we normally aim to screen 8000–16,000 genomes, a maximum of P_0 800 worms is required (10 P_0 worms will give rise to approximately 100 F1 progeny worms when following this protocol). Using the typical concentration of 50 mM EMS, 2000 copies of a gene need to be screened to find a mutation [9]. Because each plate of worms descends from 100 F_1s, 200 copies of a particular locus can be assayed for mutations, so 200 genomes are screened per plate.
4. Plating the worms late in the afternoon works for our specific worm models, as they are L4 worms at an appropriate time for us. However, always make sure to check in advance how long it takes before a specific model develops from an L1 to an L4.
5. If worms are not in the L4 stage yet, wait. If they are too old, cancel the experiment as the efficiency of EMS diminishes with increasing age.
6. EMS is highly toxic! Make sure to wear protective gear and always parafilm tubes. In addition, make sure to collect the pipette tips that are used for EMS purposes in separate Falcon tubes. Add water and pellets of KOH and incubate overnight before this waste can be discarded.
7. *See* **Note 3**. Ten worms are transferred to each plate. Concerning the temperature, we normally keep worms at 20 °C, but for specific experiments it is also possible to keep the worms at 15 or 25 °C. You should be aware of the fact that worms develop faster when the temperature increases (up till a specific point). Adjust the timing of the next steps in the protocol according to that.
8. Just follow the worms over time. When they have laid about 100 eggs per plate, the P_0 can be removed.

9. It is also possible to wash off the F_1s with M9. However, in terms of contamination, we prefer removing them with a vacuum pump and sterile tips.
10. The remaining eggs will be from a ~14-h cohort.
11. When you decide to screen worms at an older age (i.e., from adulthood D2 and older), make sure to transfer worms to FUDR plates at adulthood D1. To avoid offspring or eggs from hatching, and thus to maintain synchronized populations, FUDR is commonly used in the worm field [47].
12. Only one individual mutant per plate is kept for further analysis. It is possible that there are multiple mutants per plate, but one cannot ensure that they are independent from each other. It is always wise to keep those mutants. Always freeze the mutants as soon as possible; there is often genetic drift which may change phenotypes on the long term.
13. Our experience (when using automated screening or phenotypes that are easily observed by eye) is that screening 1 or 2 384-well plates per day (~35 to 70 12-well plates when including a control condition at each plate) is workable. Make sure to run some small trials before starting the real screen. In this way one can estimate how labor intensive the phenotyping actually is. In addition, if the phenotype is easy to screen for, one can also grow and test worms in a 96-well plate format.
14. When you decide to screen worms at an older age (i.e., from adulthood D2 and older), make sure to also prepare NGM plates with FUDR (12.5 μg/mL) (and IPTG, ampicillin, and the right RNAi clones). Worms should be transferred at D1 to these plates. To avoid offspring or eggs from hatching, and thus to maintain synchronized populations, FUDR is commonly used in the worm field [47].
15. It is possible to reduce or increase the volume of LB that is added to each well. As 40 μL of each concentrated RNAi clone is required, the lower limit is approximately 400 μL (10× concentrating this volume leaves exactly 40 μL).
16. Increasing the number of worms to higher numbers may increase the risk of starvation. 20–30 worms are fine when screening at adulthood D1. Interestingly, it is also possible to transfer L1 worms first to normal NGM plates (without RNAi clones, but with normal OP50) and later—at a specific age—move them to RNAi plates. In this way, knockdowns of specific genes can be assessed at different ages and thus different stages of a disease.
17. It is important to realize that RNAi works quite well in specific tissues, like the gut and in the muscles. When a model is used

that expresses a specific protein in neuronal tissue, it is harder to investigate a cell-autonomous effect with RNAi, as dsRNA penetrates neuronal tissues less efficiently. There are however certain *C. elegans* strains available with a higher sensitivity for neuronal RNAi (e.g., TU3311). By crossing your mutant of interest with these strains, one can also use RNAi for knock-downs in neuronal tissue. However, do realize that it does not always work so well. Muscle tissue is superior to neuronal tissue when doing RNAi screens.

18. When performing manual thrashing assays during a time course (e.g., D1, D4, D8, and D12) it is better to reduce the number of worms per plate, as their time on the plate will increase (D8 worms are normally 8 days on the same plate). Therefore it is not unlikely to only transfer 20–30 worms per 6 cm NGM plate.
19. Normally, worms are kept at 20 °C, but for the sake of specific experiments one can either decrease or increase the temperature to, e.g., 15 °C and 25 °C, respectively.
20. It is highly recommend to always start with the manual thrashing assay (trial) when using new worm strains. These values can be used as reference values for the automated tracking of worms. In this way, parameters in the tracking software can be optimized.
21. Plating worm in the middle of a 9 cm NGM plate allows worms to crawl from the middle to the edges of the plate. This creates in general an outstanding graphical representation of the crawling behavior of specific worm strains.
22. We have performed fractionation studies with as less as 5 × 9 cm NGM plates per condition. However, working with 10 × 9 cm NGM plates will yield a more workable nuclear fraction, as the pellet is better visible during the washing steps.
23. Fractionation can be studied at different ages of the worms. Here, we describe a protocol for worms on the first day of their adulthood (D1). However, the protocol is also feasible for L4 s and older worms. In general, it is easier to break the cuticle of larval stage worms with a pellet pestle than older worms. Thus, the number of strokes should be adjusted accordingly to the age of the worms. Importantly, when performing fractionation studies on worms that are older than D1 one should transfer worms to FUDR NGM plates. To avoid offspring or eggs from hatching, and thus to maintain synchronized populations, FUDR is commonly used in the worm field [47].
24. It is suggested to calculate the volume of "complete hypotonic buffer" that is required for the experiment. Pipette the required volume of hypotonic buffer in a tube and add DTT

and protease inhibitors, compounds that should be both stored at −20 °C.

25. As described in **Note 23**, the number of strokes depends on the age of the worms. In addition, other worm strains may also require more or less strokes. Always follow the process with a microscope.
26. It is important to change tubes after each wash, as tubulin and fatty substances tend to stick to the tube walls which may contaminate the nuclear fraction.
27. When equal volumes are loaded, one can actually compare the amount of specific protein between the different fractions within a sample. When equal protein amounts are loaded, one can look for an enrichment of a protein within specific fractions between conditions or worm strains. Importantly, always write down the volume of the total input fraction and the volume in which the nuclear fraction is dissolved. Enriching the nuclear fraction, i.e., dissolving the pellet in less volume, increases the concentration of proteins, which may be useful during immunoblotting. However, in this situation the signal cannot be compared directly to that of the input, as this fraction is not enriched in the same way. Therefore, corrections based on the described volumes should be performed afterwards.
28. 12% SDS polyacrylamide gel is commonly used. However, when needed, one can easily decrease the percentage in order to get a better separation of proteins.
29. There is no strict rule to perform electrophoresis at a specific voltage.
30. Since α-synuclein is fused to YFP in the OW40 worm strain, antibodies to one of these proteins can be used. When possible, we highly recommend to cut membranes around the 70 kDa marker, so LMN-1 can be blotted at the same time as YFP and/or tubulin. Again, these steps can be easily adjusted and the researcher should optimize the blotting process for his or her own specific research question (depending on the proteins of interest).

References

1. Morley JF, Brignull HR, Weyers JJ et al (2002) The threshold for polyglutamine-expansion protein aggregation and cellular toxicity is dynamic and influenced by aging in Caenorhabditis elegans. Proc Natl Acad Sci U S A 99:10417–10422
2. Dillin A, Hsu AL, Arantes-Oliveira N et al (2002) Rates of behavior and aging specified by mitochondrial function during development. Science 298:2398–2401
3. Lee SS, Kennedy S, Tolonen AC et al (2003) DAF-16 target genes that control C. elegans life-span and metabolism. Science 300:644–647
4. Nollen EAA, Garcia SM, van Haaften G et al (2004) Genome-wide RNA interference screen

identifies previously undescribed regulators of polyglutamine aggregation. Proc Natl Acad Sci U S A 101:6403–6408
5. Kim Y, Sun H (2007) Functional genomic approach to identify novel genes involved in the regulation of oxidative stress resistance and animal lifespan. Aging Cell 6:489–503
6. Van Ham TJ, Thijssen KL, Breitling R et al (2008) C. elegans model identifies genetic modifiers of α-synuclein inclusion formation during aging. *PLoS Genet* 4: e1000027–e1000011
7. Habchi J, Arosio P, Perni M et al (2016) An anticancer drug suppresses the primary nucleation reaction that initiates the production of the toxic A 42 aggregates linked with Alzheimer's disease. Sci Adv 2:e1501244–e1501244
8. Arvanitis M, Li DD, Lee K et al (2013) Apoptosis in C. elegans: lessons for cancer and immunity. Front Cell Infect Microbiol 3:67
9. Jorgensen EM, Mango SE (2002) The art and design of genetic screens: caenorhabditis elegans. Nat Rev Genet 3:356–369
10. Hamilton B, Dong Y, Shindo M et al (2005) A systematic RNAi screen for longevity genes in C. elegans. Genes Dev 19:1544–1555
11. Van der Goot AT, Zhu W, Vázquez-Manrique RP et al (2012) Delaying aging and the aging-associated decline in protein homeostasis by inhibition of tryptophan degradation. Proc Natl Acad Sci U S A 109:14912–14917
12. Van Ham TJ, Holmberg MA, van der Goot AT et al (2010) Identification of MOAG-4/SERF as a regulator of age-related proteotoxicity. Cell 142:601–612
13. Lai CH, Chou CY, Chang LY et al (2000) Identification of novel human genes evolutionarily conserved in *Caenorhabditis elegans* by comparative proteomics. Genome Res 10:703–713
14. Shaye DD, Greenwald I (2011) OrthoList: a compendium of C. elegans genes with human orthologs. PLoS One 6(5):e20085
15. Faber PW, Alter JR, Macdonald ME et al (1999) Polyglutamine-mediated dysfunction and apoptotic death of a *Caenorhabditis elegans* sensory neuron. Proc Natl Acad Sci U S A 96:179–184
16. Lasko M, Vartianen S, Moilanen AM et al (2003) Dopaminergic neuronal loss and motor deficits in Caenorhabditis elegans overexpressing human alpha-synuclein. J Neurosci 86:165–172
17. Wang J, Farr GW, Hall DH et al (2009) An ALS-linked mutant SOD1 produces a locomotor defect associated with aggregation and synaptic dysfunction when expressed in neurons of Caenorhabditis elegans. PLoS Genet 5: e1000350
18. Ash PE, Zhang YJ, Roberts CM et al (2010) Neurotoxic effects of TDP-43 overexpression in C. elegans. Hum Mol Genet 19:3206–3218
19. Treusch S, Hamamichi S, Goodman JL (2011) Functional links between Aβ toxicity, endocytic trafficking and Alzheimer's disease risk factors in yeast. Science 334(6060):1241–1245
20. Martinez BA, Petersen DA, Gaeta AL et al (2017) Dysregulation of the mitochondrial unfolded protein response induces non-apoptotic dopaminergic neurodegeneration in *C. elegans* models of Parkinson's disease. J Neurosci 37(46):11085–11100
21. Link CD (1995) Expression of human beta-amyloid peptide in transgenic Caenorhabditis elegans. Proc Natl Acad Sci U S A 92:9368–9372
22. Ben-Zvi A, Miller EA, Morimoto RI (2009) Collapse of proteostasis represents an early molecular event in Caenorhabditis elegans aging. Proc Natl Acad Sci U S A 106:14914–14919
23. Yano H, Baranov SV, Baranova OV et al (2014) Inhibition of mitochondrial protein important by mutant huntingtin. Nat Neurosci 17 (6):822–831
24. Wang W, Wang L, Lu J et al (2016) The inhibition of TDP-43 mitochondrial localization blocks its neuronal toxicity. Nat Med 22:869–878
25. Ryan T, Bamm VV, Stykel MG et al (2018) Cardiolipin exposure on the outer mitochondrial membrane modulates α-synuclein. Nat Commun 9:817
26. Ferreira IL, Resende R, Ferreiro E et al (2010) Multiple defects in energy metabolism in Alzheimer's disease. Curr Drug Targets 11:1193–1206
27. Jarrett SF, Lewin AS, Boulton ME (2010) The importance of mitochondria in age-related and inherited eye disorders. Ophthalmic Res 44:179–190
28. Kawamata H, Manfredi G (2010) Mitochondrial dysfunction and intracellular calcium dysregulation in ALS. Mech Ageing Dev 131:517–526
29. Ren J, Pulakat L, Whaley-Connell A et al (2010) Mitochondrial biogenesis in the metabolic syndrome and cardiovascular disease. J Mol Med 88:993–1001
30. Koopman M, Michels H, Dancy BM et al (2016) A screening-based platform for the assessment of cellular respiration in Caenorhabditis elegans. Nat Protoc 11 (10):1798–1816

31. Kamath RS, Ahringer J (2003) Genome-wide RNAi screening in *Caenorhabditis elegans*. Methods 30(4):313–321
32. Rual JF, Ceron J, Koreth J et al (2004) Toward improving *Caenorhabditis elegans* phenome mapping with an ORFeome-based RNAi library. Genome Res 14(10b):2162–2168
33. Perni M, Challa PK, Kirkegaard JB et al (2018) Massively parallel C. elegans tracking provides multi-dimensional fingerprints for phenotypic discovery. J Neurosci Methods 306:57–67. https://doi.org/10.1016/j.jneumeth.2018.02.005
34. Stiernagle T (2006) Maintenance of *C. elegans*. In: Wormbook (ed) The *C. elegans* Research Community
35. Corsi AK, Wightman B, Chalfie M (2015) A Transparent window into biology: a primer on Caenorhabditis elegans. In: Wormbook (ed) The *C. elegans* Research Community
36. Brenner S (1974) The genetics of Caenorhabditis elegans. Genetics 77:71–94
37. Wicks SR, Yeh RT, Gish WR et al (2001) Rapig gene mapping in *Caenorhabditis elegans* using a high density polymorphism map. Nat Genet 28(2):160–164
38. Swan KA, Curtis DE, McKusick KB et al (2002) High-throughput gene mapping in Caenorhabditis elegans. Genome Res 12(7):1100–1105
39. Sin O, Michels H, Nollen EA (2014) Genetic screens in Caenorhabditis elegans models for neurodegenerative diseases. Biochim Biophys Acta 1842(10):1951–1959
40. Herndon LA, Schmeissner PJ, Dudaronek JM et al (2002) Stochastic and genetic factors influence tissue-specific decline in ageing C. elegans. Nature 419:808–814
41. Augustin H, Partridge L (2009) Invertebrate models of age-related muscle degeneration. Biochim Biophys Acta 1790:1084–1094
42. Pierce-Shimomura JT, Chen BL, Mun JJ (2008) Genetic analysis of crawling and swimming locomotory patterns in C. elegans. Proc Natl Acad Sci U S A 105(52):20982–20987
43. Kontopoulos E, Parvin JD, Feany MB (2006) Alpha-synuclein acts in the nucleus to inhibit histone acetylation and promote neurotoxicity. Hum Mol Genet 15(20):3012–3023
44. Barmada SJ, Skibinski G, Korb E et al (2010) Cytoplasmic mislocalization of TDP-43 is toxic to neurons and enhanced by a mutation associated with familial ALS. J Neurosci 30(2):639
45. Chen F, Hersh BM, Conradt B et al (2000) Translocation of C. elegans CED-4 to nuclear membranes during programmed cell death. Science 287(5457):1485–1489
46. La Rocca G, Burgio G, Corona DF (2007) A protein nuclear extract from D. melanogaster larval tissues. Fly (Austin) 1(6):343–345
47. Mitchell DH, Stiles JW, Santelli J et al (1979) Synchronous growth and aging of Caenorhabditis elegans in the presence of fluorodeoxyuridine. J Gerontol 34:28–36

Chapter 10

Recombinant Expression and Purification of *N*-Acetylated Alpha-Synuclein

Tara A. Eastwood and Daniel P. Mulvihill

Abstract

The majority of proteins in eukaryotic cells are subject to amino-terminal (Nt) acetylation. Recombinant protein expressed using prokaryotic expression systems such as *E. coli* would not normally be Nt-acetylated as these cells lack the appropriate *N*-α-terminal acetylation complex. Here we describe a simple protocol that allows the recombinant expression and purification of Nt-acetylated alpha-synuclein (aS) from *E. coli.*

Key words Alpha-synuclein, Recombinant protein, NatB, *N*-α-terminal acetyltransferase, Amino-terminal acetylation, *E. coli*

1 Introduction

E. coli is commonly used as a cheap and simple system for producing large quantities of protein. However this system has limitations, as a prokaryote *E. coli* lacks the necessary machinery to perform posttranslational modifications of proteins such as Nt-acetylation as would normally occur in a eukaryotic cell.

Nt-acetylation involves the transfer of an acetyl group from acetyl coenzyme A to the amino-terminal amino acid of a protein. The addition of the acetyl group to the *N*-terminal residue neutralizes positive charges which can affect the function and stability of proteins as well as regulate protein–protein and protein–membrane interactions [1] and further posttranslational modifications.

Nt-acetylation of amino-terminal amino acids in eukaryotes is catalyzed by a group of protein complexes called *N*-α-terminal acetyltransferases (NATs). The NatB complex Nt-acetylates proteins which have an *N*-terminal methionine residue which is followed by an acidic or arginine residue [2] and thus is the NAT responsible for the Nt-acetylation of aS, which has the *N*-terminal residues M–D.

Tim Bartels (ed.), *Alpha-Synuclein: Methods and Protocols*, Methods in Molecular Biology, vol. 1948,
https://doi.org/10.1007/978-1-4939-9124-2_10, © Springer Science+Business Media, LLC, part of Springer Nature 2019

Previously the purification of Nt-acetylated aS has required purification from eukaryotic cells or chemical Nt-acetylation of recombinant aS purified from *E. coli* cells.

Here we describe a simple method which uses reagents developed in this lab for producing [3] and purifying Nt-acetylated and fluorescently tagged recombinant aS from *E. coli*. For ease of purification and for use in in vitro studies using fluorescent microscopy we co-expressed the gene encoding aS-mCerulean3-His_6 together with the genes encoding for the fission yeast NatB complex. Using this method it is possible to produce and purify large quantities of Nt-acetylated aS, fluorescently tagged with Cerulean3 fluorescent protein, from *E. coli* with significant cost and time benefits over eukaryotic expression and chemical acetylation alternatives.

2 Materials

1. LB medium: 10 g Tryptone, 10 g sodium chloride (NaCl), 5 g yeast extract, and 15 g agar (if solid medium required). Make up to 1 L with distilled water and autoclave to sterilize. Antibiotics are added to the medium after autoclaving and cooling.
2. Antibiotics: 1 μL of the appropriate antibiotics per 1 mL of medium are added from the stock solutions of 100 mg/mL ampicillin in distilled water or 25 mg/mL chloramphenicol in ethanol.
3. *E. coli* expression strain BL21-DE3.
4. pNatB plasmid: pACYCduet-naa20+-naa25+ (*see* **Note 1**).
5. aS-mCerulean3-His_6 expression plasmid: pETDuet-1.aS-mCerulean3-His_6 (*see* **Note 2**).
6. Sterile calcium chloride-glycerol solution for competent cells: 0.1 M $CaCl_2$ and 10% glycerol.
7. Cobalt column for His-tag purification: 5 mL Syringe sealed with glass wool with approximately 3 mL of Co^{2+} metal affinity resin.
8. aS His-tag purification binding buffer: 50 mM Tris base, 500 mM NaCl, 1% Triton X-100, 10 mM imidazole, pH 7.8, add PMSF to 1 mM prior to use.
9. aS His-tag purification wash buffer: 50 mM Tris base, 500 mM NaCl, 1% Triton X-100, 10 mM imidazole, pH 6.0.
10. aS His-tag purification elution buffer: 50 mM Tris base, 500 mM NaCl, 150 mM imidazole, pH 6.0.
11. 5 mL HiTrapQ ion-exchange column and GE Healthcare, AKTA FPLC system.
12. aS FPLC purification buffer A: 20 mM Tris, pH 7.5.
13. aS FPLC purification buffer B: 20 mM Tris, 1 M NaCl, pH 7.5.

3 Methods

3.1 Preparation of BL21-DE3-pNatB Culture

1. Defrost an aliquot of BL21-DE3 cells on ice, then add 2 μL of pNatB plasmid to the cells, and stir gently with a pipette tip.
2. Incubate cells on ice for 30 min, and then heat shock for 90 s at 42 °C. Immediately return onto ice for 2 min.
3. Add 100 μL of LB medium to the cells and incubate in a shaking incubator (200 rpm) at 37 °C for 1 h.
4. Plate cells onto a LB agar plate supplemented with chloramphenicol and incubate at 37 °C overnight.

3.2 Preparation of BL21-DE3-pNatB Competent Cells

1. Set up a 5 mL overnight shaking (200 rpm) culture from a single colony inoculum of BL21-DE3-pNatB in LB media with chloramphenicol.
2. Next morning use 0.5 mL of this culture to inoculate 50 mL of fresh LB with chloramphenicol within a sterile 250 mL baffled flask.
3. Grow cells in a shaking (200 rpm) incubator at 37 °C until the culture reaches a 595 nm optical density of 0.6–0.8.
4. Cool the cells on ice for 10 min, before centrifuging at 1,000 rcf at 4 °C for 10 min. All steps should now be performed on ice or at 4 °C.
5. Gently resuspend the pelleted cells in 10 mL of cold, sterile solution containing 0.1 M $CaCl_2$ and 10% glycerol and cool on ice for a further 15 min.
6. Centrifuging at 1,000 rcf at 4 °C for 10 min, resuspend the pellet in 1 mL of cold, sterile solution containing 0.1 M $CaCl_2$ and 10% glycerol.
7. Dispense competent cells as 50 μL aliquots into sterile 200 μL microfuge tubes, prefrozen at −80 °C. Quickly seal tubes and drop into liquid nitrogen, before storage at −80 °C.

3.3 Transformation of aS-Encoding Vector into BL21-DE3-pNatB Competent Cells

1. Defrost an aliquot of BL21-DE3-pNatB cells on ice, then add up to 5 μL of aS-mCerulean3-His_6 plasmid to the cells, and stir gently with a pipette tip.
2. Incubate cells on ice for 30 min, and then heat shock for 90 s at 42 °C. Immediately return to ice for 2 min.
3. Add 100 μL of LB medium to the cells and incubate in a shaking incubator (200 rpm) at 37 °C for 1 h.
4. Plate cells onto a LB agar plate supplemented with both chloramphenicol and ampicillin and incubate at 37 °C overnight.

3.4 Expression of Nt-Acetylated Recombinant aS in E. coli

1. Set up a 5 mL overnight shaking (200 rpm) culture from a single-colony inoculum of BL21-DE3-pNatB transformed with aS-mCerulean3-His$_6$-expressing plasmid, in LB media with chloramphenicol and ampicillin (*see* **Note 3**).
2. Use 1 mL of the overnight culture to inoculate 1 L of LB with chloramphenicol and ampicillin in a sterile 2 L conical flask, and incubate in a shaking incubator (200 rpm) at 37 °C.
3. Once the culture reaches a 595 nm optical density of 0.4–0.5, protein production is induced by the addition of IPTG to a final concentration of 20 μg/mL (*see* **Notes 4** and **5**). The culture is then incubated with shaking (200 rpm) at 25 °C for a further 2–3 h.
4. The culture is then spun at 1800 rcf for 20 min and the resultant pellets frozen at −20 °C until protein purification is undertaken.

3.5 Cobalt Column Purification of Nt-Acetylated Recombinant aS

1. The frozen cell pellets are defrosted on ice and then resuspended in 16 mL of His-tag purification binding buffer.
2. The cells were then sonicated for a total of 2 min, made up of 20-s pulses with a 59-s gap between them.
3. The suspension is then centrifuged at 4400 rcf for 30 min and the protein-containing supernatant kept.
4. The Cobalt purification column is first equilibrated by rinsing through with approximately 5 resin volumes of His-tag purification binding buffer. Buffer is added gently by slowly running down the inside of the syringe so as not to disturb the surface of the resin. The buffer is allowed to run through the column by gravity and the flow is stopped by placing a cap on the syringe. At no time should the column be allowed to run dry.
5. The protein-containing supernatant is then loaded onto the column, and all liquid that flows through the column is collected in case not all protein has bound to the column (*see* **Note 6**). The column will be seen to change color from bright pink to orange/brown as the protein binds to it (Fig. 1).
6. Once the protein has been allowed to bind to the Cobalt column it is washed with 5 resin volumes of His-tag purification binding buffer followed by 5 resin volumes of His-tag purification wash buffer.
7. The aS-mCerulean3-His$_6$ is then eluted from the Cobalt column by the addition of His-tag purification elution buffer. The elution buffer flow through is collected in 1 mL fractions until the column has returned to a bright pink color (Fig. 1) and the flow through returns from green to colorless (Fig. 2).
8. The presence of aS-mCerulean3-His$_6$ in the fractions can be confirmed by running a small amount on a Coomassie-stained

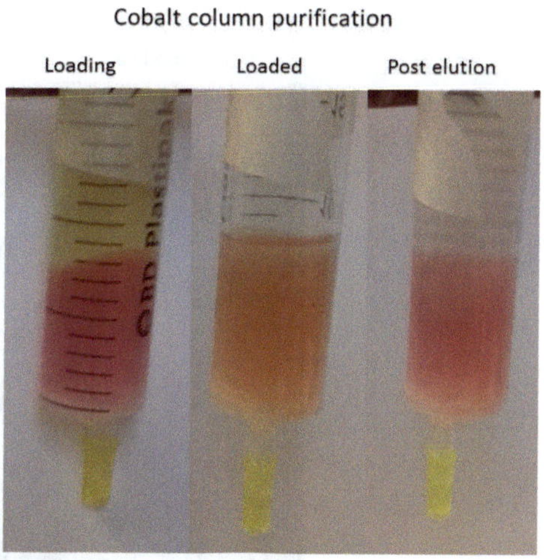

Fig. 1 Affinity column color changes during purification

Fig. 2 Affinity column elution fractions showing increase and then decrease in green color

SDS-PAGE gel; alternatively if placed in UV light the mCerulean3-containing fractions will be seen to fluoresce.

9. The aS-mCerulean3-His$_6$-containing fractions can be pooled for future use or for further purification.

3.6 Further Purification of Nt-Acetylated Recombinant aS Using FPLC

1. Due to the high salt content of the His-tag purification elution buffer, the aS-mCerulean3-His$_6$ protein obtained from the Cobalt column must first be dialyzed into aS FPLC buffer A (*see* **Note 7**).
2. The protein is placed in cellulose membrane dialysis tubing with a molecular weight cutoff of 14,000, which is securely sealed at both ends. This is then placed in 5 L of aS FPLC buffer A and placed on a magnetic stirrer at 4 °C. The aS FPLC buffer A is changed at least three times over a 24- to 48-h period.
3. The protein is then further purified by FPLC. The protein, in aS FPLC buffer A, is loaded onto the two 5 mL HiTrapQ ion-exchange columns connected in sequence into the FPLC machine. The green protein can be seen binding to the column (Fig. 3).
4. The aS-mCerulean3-His$_6$ protein is eluted using gradually increasing concentrations of aS FPLC buffer B, which are collected as fractions. The aS-mCerulean3-His$_6$ protein-

Fig. 3 Anion-exchange column purification—shown as a tight band of green-colored aS-mCerulean3-His$_6$ at the top of the column after loading, the movement of green-colored protein through the column during elution, and the absence of green protein postelution

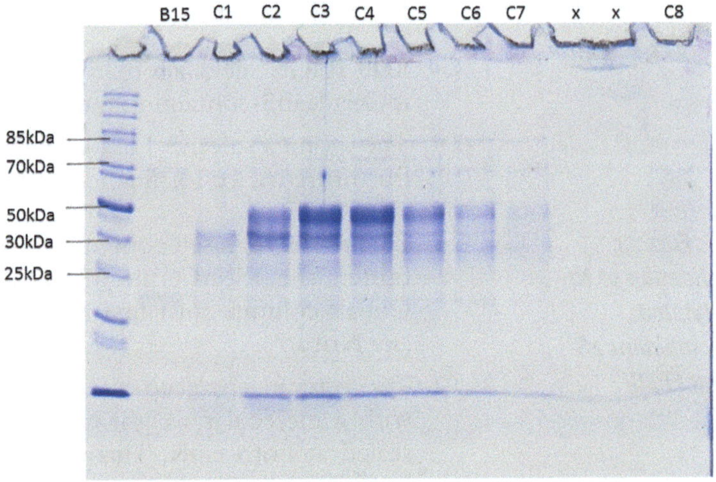

Fig. 4 Coomassie-stained SDS-PAGE gel showing the presence of protein post-anion-exchange column, at the expected size for aS-Cerulean3-His$_6$ of 42.2 kDa

containing fractions are identified by an increased absorbance at 280 nm and by the presence of green in the fractions. The purity and presence of aS-mCerulean3-His$_6$ protein is also confirmed by running a small amount of each fraction on a Coomassie-stained SDS-PAGE gel (Fig. 4); alternatively if placed in UV light the aS-mCerulean3-His$_6$-containing fractions will be seen to fluoresce (Fig. 5).

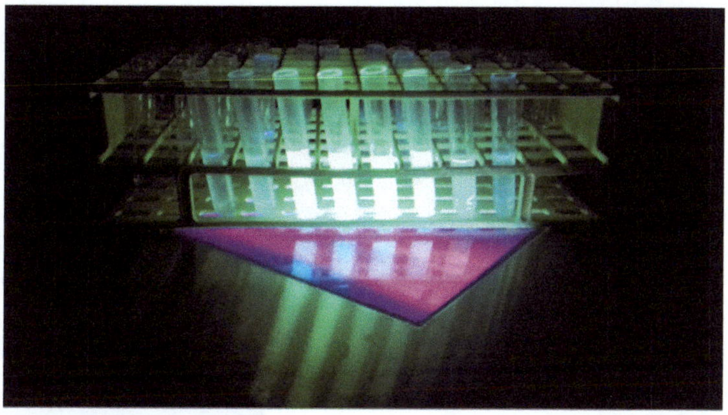

Fig. 5 Post-FPLC fractions placed in UV light. The aS-mCerulean3-His_6-containing fractions fluoresce

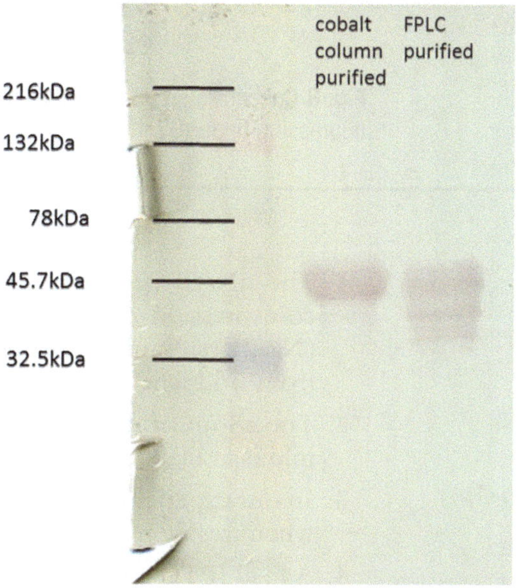

Fig. 6 Western blot of aS-Cerulean3-His_6 showing protein purified by affinity column (left) and subsequent anion exchange (right). The primary antibody used was anti-alpha-synuclein antibody (Abcam ab155038)

5. A Western blot can also be performed to confirm the presence of aS in the samples (Fig. 6).
6. The fractions confirmed to contain aS-mCerulean3-His_6 protein are then pooled and drop frozen in liquid nitrogen (Fig. 7). The frozen drops of protein are stored at −80 °C to be defrosted as needed.
7. If required the protein may be concentrated using a centrifugal concentrator (Fig. 8).

Fig. 7 Drop frozen Nt-acetylated aS-mCerulean3-His$_6$

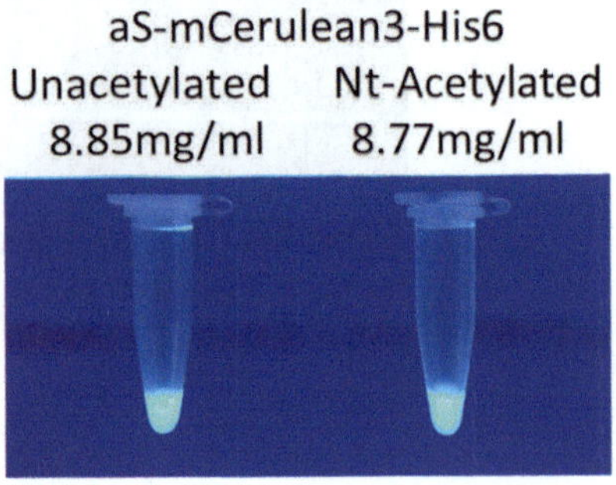

Fig. 8 Concentrated aS-mCerulean3-His$_6$ both unacetylated and Nt-acetylated, fluorescing under UV light

4 Notes

1. The pNatB (pACYCduet-naa20+-naa25+) plasmid, allowing co-expression of the catalytic (Naa20) and regulatory (Naa25) subunits of the fission yeast NatB complex, is available from Addgene.
2. The aS-mCerulean3-His$_6$ (pJC20-aS-mCerulean3-His$_6$) plasmid is available from Addgene.
3. In our experience, expression and acetylation are most efficient when freshly transformed cells are used.
4. A final IPTG concentration of 20 μg/mL is used for induction compared to the more commonly used concentration of 100 μg/mL, as the slower induction rate has less of a toxic effect on the cells and results in a higher overall level of protein being produced.
5. Simultaneous induction of the production of the NatB complex and aS-mCerulean3-His$_6$ may result in a small proportion of the aS-mCerulean3-His$_6$ not being acetylated while the initial level of NatB complex is low. If 100% Nt-acetylated aS-mCerulean3-His$_6$ is required, aS-mCerulean3-His$_6$ can be produced using a plasmid with a different promoter for sequential induction [4].
6. The protein-containing supernatant should have a bright green color due to the mCerulean3 protein present. As the protein

binds to the column it can be seen to change color from bright pink to an orange brown. If the whole column is seen to turn brown before all the supernatant is added stop and repeat the Cobalt column purification for the remainder of the supernatant. Similarly if the flow-through liquid collected is seen to have any green color it may contain aS-mCerulean3-His_6 protein and can be reapplied to further Cobalt column purification runs.

7. The aS-mCerulean3-His_6 protein must be dialyzed into aS FPLC buffer A prior to FPLC purification as the salt content of the His-tag purification elution buffer will prevent the aS-mCerulean3-His_6 protein from binding to the HiTrapQ ion-exchange columns.

Acknowledgments

Work in this lab is supported by the University of Kent, BBSRC, and the Royal Society.

References

1. Starheim KK, Gevaert K, Arnesen T (2012) Protein N-terminal acetyltransferases: when the start matters. Trends Biochem Sci 37 (4):152–161. https://doi.org/10.1016/j.tibs.2012.02.003
2. Polevoda B, Arnesen T, Sherman F (2009) A synopsis of eukaryotic N-alpha-terminal acetyltransferases: nomenclature, subunits and substrates. BMC Proc 3(Suppl 6):S2. https://doi.org/10.1186/1753-6561-3-S6-S2
3. Johnson M, Coulton AT, Geeves MA, Mulvihill DP (2010) Targeted amino-terminal acetylation of recombinant proteins in E. coli. PLoS One 5 (12):e15801. https://doi.org/10.1371/journal.pone.0015801
4. Eastwood TA, Baker K, Brooker HR, Frank S, Mulvihill DP (2017) An enhanced recombinant amino-terminal acetylation system and novel in vivo high-throughput screen for molecules affecting α-synuclein oligomerisation. FEBS Lett 591(6):833–841. https://doi.org/10.1002/1873-3468.12597

Chapter 11

Circular Dichroism and Isothermal Titration Calorimetry to Study the Interaction of α-Synuclein with Membranes

Matteo Rovere

Abstract

α-Synuclein's physiology and pathology have been linked by numerous reports to its ability to bind and remodel membranes, especially at synaptic terminals. It is therefore critical for researchers investigating the determinants of these interactions to rely on methods capable of providing an accurate and complete physicochemical snapshot of the binding events. Circular dichroism (CD) and isothermal titration calorimetry (ITC) are established techniques for the study of binding equilibria in biological systems and, especially when used in combination, allow a thorough characterization of the protein-lipid interplay.

Here we provide general guidelines and describe some common pitfalls of these experiments. This protocol describes the preparation of small unilamellar vesicles (SUVs), mimicking the curved bilayers α-synuclein normally interacts with, the CD-monitored titration of α-synuclein with SUVs, the ITC (lipid-into-protein) experiment, and the subsequent data analysis using an *n* independent binding site model.

Key words α-Synuclein, Circular dichroism, Isothermal titration calorimetry, Small unilamellar vesicles

1 Introduction

α-Synuclein (αS) has long been known to interact with curved, negatively charged lipid interfaces [1] similar to those of synaptic vesicles [2]. In fact, its N-terminal region contains seven 11-mer imperfect amino acid repeats, forming a periodic arrangement of hydrophobic and polar residues reminiscent of the lipid-binding class-A_2 apolipoproteins [3]. The N-terminus can thus fold, upon interaction with curved lipid bilayers [4] or detergent micelles [5], into amphipathic helical structures, with a predominantly apolar side embedded in the bilayer and the hydrophilic residues exposed. The binding partners (in particular the curvature of the binding interface) can determine the exact structure of the folded helix, ranging from a single helix with 11/3 periodicity, when interacting with small unilamellar vesicles (SUVs) [4], to a more curved "broken" α-helix, as described when αS is bound to SDS micelles

Tim Bartels (ed.), *Alpha-Synuclein: Methods and Protocols*, Methods in Molecular Biology, vol. 1948,
https://doi.org/10.1007/978-1-4939-9124-2_11,

[5, 6]. The presence of several lysine residues at the boundary between the hydrophilic and hydrophobic sides of the amphipathic helix can further stabilize these structures by forming ionic pairs with the negatively charged lipid heads. While its function is still debated, most recent studies have narrowed down the role of αS in neuronal physiology to the events surrounding synaptic vesicle fusion and recycling, thus confirming that its action is tightly dependent on a fine control of its interaction with membranes [7]. In addition, several of the familial PD-causing mutations in the SNCA (αS) gene lead to changes in the affinity of the protein for lipid interfaces, suggesting that this imbalance could be one of the determinants of their pathogenicity [8, 9]. It is therefore important for researchers interested in studying these equilibria to rely on accurate and sensitive techniques for the analysis of these interactions.

Circular dichroism (CD) has long been one of the methods of choice to study αS's binding equilibria [1], since the protein undergoes a coil-to-helix transition upon interaction with membranes. Among the advantages of CD are the small amount of sample required, its compatibility with vesicular dispersions, and the ability to provide real-time structural information. On the other hand, isothermal titration calorimetry (ITC), while being a well-established tool in the field of protein-ligand interactions, is not as common in protein-lipid studies [10]. Nevertheless, it has been successfully employed in the study of αS's binding to SUVs [11, 12] and, when coupled with CD, can both validate its findings and expand the picture with a complete thermodynamic characterization of the binding events.

While the applications, the instrumental design, and the physics underlying these techniques are beyond the scope of this chapter and have been described exhaustively in the literature [13–19], it is necessary to recall at least the basic principles and the instrumental output of both CD and ITC. Circular dichroism is based on the differential absorption of left- and right-handed circularly polarized light upon interaction with chiral molecules or supramolecular assemblies (as in the case of proteins and other biomolecules). When measuring in the 260–185 nm wavelength range (far-UV CD), the spectral profile will reflect the secondary structure composition of the sample, while at longer wavelengths (260–350 nm, near-UV CD) information on the tertiary structure can be obtained (e.g., disulfide bonds). In particular, predominantly helical proteins show a double minimum at 222 and 208–210 nm and a stronger maximum at 191–193 nm, while CD spectra of proteins with a disordered (random coil) structure only have a strong negative band near 200 nm (and some weak signal, either positive or negative, between 220 and 230 nm) (*see* Fig. 1). ITC, on the other hand, is not a technique which directly provides structural information; however, by monitoring heat changes upon ligand

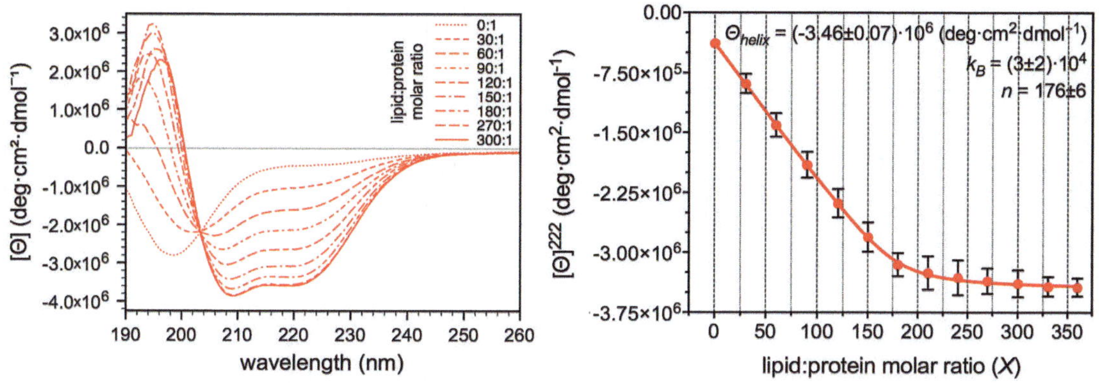

Fig. 1 (*Left*) Far-UV circular dichroism (CD) spectra of 10 μM αS titrated with 6 mM 70% POPC and 30% DOPS small unilamellar vesicles (SUVs) at 25 °C. (*Right*) Binding curve of the same titration (222 nm molar ellipticity signal plotted against the lipid:protein ratio) and the result of its fitting with an *n* independent binding site model (*solid line*). Data are obtained from three independent experiments and plotted along with their SDs. Thermodynamic and stoichiometric parameters obtained from the fitting of the binding curve, along with their SEMs, are shown in the inset

(or protein, or lipid) addition, it can yield insights into a wealth of phenomena such as conformational rearrangements and inter- or intramolecular interactions. Modern ITC instruments are almost completely automated, but the design is very similar to the first instruments of this kind [16]. They have two cells, a reference and a sample cell, embedded in a thermostating element that maintains the temperature constant and also monitors the heat exchanges. This is accomplished through a sensitive feedback system which measures the differential power (DP) that needs to be provided to the sample cell (compared to the reference cell) in order to compensate for the heat exchanges. Heat releases from the sample cell will reflect in a lower power input (negative peaks in the graph of the differential power (DP) vs. time, as in Fig. 2); vice versa, endothermic processes will produce positive DP peaks. The calorimetric unit is coupled with a mechanically controlled syringe (also providing stirring in the sample cell) which injects the titrant.

2 Materials

1. Ammonium acetate buffer: 10 mM NH_4Ac, pH 7.40.
2. Purified (>95%) αS.
3. 16:0–18:1 Phosphatidylcholine (POPC) in chloroform (850457C, Avanti Polar Lipids).
4. 18:1 Phosphatidylserine (DOPS) in chloroform (840035C, Avanti Polar Lipids).
5. Ar (or N_2) gas source.
6. Lyophilizer or vacuum desiccator.

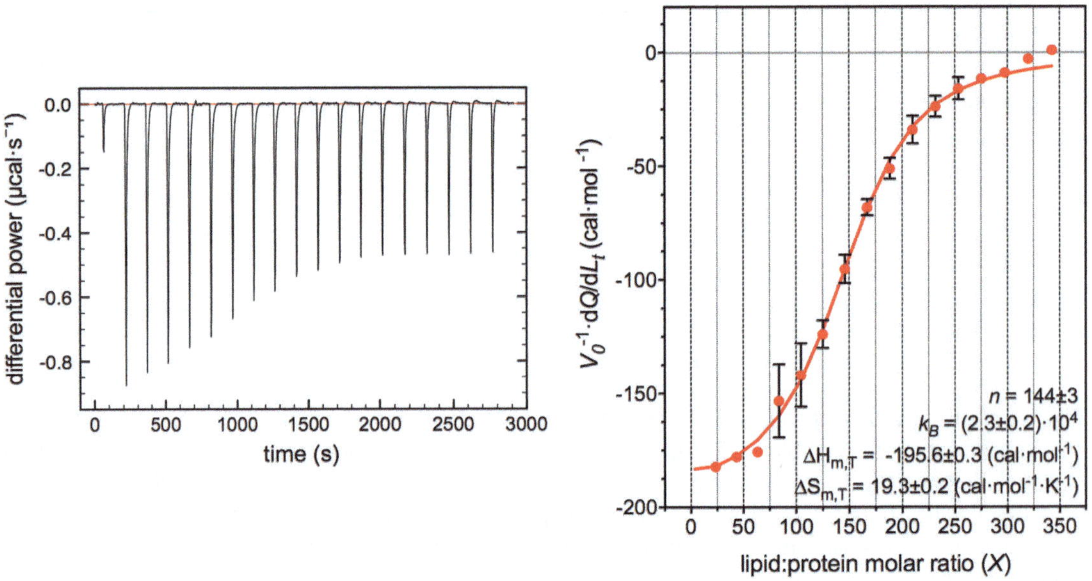

Fig. 2 (*Left*) Representative differential power (DP) trace of the isothermal titration calorimetry (ITC) experiment of 5 μM αS titrated with 10 mM 70% POPC and 30% DOPS small unilamellar vesicles (SUVs) at 25 °C. (*Right*) Binding curve of the same experiment, obtained by integrating the DP signal, and the result of its fitting with an *n* independent binding site model (*solid line*). Data are obtained from three independent experiments and plotted along with their SDs. Thermodynamic and stoichiometric parameters obtained from the fitting of the binding curve, along with their SEMs, are shown in the inset

7. Microprobe or bath (pref. programmable) sonicator.
8. Spectrophotometer (absorbance) cuvette (pref. with stopper) (e.g., 21-Q-1, Starna Cells).
9. CD spectropolarimeter equipped with a cell-thermostating Peltier device (e.g., J-815, Jasco).
10. Isothermal titration calorimeter (e.g., Microcal™ iTC$_{200}$, Malvern Panalytical).

3 Methods

While either technique could be applied independently to the study of αS-lipid interactions, we encourage the use of similar conditions in both the ITC and CD experiments, so that the results obtained can be compared and correlated.

3.1 General Notes

3.1.1 αS for CD/ITC Experiments

Protein purity is essential for the success of these experiments, as both ITC and CD are extremely sensitive to impurities and would incorporate spurious heat or ellipticity contributions in the signal recorded. In addition, protein quantification (especially if performed using the 280 nm absorbance) will be affected by other absorbing species present in the sample. Usually an αS sample at least 95% pure should be used (quantified, e.g., by CBB- or Ag-stained SDS-PAGE). For a protein prone to aggregation, like αS, care must also be taken to remove larger aggregates before use and minimize their formation during storage. The protein can be stored frozen at −80 °C (snap-freezing with liquid nitrogen or an ethanol/dry ice bath is recommended) or lyophilized at −20 °C, but upon thawing or resuspension it should be centrifuged at 20,000 × *g* for at least 20 min and the supernatant carefully transferred to a clean tube. Filtration through a 0.22 μm syringe filter can also be used in place of (or in addition to) centrifugation. If the absence of oligomers is critical for the experiment, size-exclusion chromatography (e.g., on Superdex 200, GE Healthcare) of the resuspended/thawed sample and isolation/concentration of the monomeric peak fractions should be performed prior to the measurements. Use of silanized or low-binding tubes (e.g., Eppendorf™ LoBind) can also help reducing aggregation.

The buffering system to be used can vary, but CD measurements prefer low-absorbing buffering solutions over the whole range of the measurement (for near-UV CD, usually 260–190 nm). High buffer absorbances will lead to very little transmitted light and thus low S/N, especially in the lower wavelengths (*see* **Note 8**). Chloride should be avoided when possible (fluorides or sulfates can be used to maintain the same ionic strength). Most buffers can be used at concentrations lower than 50 mM (HEPES can only be used at 5 mM or lower concentrations). Ammonium acetate (NH_4Ac) is especially convenient if the sample is lyophilized, as it decomposes in volatile components; its only drawback is its weak buffering capacity. ITC has very few limits on the buffers that can be used (but *see* **Note 15**), as long as they ensure the stability of both the protein and lipid dispersion.

Protein concentration is one last aspect that is critical for an accurate characterization of the binding equilibria. While a variety of dye-based methods are available and commonly used in lab practice [20], the spectrophotometric method is highly recommended, especially for a purified protein sample, as is the αS solution. The quantitation relies on protein absorption at 280 nm. Given the absence of Trp or Cys residues in the αS sequence, its four Tyr are the only residues contributing to the absorbance at 280 nm. Using the tabulated values from Pace et al. [21], we can estimate the extinction coefficient at 280 nm for αS to be 5890 L mol^{-1} cm^{-1} (0.412 mL mg^{-1} cm^{-1}). If an extremely accurate protein quantitation is necessary, other methods, such as amino acid analysis, should be preferred.

3.1.2 Liposomes for CD/ITC Experiments

Depending on the scope of the experiment and the specific question the researcher is seeking to answer, a variety of different lipid dispersions can be prepared. The most commonly used ones (and the most useful for the studies of αS-lipid binding) are multilamellar vesicles (MLVs), large unilamellar vesicles (LUVs, diameter ≥100 nm), and small unilamellar vesicles (SUVs, diameter <100 nm). While MLVs are a somewhat crude lipid preparation, heterogeneous in its morphology and size distribution, they have the advantage of being very easy to prepare, as they assemble readily upon hydration of dried lipids in buffer. If a more uniform and better characterized distribution is needed, further processing steps can be performed on an MLV preparation (e.g., extrusion or sonication) leading to unilamellar vesicles, with their size depending on the protocol and instrumentation used [22]. In our liposome preparation protocol we will describe the preparation of an SUV dispersion by sonication of an MLV preparation. SUVs (and LUVs) can also be obtained by extrusion of MLVs through polycarbonate membranes of defined pore size [22, 23].

Both synthetic lipids and lipid extracts can be acquired from a variety of manufacturers and distributors (e.g., Avanti Polar Lipids, Matreya, Sigma-Aldrich), either in powder or in the form of a chloroform stock solution. Using lipid powders has the advantage of easy handling and quantitation of the reagents. Nevertheless, if lipid mixtures are to be prepared, preliminary mixing of the different lipids in either chloroform or *tert*-butanol is a necessary step, which thus makes the chloroform solutions especially handy, unless small contaminations from the organic solvent are to be avoided (e.g., in NMR). Chloroform stocks are usually shipped in flame-sealed vials filled with inert gas (Ar or N_2) and care must be taken to prevent both their oxidation and solvent evaporation upon storage and manipulation, especially if insaturations are present. It is good practice to use dark-walled glass vials with chloroform-resistant caps for storage and keep the lipid solutions at −20 °C. Keeping manipulations to a minimum and always layering the top of the lipid solution with an inert gas will also prevent lipid oxidation over time (both Ar and N_2 can be used for this scope, but Ar, being heavier than air, has the advantage of not diffusing away).

The lipid composition to be used will depend on the specific question being addressed and the other techniques to be employed. αS has a preference for small (highly curved) charged vesicles [1], so some of the most common lipid mixtures are POPC/POPS (or DOPS) (16:0–18:1 phosphatidylcholine/16:0–18:1 (or 18:1) phosphatidylserine), POPG (16:0–18:1 phosphatidylglycerol), and POPG/POPC, PE/PI (phosphatidylethanolamine/phosphatidylinositol). All of these lipid classes (except POPG) also have the advantage of being prominent components of synaptic vesicles [2]. Either crude (or partially refined) lipid extracts or synthetic lipids can be used for liposome preparation. While being possibly more representative of the lipid composition of a certain tissue or organism, lipid extract can vary significantly from batch to batch and manufacturer to manufacturer, making it sometimes challenging to reproduce results. In addition, the presence of impurities and their heterogeneous nature can lead to unpredictable lipid rearrangement or phase transition processes, all of which provide strong background signals in ITC (CD usually does not suffer from these artifacts), sometimes masking the binding signal from the protein. If lipid extracts are to be used, special care must be taken to avoid artifacts and ensure the consistency of results. In the liposome preparation protocol below, a binary synthetic lipid mixture of 70% POPC (16:0–18:1 phosphatidylcholine) and 30% DOPS (18:1 phosphatidylserine) will be used.

3.2 Liposome Preparation (SUVs)

1. Weigh (if provided as a powder) the amounts of POPC/DOPS needed to prepare a 70:30 mixture at the concentration desired (*see* **Notes 1** and **2**), resuspend the powders in chloroform (or *tert*-butanol, or another organic solvent of choice), and mix well (if already available as chloroform stocks, the lipid solutions can be simply mixed together).
2. Dry the lipid film under a gentle stream of argon or nitrogen gas (~20 min), and transfer the (open) tube to a vacuum desiccator or a lyophilizer for further removal of organic solvent traces (from a few hours to overnight).
3. Resuspend the lipids in the chosen buffer (NH_4Ac 10 mM, pH 7.40 in our case) and vortex vigorously; the dispersion should have, at this point, a cloudy, opaque, whitish appearance due to robust scattering by the lipid slabs and nascent MLVs.
4. Hydrate the lipids at a temperature above their phase transition for 30–60 min (when using unsaturated lipids this can be usually done at RT or 37 °C, *see also* **Notes 3** and **4**) [22].
5. To prepare SUVs, the MLV dispersion has to be sonicated extensively in a bath or microprobe sonicator; in order to prevent overheating, a 1–5-s pulse ON/1–5-s OFF sequence can be used for 10–20 min of total active sonication time (*see* **Notes 5** and **6**).
6. After sonication, the sample is centrifuged for 10 min at >20,000 × *g* to remove large lipid particulates (and metal shavings, if a microprobe sonicator is used) and the supernatant transferred to a clean tube (while we prefer to prepare the liposomes fresh before the experiments, SUV dispersions can also be stored [22]).
7. Dynamic light scattering (DLS) can be used to quickly characterize the size distribution of the liposome preparation, requiring only minimal amounts of sample; TEM with negative staining is also a valuable tool to evaluate the homogeneity of the dispersion (but collapse and drying artifacts make it less suitable to provide accurate size estimates) [22, 24] (*see* **Note 7**).

3.3 Protein Preparation and Quantitation

1. Thaw on ice or resuspend (if lyophilized) the αS aliquot to be used; centrifuge it at 20,000 × *g* for 20 min, to remove larger aggregates, then carefully remove the supernatant, and transfer it to a clean tube.
2. Measure the αS concentration using the method of choice (e.g., 280 nm absorbance); if a lengthy quantitation technique (e.g., amino acid analysis or micro-Kjeldahl) is to be used, an aliquot of the protein stock solution can be removed for later processing.
3. Both for CD and ITC a solution with the required concentration (5–15 μM) can at this point be prepared and stored on ice.

3.4 CD-Monitored SUV Titration

(Although the procedures described here are for a Jasco J-815 spectropolarimeter, similar ones should be carried out for instruments from different manufacturers as well, while always following the producers' specifications.)

1. The instrument, the water bath, and the Peltier element should be switched on and nitrogen gas should be fluxed for >3 min through the machine in order to purge it from oxygen (as the ozone generated through UV irradiation would damage the optics and absorb in the 190–260 nm range); titrations can usually be measured at RT (20–25 °C) and the Peltier element ensures that the sample does not overheat because of irradiation.
2. After purging the instrument, while always keeping the nitrogen flux running, the lamp can be switched on (on Jasco instruments this is done automatically after the purging countdown is over) and should be allowed to warm up for about 30 min before use, in order to minimize drift.
3. A buffer blank should be measured beforehand (ideally blanks of the SUV dispersion can also be measured at different concentrations, *see* **Notes 8** and **9**), ensuring that there is minimal absorbance from the buffer, especially at low wavelengths, and that the cuvette used is clean (*see* **Notes 10** and **12**).
4. After transferring the 5–15 μM stock solution of αS in the cuvette and measuring the spectrum of the unfolded protein sample, additions can be made from the SUV solution and spectra measured for each of the points of the titration curve (care must be taken to mix well the sample after the lipid addition; this can be done easily by inverting a few times in a stoppered cuvette, or by pipetting, avoiding bubbling).
5. The range of lipid:protein molar ratios to be used (which will determine the optimal concentration of the lipid stock solution and the protein sample as well) has to be optimized depending

on the lipid and the liposome size/morphology (usually for SUVs saturation is reached at lipid:protein ratios ≤ 1000).

6. The objective is to reach saturation (thus a helical CD signal that is stable upon lipid addition) after measuring 12–17 data points, with at least 4 points in the plateau region, in order to have a robust dataset for the subsequent fitting and data analysis (*see* **Note 11**) (Fig. 1).
7. When the endpoint of the titration has been reached (having at least four approximately superimposable spectra is usually a good rule of thumb), the cuvette can be emptied and cleaned, if another titration has to be measured with the same cell (at least three titrations should usually be measured for each condition, in order to ensure reproducibility and liposome stability; day-to-day repeats should also be performed from time to time) (*see* **Note 12**).
8. The data should be exported in .txt, .asc, or .csv format for further manipulation; it is usually advised to convert the ellipticity signal to molar (or residue) ellipticity, subtract the buffer blank, and correct for the dilution caused by sample addition before exporting (*see* **Note 13**).
9. Before leaving the instrument, nitrogen should also be fluxed for about 5 min after shutting down the spectropolarimeter's lamp.

3.5 Isothermal Titration Calorimetry of the αS/SUV Interaction

(The procedures described here are referred to a Malvern Panalytical Microcal™ iTC$_{200}$, but similar ones should be carried out for instruments from different manufacturers as well while always following the producers' specifications).

1. Before loading the sample cell and the syringe it is recommended to perform a wash cycle on both of them (most commercial calorimeters have built-in automated washing routines) (*see* **Note 14**).
2. Defining an optimal injection sequence (number of injections, volume, and equilibration time) is likely to require some optimization depending on the lipid preparation and the temperature at which the experiment is run; nevertheless, a few guidelines are to be observed:
   (a) The first injection should be set at a smaller volume than the rest (e.g., 0.4 μL for an iTC$_{200}$) and discarded, since it is usually imprecise (Fig. 2).
   (b) αS binding to SUVs in most cases provides robust negative (exothermic) heat signals, but care should be taken to work in a concentration range and use injection parameters providing both a good number of data points and a strong DP signal.

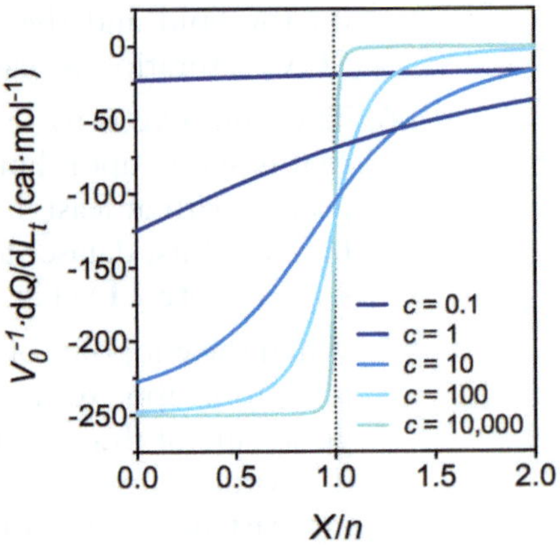

Fig. 3 Simulated ITC binding curves of 2.5 μM αS (P_t) with varying *c* (Wiseman *c* parameter [16]). *See* Subheading 3.6

(c) The equilibration time between injections should be long enough for the DP signal to return to its baseline, as the enthalpic contributions are measured by integrating each DP peak.

(d) A composite parameter that can help define the optimal range of concentrations for an ITC experiment is *c*, defined by Wiseman et al. as [16]

$$c = nk_{\mathrm{B}}P_{\mathrm{t}}$$

c should fall between 10 and 1000 [19], or else the binding sigmoid will be either too shallow or too steep to fit with acceptable accuracy some of the binding parameters (n is the binding stoichiometry, k_{B} the microscopic binding constant, and P_{t} the protein concentration; tentative values for n and k_{B} can be found in the literature or estimated; *see* Subheading 3.6 and Fig. 3).

(e) A good initial set of parameters (for an iTC_{200} with a syringe volume of about 40 μL) is 19 injections (2 μL each), with 150 s of equilibration time.

3. Once the conditions for the experiments have been defined, the syringe can be filled with titrant (SUVs, in our case); in some commercial instruments (e.g., iTC_{200}) syringe filling is automated (instrument-specific guidelines can be obtained from the manufacturers).

4. After filling the syringe, the cell can be manually filled with the αS stock solution using the Hamilton syringe provided with the instrument; care must be taken not to introduce bubbles when filling it, which is usually accomplished by slowly injecting the sample while gently retracting the needle and then aspirating and dispensing a small volume of solution a few times, in order

to dislodge possible air bubbles attached to the walls (*see* **Note 15**).

5. Once both cell and syringe are full, the syringe assembly can be moved into the cell and the experiment is started; after performing an internal power calibration and stabilizing the cell temperature and the DP baseline, the instrument will perform the titration and record the DP signal over the course of the injections (*see* **Note 16**).
6. After the titration is complete, both the cell and the syringe are washed before performing either the repeats (at least three titration curves for each condition should be measured) or the blank titration (lipid-into-buffer; a blank titration of buffer-into-protein could also be measured, but the heat contributions from protein dilution are usually so small that they are not corrected for) (*see* **Note 17**).
7. At the end of the day of experiments, it is good practice to perform a water-in-water (using filtered, degassed, deionized water) titration to verify whether the cell is clean; the heat peaks from the injections should be smaller than 0.02 μcal and the DP baseline linear (if this is not the case, the sample cell should be washed again; *see* **Note 17**).

3.6 Data Analysis

Before proceeding to the data analysis, some processing needs to be performed on the raw data for both CD and ITC experiments. For CD-monitored titrations, after subtracting the blank, normalizing the spectra, and correcting for the dilution due to titrant addition, the molar (or residue) ellipticity at 222 nm is plotted against the lipid:protein ratio for each titration point (Fig. 1). While other wavelengths can be employed, 222 nm is the usual choice when monitoring helix-to-coil transitions, due to the presence of one of the two minima characteristic of the helical CD spectrum and the minimal contribution arising from coils at such wavelength. ITC raw data require a slightly more complex processing step, which is usually handled using the software package provided with the instrument. First, the DP data must be integrated; the software calculates a baseline and evaluates the peak area for each of the injections (*see* **Note 18**), after which it plots the integrated data points as noncumulative heats versus the ligand:protein ratio (lipid:αS, in our case). If a lipid-into-buffer titration was performed, at this point (after integration of the raw DP signal of the blank titration) the data can be subtracted and further spurious contributions in the second plateau region of the binding sigmoid (at high lipid:protein ratios) can be eliminated in order to obtain a null signal upon αS saturation (*see* **Note 19**) (Fig. 2).

Although the software packages of most commercial ITC instruments have built-in data analysis and fitting routines and several commercial data processing programs (e.g., Origin,

GraphPad Prism) allow sigmoidal and ligand-binding fitting, it is important to explore the equations used and understand the assumptions and limitations of different models. These aspects become critical when comparing and integrating data from different techniques, since it is important to confirm that the model used is the same in both cases and how the fitting parameters are defined. In addition, modifications of the equations in order to better describe the data or address specific questions are only possible if one is familiar with the underlying model.

Here we briefly discuss the derivation of the equations describing the binding of αS to SUVs with an *n* independent binding site model, the simplest that can be used to describe binding interactions. Despite its simplicity, it has been shown to be able to describe the binding of αS to a variety of membranes [11, 12, 25]. When fitting ITC data, it is advisable to be conservative in the choice of the model to be used and only transition to more complex interaction modes (e.g., multiple binding sites [19, 26]) if strong evidence warrants the use of more fitting parameters and the underlying assumptions are convincingly verified.

Classically the *n* independent binding site model has been used to describe the binding of ligands (e.g., ions or small molecules) to macromolecules bearing *n* sites independent from each other (i.e., the microscopic binding constant for each of these sites does not depend on the occupancy of the other sites). The expression of the microscopic binding constant as a function of the fractional saturation ϑ (ratio of occupied sites to total sites) can be easily derived [27]:

$$k_B = \frac{\vartheta}{(1-\vartheta)[\mathrm{L}]} \tag{1}$$

In our case, although the "ligand" (titrant) L is the dispersion of SUVs and the macromolecule is αS, the same equations can be used. Since both the CD and the ITC data are plotted as a function of the lipid:protein molar ratio, the binding event can be seen as one between *n* molecules of lipid (forming a "binding site" on the vesicular surface) and a molecule of αS ($P + nL \rightleftharpoons PL_n$). Therefore, the somewhat artificial (in this case) microscopic binding constant k_B will describe the binding equilibrium of a single lipid molecule to αS and the fractional saturation ϑ will be the ratio of SUV-bound αS molecules to total αS. If we call [P] and [L] the concentrations of free protein (αS) and lipid, P_t and L_t the total concentrations of protein and lipid, and $[\mathrm{PL}_n]$ the "concentration" of lipid-bound αS, we have

$$\vartheta = \frac{[\mathrm{PL}_n]}{P_t} \tag{2}$$

and

$$\begin{cases} L_t = [L] + n[PL_n] \\ P_t = [P] + [PL_n] \end{cases} \quad (3)$$

By substituting Eqs. (3) and (2) in (1), we can write a quadratic equation in the unknown ϑ:

$$nk_B P_t \vartheta^2 - (nk_B P_t + k_B L_t + 1)\vartheta + k_B L_t = 0 \quad (4)$$

and solve it ($0 \le \vartheta \le 1$):

$$\vartheta = \frac{1}{2}\left[1 + \frac{L_t}{nP_t} + \frac{1}{nk_B P_t} - \sqrt{\left(1 + \frac{L_t}{nP_t} + \frac{1}{nk_B P_t}\right)^2 - \frac{4L_t}{nP_t}}\right] \quad (5)$$

If we call the total lipid:protein molar ratio X and define c as $nk_B P_t$, Eq. (5) becomes

$$\vartheta = \frac{1}{2}\left[1 + \frac{X}{n} + \frac{1}{c} - \sqrt{\left(1 + \frac{X}{n} + \frac{1}{c}\right)^2 - \frac{4X}{n}}\right] \quad (6)$$

c is important for ITC measurements, as it serves as a preliminary measure of the sharpness of the sigmoid during the experimental design (*see* Fig. 3 and ref. 16). Both CD and ITC titration curves can now be described in terms of the fractional saturation ϑ.

3.6.1 CD Data Analysis

In CD-monitored titration the signal at 222 nm comes from the (weak) coil signal from unbound αS and the (strong) helical signal of lipid-bound αS, each proportional to the molar ratios of the two species:

$$[\Theta]_i^{222} = \frac{[PL_n]}{P_t}[\Theta]_{\text{helix}}^{222} + \frac{[P]}{P_t}[\Theta]_{\text{coil}}^{222} \quad (7)$$

ϑ (at each point of the titration) will then be (using Eq. (3))

$$\frac{[PL_n]}{P_t} = \vartheta = \frac{[\Theta]_i^{222} - [\Theta]_{\text{coil}}^{222}}{[\Theta]_{\text{helix}}^{222} - [\Theta]_{\text{coil}}^{222}} \quad (8)$$

where $[\Theta]_i^{222}$ is the molar (or residue) ellipticity at each point of the titration. Combining Eqs. (5) and (8) we can write $[\Theta]_i^{222}$ as a function of X, obtaining the final form of the titration curve:

$$[\Theta]_i^{222} = [\Theta]_{\text{coil}}^{222} + \frac{[\Theta]_{\text{helix}}^{222} - [\Theta]_{\text{coil}}^{222}}{2}\left[1 + \frac{X}{n} + \frac{1}{nk_B P_t} - \sqrt{\left(1 + \frac{X}{n} + \frac{1}{nk_B P_t}\right)^2 - \frac{4X}{n}}\right] \quad (9)$$

This equation can be used to fit CD titration curves using an n independent site binding model: n and k_B can be left as unconstrained parameters to be fitted, $[\Theta]_{\text{coil}}^{222}$ is the value of the 222 nm signal from the free protein before the first addition, and $[\Theta]_{\text{helix}}^{222}$ is

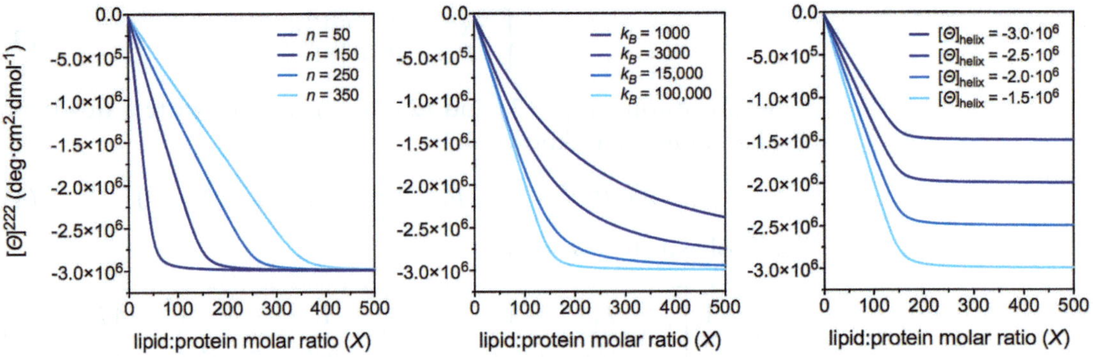

Fig. 4 Simulated CD binding curves of 10 μM αS (P_t) with varying (*left*) *n*, (*center*) k_B, (*right*) $[\Theta]^{222}_{helix}$. *See* Subheading 3.6.1

the 222 nm molar (or residue) ellipticity at the end of the titration, in a condition of lipid saturation (Figs. 1 and 4).

3.6.2 ITC Data Analysis

(The ITC data analysis procedures and equations are described for a lipid-into-protein titration, but they can be easily adapted to protein-into-lipid experiments or other designs; *see also* refs. 18, 19, 28).

While in CD the signal is cumulative (and the shape of the titration curve reflects this), in ITC the heat release from each injection is usually plotted, thus measuring the ΔH contributions arising from the transition of the system from the equilibrium state reached at injection i to the new equilibrium after injection $i + 1$. For that reason, we observe a sigmoidal curve, where the early injections provide a robust and almost constant heat release, while the last ones, upon lipid saturation, cause little or no exchange. In order to describe the shape of this titration curve with our model, we must first express the cumulative heat (the total heat exchanged by our sample, initially lipid free) as a function of the fractional saturation ϑ [18]:

$$Q = n\vartheta P_t \cdot \Delta H_{m,T} \cdot V_0 \qquad (10)$$

$\Delta H_{m,T}$ is the binding enthalpy per mole of ligand at the titration temperature T and V_0 the cell volume. While in CD-monitored titrations each injection can be manually corrected beforehand to account for dilution, dilution effects in ITC are usually included in the expression of the fitting equation, but are neglected here for simplicity (*see* [26]). The final form of the ITC titration curve is a plot of the inter-addition differences in heat release against lipid: protein molar ratios (X). Although the additions are discrete ($\Delta Q/\Delta L_t$), we can imagine to be working in the limiting case of infinitesimal titrant additions and obtain the functional form of the curve by deriving the heat release by the lipid concentration (dQ/dL_t):

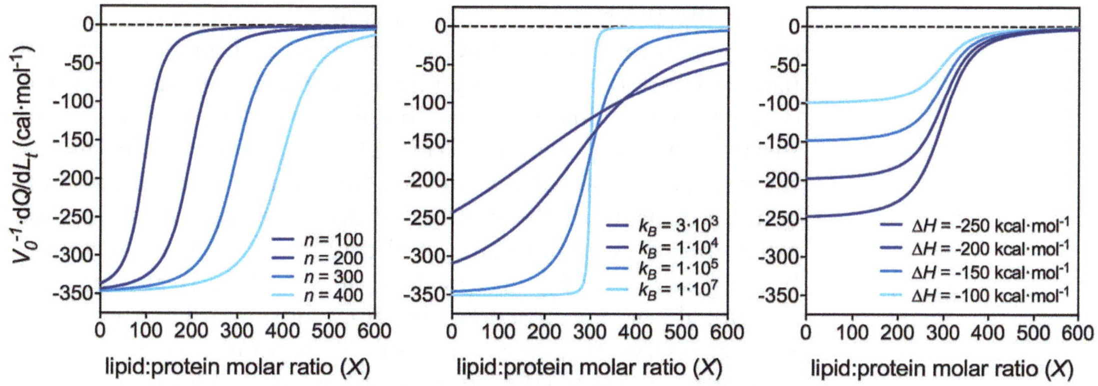

Fig. 5 Simulated ITC binding curves of 2.5 μM αS (P_t) with varying (*left*) n, (*center*) k_B, (*right*) ΔH. *See* Subheading 3.6.2

$$\frac{dQ}{dL_t} = \frac{\Delta H_{m,T} V_0}{2} + \Delta H_{m,T} V_0 \cdot \frac{1 - \frac{L_t}{nP_t} - \frac{1}{nk_B P_t}}{2\sqrt{\left(1 + \frac{L_t}{nP_t} + \frac{1}{nk_B P_t}\right)^2 - \frac{4L_t}{nP_t}}} \quad (11)$$

Equation (11) can be rewritten as a function of X:

$$\frac{dQ}{dL_t} = \frac{\Delta H_{m,T} V_0}{2} + \Delta H_{m,T} V_0 \cdot \frac{1 - \frac{X}{n} - \frac{1}{nk_B P_t}}{2\sqrt{\left(1 + \frac{X}{n} + \frac{1}{nk_B P_t}\right)^2 - \frac{4X}{n}}} \quad (12)$$

Equation (12) is used for the fitting of ITC binding data with an n independent site model [16, 19] (Figs. 2 and 5). V_0 and P_t are known instrumental and experimental parameters, while n, k_B, and $\Delta H_{m,T}$ are the unconstrained fitting parameters to be obtained from the binding data. By combining k_B and $\Delta H_{m,T}$, the entropic component of the Gibbs free energy can also be obtained:

$$\Delta H_{m,T} - T\Delta S_{m,T} = \Delta G_{m,T} = -RT \ln k_B \quad (13)$$

Since both Eqs. (9) and (12) were derived from the same binding model, they should provide comparable thermodynamic and binding parameters, allowing for cross-technique validation of the data obtained. This, naturally, relies on the assumption that the phenomena observed (binding of αS to lipid membrane and coil-helix transition of αS) are closely correlated, which has been shown to be the case for a number of model membranes [11, 12, 29]. Even so, one could imagine particular membrane compositions (e.g., bearing a strong positive surface charge) with which αS could interact in its unfolded form as well, e.g., through mainly electrostatic interactions with its negatively charged C-terminus. For this and other reasons, the imperfect correlation between ITC and CD measurements should not necessarily be considered an artifact, but could just be the result of multiple different binding modes.

In the absence of the afore described complications, the data obtained from CD and ITC will provide a complete thermodynamic picture of the binding phenomenon, by measuring stoichiometry (n), microscopic binding constants (k_B), and thermodynamic potentials (ΔH, ΔG, $\Delta S_{m,T}$). Often, because of the peculiar nature of the binding interaction (the stoichiometry is not a measure of the number of lipid-binding sites on αS, but rather provides an estimate of the number of lipids forming a binding site for αS), the product of n and k_B is also reported (a "global" or "macroscopic" binding constant). For this reason, n can also serve by itself as a measure of the number of binding sites present on a vesicle (the number of lipid molecules in an average vesicle can be assumed constant): a lower n can be thought of as a higher "avidity" (vs. the affinity described by k_B), even when k_B does not change by much. Finally, because of the complexity of the phenomenon observed (a combination of binding, folding, and membrane remodeling), it is often hard to interpret straightforwardly, in terms of molecular mechanisms, the signs and entities of the thermodynamic potentials [10, 28]. Yet, through a combination of carefully designed experiments, using, e.g., mutants or different membrane compositions (and most likely with the help of other techniques as well), it is possible to uncouple and characterize aspects of αS's binding and folding pathway [11, 29, 30].

4 Notes

1. Lipid quantitation in this protocol is based on the accurate weighing of the lipid powders or the careful aliquoting of the chloroform stock solution, both of which lead to errors (albeit generally small) in the estimates of the stock solution concentration and composition, because of the difficulties in the manipulation of powders (often waxy or hygroscopic) and the pipetting of organic solvents. If the accuracy of the thermodynamic potentials and constants to be measured is critical, an aliquot of the lipid solution can be kept and analyzed later, using, e.g., a spectrophotometric phosphate micro-assay [31].
2. For CD-monitored αS titrations a good starting concentration for the SUV stock solution is 10 mM (the protein concentration can be between 5 and 15 μM); for ITC, when performing a lipid-into-protein titration with an αS concentration of 5 μM in the cell, a 20 mM lipid stock solution should be prepared. Generally, while the stability of the SUV dispersions varies wildly with temperature and lipid composition, it is challenging to obtain homogeneous and stable dispersions that are more than 40 mM in lipids (*see also* **Note 6**).

3. The phase transition temperature (T_m) for most lipids and lipid mixtures can be obtained from the literature, manufacturer, or lipid data collections [32]. It must be noted that the T_m will also change slightly with the lipid assembly being studied (e.g., SUVs, having a more strained packing, will undergo their phase transition at temperatures slightly lower than a planar bilayer with the same composition).
4. While the time, temperature, and mode of hydration could be considered minor parameters in the protocol, they affect the distribution of sizes and morphologies of the MLV dispersion produced after the hydration step [22]. If the MLVs are then to be sonicated these parameters become less critical, but it is nevertheless advisable to keep them constant throughout the experiments.
5. As with the other parameters in the liposome preparation protocol, the pulse sequence, total sonication time, and pulse parameters (e.g., power/amplitude) should be kept constant to ensure the consistency of the lipid dispersions prepared over time. Both bath and microprobe setups can be used [22]. Given the many variables involved in the preparation of liposomes, a lab setting up a new liposome preparation protocol should characterize the morphology and size of their liposome dispersions (*see* **Note** 7) and ensure minimal run-to-run variability.
6. For lipid samples more concentrated than 20 mM or rich in cholesterol, longer sonication times are advised (30 min or longer). TEM and DLS can confirm the homogeneity of the sonicated sample, if doubts should persist (*see* **Note** 7).
7. Liposomal samples for TEM can be deposited on glow-discharged carbon-coated grids and stained with neutral phosphotungstic acid or ammonium molybdate (1–2%). Uranyl formate and acetate, although they provide better contrast, should be used with caution and their micrographs compared to the ones obtained using neutral negative stains, since the low pH of their staining solutions often generates artifacts [22, 24].
8. As mentioned in Subheading 3.1.1, buffers (and lipids, but often the lipid choice is dictated by the specific research question) that absorb robustly in the 260–185 nm range should be avoided. A CD spectropolarimeter measures the differential absorbance of circularly polarized light, so a strongly absorbing sample will reduce the amount of transmitted light and thus the signal detected, decreasing the S/N ratio. The instrument will compensate by increasing the voltage of its photomultiplier (HT on Jasco instruments, which should always be plotted in the output along with the ellipticity as a quality control measure), but HT values above 600 V indicate a signal quality too

low to be trusted and the corresponding data points should be discarded.

9. SUVs (and small LUVs, ~100 nm) provide minimal (and approximately constant with the increase in lipid concentration) scattering signal [33, 34]; thus a tedious measurement of a blank titration for each day of experiments can be avoided (it is advisable, however, to verify that the signal is indeed neglectable or small but constant by performing, at least once, an SUV-into-buffer titration). Larger vesicles, on the other hand, give a robust scattering signal, which both makes it challenging to reach low wavelengths and reduces the overall protein signal because of its broad, positive signal in the region 200–260 nm.
10. A good set of starting parameters for a CD spectrum are a scanning speed of 50–100 nm/min., bandwidth of 1 nm (data pitch 0.5 nm), integration time (measurement time) of 2–4 s (D.I.T. on Jasco instruments), and a wavelength range of 260–190 nm. Three or more accumulations can be recorded, depending on the quality of the signal (or the scanning speed can be lowered to, e.g., 50 nm/min).
11. A common artifact which can help recognize the presence of robust (non-neglectable) scattering contributions is the lack of a plateau and the increase of the signal at high lipid:protein ratios (consequence of the fact that the helical signal will tend to stabilize toward the end of the titration, while the scattering signal will keep increasing with the increase in lipid content).
12. Cleanliness of the cell is paramount to the measurement of high-quality and consistent data. Commercial (e.g., Hellmanex™) or homemade (e.g., 70% EtOH, 30% fuming HCl [15]) cleaning solutions can be used. Usually the cuvettes can be incubated overnight in the cleaning agent, but if multiple titrations have to be performed with the same cell 30 min will do. It is also important to rinse the cell with a generous amount of deionized water before use; a buffer should always be quickly measured beforehand. In the case of sample aggregation, harsher detergents (e.g., HNO_3 or 6 M GdnCl) should be employed.
13. When working with protein samples, the raw ellipticity output Θ (in mdeg) from the instrument can be converted to either molar or residue ellipticity:

$$[\Theta]\left(\frac{\text{deg}\cdot\text{cm}^2}{\text{dmol}}\right)=\frac{\Theta\ (\text{mdeg})}{10c\ (\text{mol}\cdot\text{L}^{-1})\ \ell\ (\text{cm})}\quad(\text{molar ellipticity})$$

$$[\Theta]_{\text{MRW}}\left(\frac{\text{deg}\cdot\text{cm}^2}{\text{dmol}}\right)=\frac{\text{MRW}\ (\text{g}\cdot\text{mol}^{-1})\cdot\Theta}{10c\ (\text{g}\cdot\text{L}^{-1})\ \ell\ (\text{cm})}\quad(\text{mean residue ellipticity})$$

where MRW (mean residue weight, the average MW of the protein's amino acids) is defined as (N is the number of amino acids)

$$\text{MRW}=\frac{\text{MW}}{N-1}\sim 110\ \text{g}\cdot\text{mol}^{-1}\quad(\text{for most proteins})$$

14. The reference cell should be refilled approximately once a week with filtered, degassed, deionized (or Milli-Q) water, taking special care not to introduce air bubbles (*see* Subheading 3.5).
15. A critical requirement of ITC is that the buffer in the cell and the buffer in the syringe are matched; otherwise spurious heat contributions from buffer mixing and dilution will be measured and could easily mask the protein-lipid heat signal. For proteins that are dialyzed, it is recommended to use the last dialyzate as a matching buffer. Buffers should be preferably degassed (e.g., by incubation overnight in a vacuum desiccator), especially if titrations are to be performed at different temperatures.
16. It is advisable to visually inspect the DP versus time plot during the titration in order to recognize eventual problems or artifacts (e.g., the presence of air bubbles, marked by spikes in the DP profile) and correct them before repeating the experiment.
17. If protein or lipid precipitation occurs, both cell and syringe should be soaked in a mild detergent (e.g., 10–20% Contrad 70) for ≥60 min (ideally overnight) and subsequently emptied and washed thoroughly.
18. While commercial software packages can usually draw a good baseline and easily identify peaks, it is advisable to visually inspect the raw data and eventually manually correct the baseline, if needed, especially if the heat signal is weak and the baseline noisy. Inspection of the raw data is also necessary to evaluate whether a point from the binding curve can be excluded as artifactual (e.g., because of spikes due to air bubbles in the cell).
19. Even after subtraction of the lipid-into-buffer blank titration, the integrated heats in the final part of the sigmoid often do not plateau at 0 kcal/mol and therefore a constant must be manually subtracted. Care must be taken to ensure that no binding or other sigmoidal processes are present in the high lipid:protein range (e.g., by performing a titration with a higher titrant concentration) and that manual baseline subtraction is therefore warranted.

Acknowledgments

M.R. thanks Kelly L. Arnett and the Center for Macromolecular Interactions at the Harvard Medical School Department of Biological Chemistry and Molecular Pharmacology for assisting with CD and ITC measurements, revising this manuscript, and many helpful discussions; Alex E. Powers for helping with data collection and analysis; and Martina Astrid Rodda for editing and revising the manuscript.

References

1. Davidson WS, Jonas A, Clayton DF, George JM (1998) Stabilization of alpha-synuclein secondary structure upon binding to synthetic membranes. J Biol Chem 273:9443–9449. https://doi.org/10.1074/jbc.273.16.9443
2. Takamori S, Holt M, Stenius K et al (2006) Molecular anatomy of a trafficking organelle. Cell 127:831–846. https://doi.org/10.1016/j.cell.2006.10.030
3. George JM, Jin H, Woods WS, Clayton DF (1995) Characterization of a novel protein regulated during the critical period for song learning in the zebra finch. Neuron 15:361–372. https://doi.org/10.1016/0896-6273(95)90040-3
4. Jao CC, Hegde BG, Chen J et al (2008) Structure of membrane-bound alpha-synuclein from site-directed spin labeling and computational refinement. Proc Natl Acad Sci USA 105:19666–19671. https://doi.org/10.1073/pnas.0807826105
5. Chandra S, Chen X, Rizo J et al (2003) A broken alpha-helix in folded alpha-Synuclein. J Biol Chem 278:15313–15318. https://doi.org/10.1074/jbc.M213128200
6. Ulmer TS, Bax A, Cole NB, Nussbaum RL (2005) Structure and dynamics of micelle-bound human alpha-synuclein. J Biol Chem 280:9595–9603. https://doi.org/10.1074/jbc.M411805200
7. Lautenschläger J, Kaminski CF, Kaminski Schierle GS (2017) α-Synuclein—regulator of exocytosis, endocytosis, or both? Trends Cell Biol 27:468–479. https://doi.org/10.1016/j.tcb.2017.02.002
8. Jo E, Fuller N, Rand RP et al (2002) Defective membrane interactions of familial Parkinson's disease mutant A30P alpha-synuclein. J Mol Biol 315:799–807. https://doi.org/10.1006/jmbi.2001.5269
9. Choi W, Zibaee S, Jakes R et al (2004) Mutation E46K increases phospholipid binding and assembly into filaments of human alpha-synuclein. FEBS Lett 576:363–368. https://doi.org/10.1016/j.febslet.2004.09.038
10. Blume A, Garidel P (1999) Lipid model membranes and biomembranes. In: Kemp R (ed) Handbook of thermal analysis and calorimetry, From macromolecules to man, vol 4. Elsevier Science B.V, Amsterdam, pp 109–173
11. Nuscher B, Kamp F, Mehnert T et al (2004) Alpha-synuclein has a high affinity for packing defects in a bilayer membrane: a thermodynamics study. J Biol Chem 279:21966–21975. https://doi.org/10.1074/jbc.M401076200
12. Bartels T, Kim NC, Luth ES, Selkoe DJ (2014) N-alpha-acetylation of α-synuclein increases its helical folding propensity, GM1 binding specificity and resistance to aggregation. PLoS One 9:1–10. https://doi.org/10.1371/journal.pone.0103727
13. Fasman GD (1996) Circular dichroism and the conformational analysis of biomolecules. Plenum Press, New York, NY
14. Kelly SM, Jess TJ, Price NC (2005) How to study proteins by circular dichroism. Biochim Biophys Acta 1751:119–139. https://doi.org/10.1016/j.bbapap.2005.06.005
15. Greenfield NJ (2006) Using circular dichroism spectra to estimate protein secondary structure. Nat Protoc 1:2876–2890. https://doi.org/10.1038/nprot.2006.202
16. Wiseman T, Williston S, Brandts JF, Lin LN (1989) Rapid measurement of binding constants and heats of binding using a new titration calorimeter. Anal Biochem 179:131–137. https://doi.org/10.1016/0003-2697(89)90213-3
17. Freire E, Mayorga OL, Straume M (1990) Isothermal titration calorimetry. Anal Chem 62:950A–959A. https://doi.org/10.1021/ac00217a002

18. Fisher HF, Singh N (1995) Calorimetric methods for interpreting protein-ligand interactions. Methods Enzymol 259:194–221. https://doi.org/10.1016/0076-6879(95)59045-5
19. Indyk L, Fisher HF (1998) Theoretical aspects of isothermal titration calorimetry. Methods Enzymol 295:350–364. https://doi.org/10.1016/S0076-6879(98)95048-0
20. Noble JE, Bailey MJA (2009) Quantitation of protein. Methods Enzymol 463:73–95. https://doi.org/10.1016/S0076-6879(09)63008-1
21. Pace CN, Vajdos F, Fee L et al (1995) How to measure and predict the molar absorption coefficient of a protein. Protein Sci 4:2411–2423. https://doi.org/10.1002/pro.5560041120
22. Szoka F, Papahadjopoulos D (1980) Comparative properties and methods of preparation of lipid vesicles (liposomes). Annu Rev Biophys Bioeng 9:467–508. https://doi.org/10.1146/annurev.bb.09.060180.002343
23. MacDonald RC, MacDonald RI, Menco BP et al (1991) Small-volume extrusion apparatus for preparation of large, unilamellar vesicles. Biochim Biophys Acta 1061:297–303. https://doi.org/10.1016/0005-2736(91)90295-J
24. Bangham A, Hill M, Miller N (1974) Preparation and use of liposomes as models of biological membranes. In: Korn ED (ed) Methods in membrane biology, vol 1. Plenum Press, New York, NY, pp 1–68
25. Galvagnion C, Buell AK, Meisl G et al (2015) Lipid vesicles trigger α-synuclein aggregation by stimulating primary nucleation. Nat Chem Biol 11:229–234. https://doi.org/10.1038/nchembio.1750
26. MicroCal ITC analysis software using Origin™ MAN0577-02-EN-00 (May 2015)
27. Cantor CR, Schimmel PR (1980) Biophysical chemistry, Part III: The behavior of biological macromolecules. W.H. Freeman & Co., New York, NY
28. Wieprecht T, Seelig J (2002) Isothermal titration calorimetry for studying interactions between peptides and lipid membranes. Curr Top Membr 52:31–56
29. Bartels T, Ahlstrom LS, Leftin A et al (2010) The N-terminus of the intrinsically disordered protein α-synuclein triggers membrane binding and helix folding. Biophys J 99:2116–2124. https://doi.org/10.1016/j.bpj.2010.06.035
30. Ferreon ACM, Deniz AA (2007) Alpha-synuclein multistate folding thermodynamics: implications for protein misfolding and aggregation. Biochemistry 46:4499–4509. https://doi.org/10.1021/bi602461y
31. Chen PS, Toribara TY, Huber W (1956) Microdetermination of phosphorus. Anal Chem 28:1756–1758. https://doi.org/10.1021/ac60068a036
32. Marsh D (2013) Handbook of lipid bilayers, 2nd edn. CRC Press, Boca Raton, FL
33. Litman BJ (1972) Effect of light scattering on the circular dichroism of biological membranes. Biochemistry 11:3243–3247. https://doi.org/10.1021/bi00767a018
34. Mao D, Wallace BA (1984) Differential light scattering and absorption flattening optical effects are minimal in the circular dichroism spectra of small unilamellar vesicles. Biochemistry 23:2667–2673. https://doi.org/10.1021/bi00307a020

Chapter 12

Yeast-Based Screens to Target Alpha-Synuclein Toxicity

Inês Caldeira Brás, Blagovesta Popova, Gerhard H. Braus, and Tiago F. Outeiro

Abstract

The budding yeast *Saccharomyces cerevisiae* (*S. cerevisiae*) has been a remarkable experimental model for the discovery of fundamental biological processes. The high degree of conservation of cellular and molecular processes between the budding yeast and higher eukaryotes has made it a valuable system for the investigation of the molecular mechanisms behind various types of devastating human pathologies. Genetic screens in yeast provided important insight into the toxic mechanisms associated with the accumulation of misfolded proteins. Thus, using yeast genetics and high-throughput screens, novel molecular targets with therapeutic potential have been identified. Here, we describe a yeast screen protocol for the identification of genetic modifiers of alpha-synuclein (aSyn) toxicity, thereby accelerating the identification of novel potential targets for intervention in Parkinson's disease (PD) and other synucleinopathies.

Key words *Saccharomyces cerevisiae*, Yeast, Parkinson's disease, Synucleinopathies, Alpha-synuclein, Genetic screens

1 Introduction

1.1 The Yeast S. cerevisiae: A Versatile System for Genetic Screens

S. cerevisiae has significantly contributed to the understanding of fundamental cellular processes in eukaryotes, including cell division, replication, metabolism, protein folding, and intracellular transport. Among the various features that make yeast a powerful model system, the "awesome power of yeast genetics" is chief [1, 2]. The simple mating process of haploid strains to sporulate diploid strains, along with the accessibility to mutant collections and open reading frame (ORF) libraries, allows the rapid isolation of genetic modifiers of specific processes or phenotypes. *S. cerevisiae* is among the simplest systems and is one of the best characterized eukaryotic organisms. In addition, yeast was the first eukaryote to have its genome fully sequenced [3]. In line with the evolutionary conservation of genes in eukaryotes [4–6], the molecular machinery necessary for many cellular functions is highly conserved between yeast and humans [7, 8]. This includes components of

Tim Bartels (ed.), *Alpha-Synuclein: Methods and Protocols*, Methods in Molecular Biology, vol. 1948,
https://doi.org/10.1007/978-1-4939-9124-2_12,

vesicular trafficking [9], mitochondrial biogenesis [10], unfolded protein response [11], autophagy [12], and protein quality control systems [13]. These features make *S. cerevisiae* a unique tool to study disease-associated cellular processes, and to aid in the identification of novel therapeutic targets, using genetic screens [14–19].

The study of the molecular basis of human diseases using yeast models can be accomplished via different approaches. One approach is the study of the function of yeast homologs of the genes of interest by assessing their effect on relevant phenotypes, through either deletion or overexpression screens [2, 20–22]. Genes that enhance or suppress toxicity can be identified by assessing their effect on cell viability, measuring, for example, colony formation. In this manner, the screens provide insight into the function of disease-associated genes. Different factors should be considered when designing the screening strategy and the phenotype used in the screen. These include, for example, the type and strength of the promoter, plasmid copy number, selection markers, and use of epitope tags.

Several collections of yeast strains and plasmids have been assembled over the years and are widely used by the yeast community [23]. For convenience, yeast strains can be arrayed in microtiter plates (96- or 384-well plates) and the entire library can be manipulated systematically either manually or using liquid-handling robots. The screens can be performed by transforming the collection of strains with a plasmid encoding for the gene of interest. In another type of screen, such as an overexpression screen, the screening strain of interest can be transformed with a collection of plasmids encoding all or selected sets of yeast genes. The protocol for both approaches is quite similar, and involve, in these cases, simple transformation procedures that are widely established. After transformation, cells can be plated onto solid media. If the protein of interest is toxic to the cells, reduced growth is to be expected, in comparison to a control. The identification of enhancers or suppressors of the phenotypes of interest can then provide insight into the cellular pathways affected by the gene of interest.

1.2 S. cerevisiae as a Model of Parkinson's Disease

Parkinson's disease (PD) is a neurodegenerative disorder that affects an increasing number of people worldwide. Yeast models reproduce important molecular mechanisms associated with PD, such as aSyn cytotoxicity and aggregation. The rationale for developing yeast models of human neurodegenerative disorders, including PD, is the fact that several disease-associated cellular pathways are highly conserved from yeast to humans [24, 25], enabling the study of the function of mammalian gene products and the fundamental molecular mechanisms associated with PD [26].

The yeast models of PD consist of the heterologous expression of human aSyn [14] or other PD-related genes [27], such as ATP13A2 [28], DJ-1 [29, 30], EIF4G1 [31], LRRK2 [32, 33],

Parkin [34], VPS35 [31], Pink1 [34], Synphilin-1 [35], and beta-synuclein [36]. Human aSyn has been expressed in yeast using different expression systems. Since aSyn expression in yeast is toxic, the use of multicopy plasmids yields more variable results, as the cells have the capacity to reduce the average plasmid copy number to reduce aSyn expression and toxicity [14]. To avoid this, stable expression of aSyn can be achieved by integrating the cDNA encoding for aSyn in the yeast genome. With this approach, the levels of toxicity can be altered by varying the number of aSyn copies integrated [14, 37, 38], thereby mimicking the dose-dependent toxicity observed in familial forms of the disease-carrying duplications or triplications of the aSyn gene [39, 40].

The expression of human aSyn in yeast results in the localization of the protein at the plasma membrane, at lower levels of expression, or in cytosolic inclusions, at higher levels of expression [14, 41, 42]. Other consequences of aSyn overexpression in yeast include impairment of vesicular trafficking, endocytosis and vacuolar degradation [19, 43–45], impairment of proteasome activity and accumulation of cytoplasmic lipid droplets [14], ER stress [46], activation of the heat-shock response [47], mitochondrial dysfunction [37], reduced life span and induction of autophagy and mitophagy [14, 19, 48, 49], impairment of endocytosis [14], and oxidative stress and apoptosis induction [17, 18, 37, 41].

The use of yeast to model basic molecular aspects of PD enabled the discovery of several cellular processes and the identification of novel players involved in aSyn-mediated cytotoxicity. Together, these results demonstrate the potential of genome-wide screens in yeast for providing insight into the molecular underpinnings of various neurodegenerative diseases, which is essential for the development of novel therapeutic strategies for these devastating conditions.

The protocols described below can be used for large-scale overexpression screens in yeast. Furthermore, we provide an additional protocol to the identification and validation of genetic modifiers of aSyn toxicity. The first protocol enables the replication of the collections of strains in 96-well plates (*see* Subheading 3.1). The second protocol is a simple method for performing up to 96 transformations at a time, enabling the rapid testing of candidate modifiers (*see* Subheading 3.2). To confirm the candidates identified, we suggest the use of the regular yeast transformation protocol (*see* Subheading 3.3). This enables the elimination of false positives that can occur in any large-scale screens. These protocols can be performed using yeast genome-wide libraries such as the *Saccharomyces* gene deletion collection [50, 51], the Tet-promoter collection (yTHC) [52], the DaMP (decreased abundance by mRNA perturbation) collection [53], the library of temperature-sensitive alleles of essential genes [54], as well as other collections of yeast mutants available in array format [55, 56].

2 Materials

1. Manual pin tool: 96 Solid Pin Multi-Blot Replicators or Singer ROTOR HDA bench robot (Singer Instruments, Somerset, UK).
2. Large-capacity refrigerated centrifuge equipped with microtiter plate rotor and Falcon tube rotor.
3. Tabletop centrifuge.
4. 96-Well round-bottom microtiter plates.
5. Gas-permeable adhesive seals.
6. Yeast peptone (YPD) liquid media: 10 g/L Yeast extract, 20 g/L peptone, 20 g/L glucose.
7. YPD media plates: 10 g/L Yeast extract, 20 g/L peptone, 20 g/L glucose, 20 g/L agar, 200 μg/mL geneticin (G418).
8. Synthetic complete dropout selection plates: 6.7 g/L Yeast nitrogen base (YNB) without amino acids, 20 g/L raffinose or galactose, approximately 0.7 g/L of complete supplement mixture (CSM) of amino acids (the exact amount of CSM dropout to add is on the container), 20 g/L agar.
9. 10× TE buffer: 0.1 M Tris-HCl pH 7.5, 0.01 M EDTA pH 7.5.
10. Carrier DNA: 10 g/mL Salmon sperm DNA.
11. Lithium acetate buffer (LiOAc): 1 M LiOAc pH 7.5 (adjusted with acetic acid).
12. Polyethylene glycol (PEG): 50% (w/v) PEG MW 3350.
13. Sterile 30% (v/v) glycerol solution.
14. Plasmid DNA.

3 Methods

3.1 Preparation of Frozen Stocks

1. Prepare the YPD media plates to inoculate the cells from the stocks the day prior to the screen.
2. Peel off the foil of the frozen 96-well microtiter plates and let them thaw completely.
3. Sterilize the 96-pin replicator by soaking in ethanol and flaming. After cooling down, immerse the pins into the glycerol stock until the bottom of the wells and rotate gently several times.
4. Replicate the glycerol stocks by pinning the small amount of liquid onto the surface of YPD media plate, and immediately return the stocks to −80 °C. to ensure that cross-contamination of strains does not occur, be very careful when handling the plates. Record the orientation of the plate.

5. Incubate the plate at 30 °C for 24–48 h (25 °C for temperature-sensitive mutants).
6. When the clones are grown, prepare sterile 96-well microtiter plates with 100 μL of YPD liquid media in each well.
7. After sterilizing the pinning tool, position it carefully over the colony spots. Pick up the cells by gently swirling the replicator in order to transfer them to the pins. Try to pick as many cells as possible.
8. Dip the replicator into the microtiter plate containing 100 μL of YPD liquid media. Mix by gentle rotations to resuspend the cells. Take care not to cross-contaminate adjacent wells.
9. Transfer the cell suspensions to 1.5 mL cryovials or sterile Eppendorf tubes and add 100 μL 30% (v/v) glycerol.
10. Invert the tubes to mix the cells and glycerol. Store at −80 °C.

3.2 High-Throughput Yeast Transformation

This protocol is designed for the transformation of four 96-well plates. It can be scaled up or down, according to the number of plates needed. We do not advise to process more than twelve 96-well plates (three 384 plates) in one round of transformation, to ensure optimal results. Most collections of yeast mutants contain the KanMX cassette that confers resistance against G418. Thus, this antibiotic can be supplemented into the YPD plates at a concentration of 200 μg/mL to maintain selective pressure.

1. Repeat the **steps 1–5** from Subheading 3.1.
2. Dispense 100 μL of sterile water into each well of the four sterile 96-well microtiter plates using 8-multichannel pipette.
3. After sterilizing the pinning tool, position it carefully over the colony spots. Pick up the cells by gently swirling the replicator in order to transfer them to the pins. Try to pick as many cells as possible.
4. Dip the replicator into the microtiter plate containing 100 μL of sterile water. Mix by using gentle rotations in order to resuspend the cells. Take care not to cross-contaminate adjacent wells.
5. Centrifuge the plates at 1500 *g* for 5 min at 4 °C using a centrifuge with microtiter plate rotor.
6. Meanwhile, prepare 22 mL of transformation mixture according to Table 1. The amount of plasmid DNA per well can differ between 0.1 and 1 μg. Boil the salmon sperm DNA for 15 min and then chill on ice before using.
7. Discard the supernatant with a single shake of the plate over a waste container. Immediately put the plate upside down on a clean paper tissue in order to absorb residual supernatant that might adhere on the ridges between the wells. Do not drag the

Table 1
Composition of the transformation mixture

Component	Volume (mL)
1 M LiOAc buffer	6.6
Salmon sperm DNA (10 mg/mL) (boiled)	2.2
Plasmid DNA (0.1–1 μg/μL)	0.440
Sterile H_2O	12.76

plate over the tissue in order to avoid cross-contaminations. It is advisable to practice this step before the experiment.

8. Pour the transformation buffer into a sterile tray or empty sterile Petri dish. Pipette 50 μL transformation mixture into each well of the plate using a multichannel pipette. Cover the microtiter plates with the lids, secure them on a rotary shaker, and shake for 5 min at 500 rpm to resuspend the cells.
9. Pipette 100 μL of 50% PEG solution into each well using multichannel pipette.
10. Seal the plates with gas-permeable adhesive seals and shake again for 5 min at 500 rpm.
11. For temperature-sensitive mutants, incubate the plates for 45 min at 30 °C or 25 °C.
12. Ensure that the plastic seal adheres well at the edges of the microtiter plate and put the plates into a 42 °C water bath. The plates will float in the water. Incubate for 40 min. For transforming temperature-sensitive mutants, incubate for 15 min.
13. Spin down the microtiter plates at 1500 *g* for 10 min and discard the supernatant as described in **step 3**.
14. Pipette 20 μL of sterile water into each well using a multichannel pipette and resuspend the cell pellet.
15. Use the sterile 96-pin replicator and mark the pins onto four plates with synthetic complete dropout selection medium. This will mark the position of the spots.
16. To perform the spotting assays, pipette 10 μL cell suspension with the multichannel pipette onto the markings left on the selection plates. Pipette carefully to avoid merging of the spots into one another. Allow the spots to dry before you transfer the plates to the incubator.
17. Incubate the plates at 30 °C (25 °C for temperature-sensitive mutants) for three to 4 days, until colonies are visible.
18. Inspect the plates visually. Strains with lower transformation efficiency may display single transformants at the edge of the

spot (*see* **Notes 1** and **2**). Use sterile inoculating needles or sterile toothpicks to distribute the single transformants through the whole area of the spot. This step will allow the subsequent transfer of colonies with the 96-well replicator for spotting tests even for strains with lower transformation efficiency. Incubate the plates for one more day at 30 °C or 25 °C.

19. Compare the initial 96-well plate with the transformation plates and determine the transformation efficiency (*see* **Notes 3** and **4**).

3.3 Regular Yeast Transformation for Validation of the Results

This protocol is designed for the confirmation of the candidate genes identified in the screen. We describe the protocol for one transformation, but it can be scaled up according to the number of genes to be tested. For yeast transformation, variations of the LiOAc TRAFO method have been used in several protocols [57].

1. Prepare a pre-inoculum of the yeast strain of interest in 10 mL YPD liquid media by incubating overnight in a shaking incubator (220 rpm) at 30 °C.
2. On the next day, dilute the culture to OD_{600nm} of 0.1 (10 mL culture is enough for two transformations).
3. Grow the inoculum for several hours, until the culture reaches an OD_{600nm} of 0.5–0.7.
4. Harvest the inoculum at 1000 *g* for 5 min at 4 °C.
5. Poor off the media and wash the cells using 30 mL of sterile water. Repeat the centrifugation at 1000 *g* for 5 min at 4 °C. After this step, maintain the cells in ice.
6. Meanwhile, prepare the following solutions for one reaction (Table 2). This can be scaled up by proportionality increasing the volumes of the components. Boil the salmon sperm DNA for 15 min and then chill on ice before use.
7. Resuspend the cells in 1 mL of sterile water and transfer them into a 1.5 mL Eppendorf tube. Centrifuge the cells at maximum speed in a tabletop centrifuge for 30 s.
8. Wash the pellet in 300 μL of LiOAc/TE solution. Repeat the centrifugation at maximum speed in a tabletop centrifuge for 30 s. Resuspend the pellet in the same solution once again.
9. For each individual transformation, add the following components, in the following order:
   (a) 50 μL of cell suspension.
   (b) 5 μL of salmon sperm DNA (10 mg/mL).
   (c) 0.5–1 μg of transformation DNA.
   (d) 300 μL of LiOAc/TE/PEG solution.

Table 2
Composition of transformation solutions for one transformation

	Component	Volume (μL)
LiOAc/TE solution	1 M LiOAc buffer pH 7.5	60
	10× TE buffer	60
	Sterile H_2O	480
LiOAc/TE/PEG solution	1 M LiOAc buffer pH 7.5	30
	10× TE buffer	30
	PEG 50% (W/V)	240

10. Vortex the mixture vigorously to ensure the resuspension of the cell pellet.
11. Incubate the mixture at 30 °C in a shaking incubator (220 rpm) for 30 min.
12. Heat shock at 42 °C for 20 min in a water bath.
13. Pellet the cell suspension at maximum speed in a tabletop centrifuge for 30 s.
14. Remove the transformation mixture and resuspend the pellet in 1 mL of sterile water.
15. Pipette 100 μL of the cell suspension onto plates of appropriate SC selection medium. Incubate at 30 °C for 2–3 days.
16. Isolate four transformants in a new SC selection medium plate for further experiments. Incubate at 30 °C for 2–3 days.
17. To perform spotting assays, inoculate the transformants in 5 mL of YPD liquid medium overnight at 30 °C.
18. On the next day, dilute the culture to OD_{600nm} of 0.1 and prepare several dilutions (10^1–10^6) from the initial culture using sterile water.
19. Drop 5 μL of each dilution onto appropriate selection plates using a multipipette, creating a grid of descending cell concentrations.
20. Let the plates dry completely (otherwise the drops will mix and it will not be possible to distinguish the cell growth).
21. Incubate the plates at 30 °C for 2–3 days, until colonies are visible.
22. Inspect the plates visually. Strains with lower transformation efficiency may display single transformants at the edge of the spot (*see* **Notes 1** and **2**).
23. Compare the transformation plates (*see* **Note 4**).

4 Notes

1. In case of low transformation efficiency, increase the number of cells transferred. Furthermore, the time for the heat shock can be optimized. This can take from 15 min to 1 h, depending on the yeast strain. We recommend the use of fresh salmon sperm DNA.
2. In case of low number of transformants *per* strain, increase the amount of plasmid DNA to 1 μg per reaction. Improve the shaking of the cell suspension before and after adding the transformation mixture. For the spotting assays, repeat the addition of the cell suspension on the selection plates. Wait for the spots from the first transfer to dry completely and make a second transfer on the same spots.
3. Detect carefully the presence of cross-contaminations. Compare your transformation plate with the initial plate and look for extra clones at originally empty positions. They are indicative for cross-contaminations. Take care not to touch neighboring wells while pipetting. Removing the seal while the plates are frozen minimizes the risk of cross-contamination.
4. Examples of yeast plates from an overexpression screen for the identification of suppressors and enhancers of aSyn toxicity are shown in Figs. 1 and 2. Yeast mutant strains were transformed with a high-copy plasmid harboring the *GAL1*-driven aSyn gene. The 96 transformed strains using the Subheading 3.2 protocol were spotted in quadruplicate, resulting in an array of 384 colonies. The spotting can be done using manual pin tools or using the Singer ROTOR HAD. Figure 2 is a representative image of human aSyn overexpression in the BY4741 wild-type

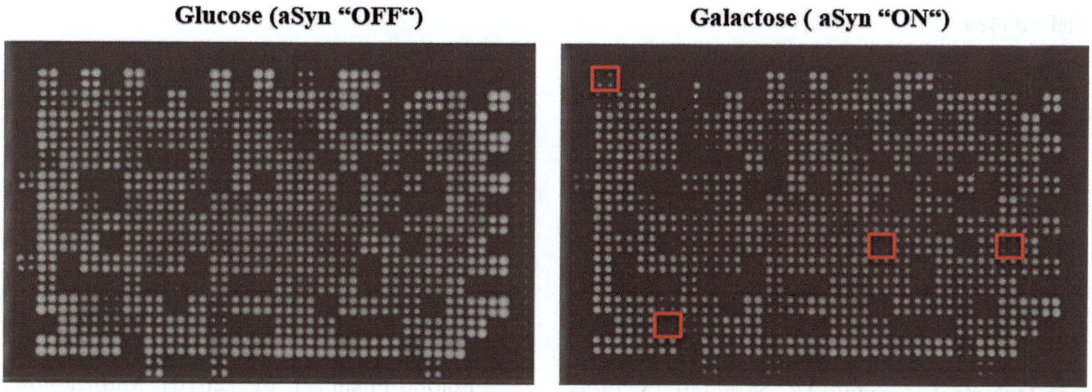

Fig. 1 Yeast overexpression screen for the identification of modifiers of aSyn toxicity. Representative plates from an aSyn modifier screen. Colonies were spotted onto SC-URA plates, supplemented with glucose, in order to repress the expression of aSyn (aSyn "OFF"), or galactose, in order to induce the expression of aSyn (aSyn "ON"). Red squares indicate yeast mutant strains that significantly enhance the toxicity of aSyn

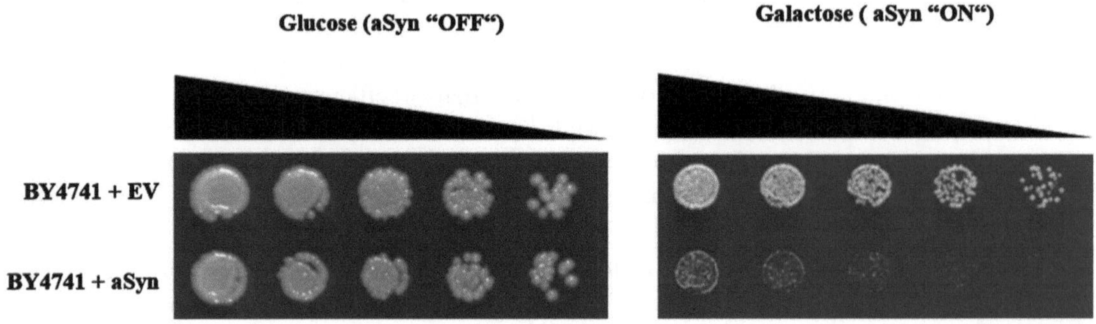

Fig. 2 Representative image of the BY4741 strain overexpressing human aSyn. Spotting assays of yeast cultures expressing aSyn under the control of a *GAL1*-inducible promoter or carrying an empty vector as a control. Cell suspensions were adjusted to the same OD_{600nm}, serially diluted, and spotted onto the surface of SC-URA plates, supplemented with glucose (aSyn "OFF"), or galactose (aSyn "ON")

strain, performed using the protocol in Subheading 3.3. Cells were spotted onto selection plates containing glucose, to repress the expression of aSyn, or galactose, to induce the expression of aSyn. The plates were imaged after 3 days of growth at 30 °C and scored for fitness (*see* Fig. 2 for an example).

Acknowledgments

T.F.O. and G.H.B. are supported by the DFG Center for Nanoscale Microscopy and Molecular Physiology of the Brain. This project has received funding from the European Union's Horizon 2020 research and innovation programme under grant agreement No. 721802.

References

1. Giorgini F, Muchowski PJ (2006) Screening for genetic modifiers of amyloid toxicity in yeast. Methods Enzymol 412:201–222
2. Tenreiro S, Outeiro TF (2010) Simple is good: yeast models of neurodegeneration. FEMS Yeast Res 10:970–979
3. Goffeau A, Barrell BG, Bussey H, Davis RW, Dujon B, Feldmann H et al (1996) Life with 6000 genes. Science 274:563–567
4. Coller J, Parker R (2004) Eukaryotic mRNA decapping. Annu Rev Biochem 73:861–890
5. Tsukuda T, Fleming AB, Nickoloff JA, Osley MA (2005) Chromatin remodelling at a DNA double-strand break site in Saccharomyces cerevisiae. Nature 438:379–383
6. Botstein D, Fink GR (2011) Yeast: an experimental organism for 21st century biology. Genetics 189:695–704
7. Mager WH, Winderickx J (2005) Yeast as a model for medical and medicinal research. Trends Pharmacol Sci 26:265–273
8. Carmona-Gutierrez D, Eisenberg T, Buttner S, Meisinger C, Kroemer G, Madeo F (2010) Apoptosis in yeast: triggers, pathways, subroutines. Cell Death Differ 17:763–773
9. Bonifacino JS, Glick BS (2004) The mechanisms of vesicle budding and fusion. Cell 116:153–166
10. Rinaldi T, Dallabona C, Ferrero I, Frontali L, Bolotin-Fukuhara M (2010) Mitochondrial diseases and the role of the yeast models. FEMS Yeast Res 10:1006–1022
11. Goeckeler JL, Brodsky JL (2010) Molecular chaperones and substrate ubiquitination control the efficiency of endoplasmic reticulum-

associated degradation. Diabetes Obes Metab 12:32–38
12. Klionsky DJ, Abeliovich H, Agostinis P, Agrawal DK, Aliev G, Askew DS et al (2008) Guidelines for the use and interpretation of assays for monitoring autophagy in higher eukaryotes. Autophagy 4:151–175
13. Brodsky JL, Skach WR (2011) Protein folding and quality control in the endoplasmic reticulum: recent lessons from yeast and mammalian cell systems. Curr Opin Cell Biol 23:464–475
14. Outeiro TF, Lindquist S (2003) Yeast cells provide insight into α-synuclein biology and pathobiology. Science 302:1772–1775
15. Willingham S, Outeiro TF, DeVit MJ, Lindquist SL, Muchowski PJ (2003) Yeast genes that enhance the toxicity of a mutant huntingtin fragment or α-synuclein. Science 302:1769–1772
16. Dixon C, Mathias N, Zweig RM, Davis DA, Gross DS (2005) α-Synuclein targets the plasma membrane via the secretory pathway and induces toxicity in yeast. Genetics 170:47–59
17. Flower TR, Chesnokova LS, Froelich CA, Dixon C, Witt SN (2005) Heat shock prevents α-synuclein-induced apoptosis in a yeast model of Parkinson's disease. J Mol Biol 351:1081–1100
18. Buttner S, Bitto A, Ring J, Augsten M, Zabrocki P, Eisenberg T et al (2008) Functional mitochondria are required for α-synuclein toxicity in aging yeast. J Biol Chem 283:7554–7560
19. Zabrocki P, Bastiaens I, Delay C, Bammens T, Ghillebert R, Pellens K et al (2008) Phosphorylation, lipid raft interaction and traffic of α-synuclein in a yeast model for Parkinson. Biochim Biophys Acta 1783:1767–1780
20. Puccio H, Koenig M (2000) Recent advances in the molecular pathogenesis of Friedreich ataxia. Hum Mol Genet 9:887–892
21. Outeiro TF, Muchowski PJ (2004) Molecular genetics approaches in yeast to study amyloid diseases. J Mol Neurosci 23:49–60
22. Khurana V, Lindquist S (2010) Modelling neurodegeneration in Saccharomyces cerevisiae: why cook with baker's yeast? Nat Rev Neurosci 11:436–449
23. Entian KD, Kötter P (2007) Yeast genetic strain and plasmid collections. In: Stansfield I, MJR S (eds) Yeast gene analysis, Methods in microbiology, vol 36, 2nd edn. Academic Press, New York, pp 629–666
24. Miller-Fleming L, Giorgini F, Outeiro TF (2008) Yeast as a model for studying human neurodegenerative disorders. Biotechnol J 3:325–338
25. Karathia H, Vilaprinyo E, Sorribas A, Alves R (2011) Saccharomyces cerevisiae as a model organism: a comparative study. PLoS One 6: e16015
26. Tenreiro S, Munder MC, Alberti S, Outeiro TF (2013) Harnessing the power of yeast to unravel the molecular basis of neurodegeneration. J Neurochem 127:438–452
27. Menezes R, Tenreiro S, Macedo D, Santos CN, Outeiro TF (2015) From the baker to the bedside: yeast models of Parkinson's disease. Microb Cell 2:262–279
28. Gitler AD, Chesi A, Geddie ML, Strathearn KE, Hamamichi S, Hill KJ et al (2009) Alpha-synuclein is part of a diverse and highly conserved interaction network that includes PARK9 and manganese toxicity. Nat Genet 41:308–315
29. Miller-Fleming L, Antas P, Pais TF, Smalley JL, Giorgini F, Outeiro TF (2014) Yeast DJ-1 superfamily members are required for diauxic-shift reprogramming and cell survival in stationary phase. Proc Natl Acad Sci U S A 111:7012–7017
30. Zondler L, Miller-Fleming L, Repici M, Goncalves S, Tenreiro S, Rosado-Ramos R et al (2014) DJ-1 interactions with alpha-synuclein attenuate aggregation and cellular toxicity in models of Parkinson's disease. Cell Death Dis 5:e1350
31. Dhungel N, Eleuteri S, Li LB, Kramer NJ, Chartron JW, Spencer B et al (2015) Parkinson's disease genes VPS35 and EIF4G1 interact genetically and converge on alpha-synuclein. Neuron 85:76–87
32. Xiong Y, Coombes CE, Kilaru A, Li X, Gitler AD, Bowers WJ et al (2010) GTPase activity plays a key role in the pathobiology of LRRK2. PLoS Genet 6:e1000902
33. Pereira C, Miguel ML, Saraiva L (2014) LRRK2, but not pathogenic mutants, protects against H2O2 stress depending on mitochondrial function and endocytosis in a yeast model. Biochim Biophys Acta 1840:2025–2031
34. Pereira C, Costa V, Martins LM, Saraiva L (2015) A yeast model of the Parkinson's disease-associated protein Parkin. Exp Cell Res 333:73–79
35. Buttner S, Delay C, Franssens V, Bammens T, Ruli D, Zaunschirm S et al (2010) Synphilin-1 enhances alpha-synuclein aggregation in yeast and contributes to cellular stress and cell death in a Sir2-dependent manner. PLoS One 5: e13700

36. Tenreiro S, Rosado-Ramos R, Gerhardt E, Favretto F, Magalhaes F, Popova B et al (2016) Yeast reveals similar molecular mechanisms underlying alpha- and beta-synuclein toxicity. Hum Mol Genet 25:275–290
37. Su LJ, Auluck PK, Outeiro TF, Yeger-Lotem E, Kritzer JA, Tardiff DF et al (2010) Compounds from an unbiased chemical screen reverse both ER-to-Golgi trafficking defects and mitochondrial dysfunction in Parkinson's disease models. Dis Model Mech 3:194–208
38. Petroi D, Popova B, Taheri-Talesh N, Irniger S, Shahpasandzadeh H, Zweckstetter M, Outeiro TF, Braus GH (2012) Aggregate clearance of α-synuclein in Saccharomyces cerevisiae depends more on autophagosome and vacuole function than on the proteasome. J Biol Chem 287: 27567–27579
39. Singleton AB, Farrer M, Johnson J, Singleton A, Hague S, Kachergus J et al (2003) Alpha-Synuclein locus triplication causes Parkinson's disease. Science 302:841
40. Ibanez P, Bonnet AM, Debarges B, Lohmann E, Tison F, Pollak P et al (2004) Causal relation between alpha-synuclein gene duplication and familial Parkinson's disease. Lancet 364:1169–1171
41. Franssens V, Boelen E, Anandhakumar J, Vanhelmont T, Buttner S, Winderickx J (2010) Yeast unfolds the road map toward α-synuclein-induced cell death. Cell Death Differ 17:746–753
42. Franssens V, Bynens T, Van den Brande J, Vandermeeren K, Verduyckt M, Winderickx J (2013) The benefits of humanized yeast models to study Parkinson's disease. Oxidative Med Cell Longev 2013:760629
43. Flower TR, Clark-Dixon C, Metoyer C, Yang H, Shi R, Zhang Z et al (2007) YGR198w (YPP1) targets A30P α-synuclein to the vacuole for degradation. J Cell Biol 177:1091–1104
44. Gitler AD, Bevis BJ, Shorter J, Strathearn KE, Hamamichi S, Su LJ et al (2008) The Parkinson's disease protein α-synuclein disrupts cellular Rab homeostasis. Proc Natl Acad Sci U S A 105:145–150
45. Soper JH, Roy S, Stieber A, Lee E, Wilson RB, Trojanowski JQ et al (2008) α-Synuclein-induced aggregation of cytoplasmic vesicles in Saccharomyces cerevisiae. Mol Biol Cell 19:1093–1103
46. Cooper AA, Gitler AD, Cashikar A, Haynes CM, Hill KJ, Bhullar B et al (2006) Alpha-synuclein blocks ER-Golgi traffic and Rab1 rescues neuron loss in Parkinson's models. Science 313:324–328
47. Yeger-Lotem E, Riva L, Su LJ, Gitler AD, Cashikar AG, King OD et al (2009) Bridging high-throughput genetic and transcriptional data reveals cellular responses to alpha-synuclein toxicity. Nat Genet 41:316–323
48. Zabrocki P, Pellens K, Vanhelmont T, Vandebroek T, Griffioen G, Wera S et al (2005) Characterization of α-synuclein aggregation and synergistic toxicity with protein tau in yeast. FEBS J 272:1386–1400
49. Sampaio-Marques B, Felgueiras C, Silva A, Rodrigues M, Tenreiro S, Franssens V et al (2012) SNCA (alpha-synuclein)-induced toxicity in yeast cells is dependent on sirtuin 2 (Sir2)-mediated mitophagy. Autophagy 8:1494–1509
50. Winzeler EA, Shoemaker DD, Astromoff A, Liang H, Anderson K, Andre B et al (1999) Functional characterization of the S. cerevisiae genome by gene deletion and parallel analysis. Science 285:901–906
51. Giaever G, Chu AM, Ni L, Connelly C, Riles L, Véronneau S, Dow S et al (2002) Functional profiling of the Saccharomyces cerevisiae genome. Nature 418:387–391
52. Mnaimneh S, Davierwala AP, Haynes J, Moffat J, Peng WT, Zhang W, Yang X et al (2004) Exploration of essential gene functions via titratable promoter alleles. Cell 118:31–44
53. Schuldiner M, Collins SR, Thompson NJ, Denic V, Bhamidipati A, Punna T, Ihmels J et al (2005) Exploration of the function and organization of the yeast early secretory pathway through an epistatic miniarray profile. Cell 123:507–519
54. Li Z, Vizeacoumar FJ, Bahr S, Li J, Warringer J, Vizeacoumar FS, Min R et al (2011) Systematic exploration of essential yeast gene function with temperature-sensitive mutants. Nat Biotechnol 29:361–367
55. Huh WK, Falvo JV, Gerke LC, Carroll AS, Howson RW, Weissman JS, O'Shea EK (2003) Global analysis of protein localization in budding yeast. Nature 425:686–691
56. Gavin AC, Bösche M, Krause R, Grandi P, Marzioch M, Bauer A, Schultz J et al (2002) Functional organization of the yeast proteome by systematic analysis of protein complexes. Nature 415:141–147
57. Gietz RD, Schiestl RH (1996) Transforming yeast with DNA. Methods Mol Cell Biol 5:255–269

Chapter 13

Probing Structural Changes in Alpha-Synuclein by Nuclear Magnetic Resonance Spectroscopy

Tapojyoti Das and David Eliezer

Abstract

Alpha-synuclein, the principal protein involved in the pathogenesis of Parkinson's disease, has been shown to exchange between multiple conformational states, with hitherto unclear physiological role of such conformational changes. Due to its ability to provide rich structural information for proteins in their near-native environment, nuclear magnetic resonance (NMR) spectroscopy has been a valuable tool to study these conformational states. In this review we describe the application of model systems and NMR methods to the study of membrane-bound states of alpha-synuclein. We provide a detailed description, primarily meant for someone new to the field, of how to prepare the necessary samples, perform the basic experiments, and obtain an initial interpretation of the results.

Key words Nuclear magnetic resonance, Biophysics, Protein folding, Structure, Lipid interactions

1 Introduction

1.1 Pathological and Physiological Roles of Alpha-Synuclein

Alpha-synuclein is a protein best known for its involvement in Parkinson's disease (PD), where a misfolded aggregated state of the protein is a primary component of Lewy bodies, the pathological hallmark of the disease. Although a pathological role for the misfolded, aggregated protein is strongly suspected, the physiological structure-function relationships of the protein are still not well characterized. Alpha-synuclein is a 140-amino-acid protein that remains intrinsically disordered in solution while forming stable secondary structure upon interaction with lipid membranes. Interestingly, the protein preferentially binds to highly curved membranes and its chief physiological interaction partner is thought to be the synaptic vesicle membrane. The membrane-binding domain of alpha-synuclein is comprised of the N-terminal ~100 residues, which form an amphipathic helical structure on membrane surfaces, while the C-terminal ~40 residues remain largely membrane unassociated and highly disordered under most conditions. The membrane-binding domain is further subdivided into helix-1 and

Tim Bartels (ed.), *Alpha-Synuclein: Methods and Protocols*, Methods in Molecular Biology, vol. 1948,
https://doi.org/10.1007/978-1-4939-9124-2_13,

helix-2 regions, separated by a break at residues 43–44 in the micelle-bound state [1, 2]. A current goal of our laboratory is to elucidate the structure-function relation of alpha-synuclein in its physiological state, with the hope that a better understanding of its physiology and function will assist in efforts to discover new therapeutics for combating disease.

1.2 Artificial Models to Study Alpha-Synuclein Structure and Function

Alpha-synuclein is a relatively small protein that exchanges between different conformational states, making it a good candidate for structural studies using solution-state nuclear magnetic resonance (NMR) spectroscopy. Subsequent to the discovery of its disorder-to-helix transition on interaction with membrane mimetics using circular dichroism spectroscopy, NMR spectroscopy-based methods have been the cornerstone of higher resolution structural studies on this protein. In order to study the membrane-bound conformation of alpha-synuclein, we utilize various models that mimic biological membranes. Each one of these membrane-mimetic models has its own advantages and drawbacks, but together they provide complementary structural information. In the following section we briefly review the different models used in various experiments, what types of questions they can address, and the respective experimental conditions optimized through experience, as a guide for someone newly entering the field.

1. **SDS micelles**: SDS (sodium dodecyl sulfate) is a surfactant that forms micelles above its critical micelle concentration, typically around 8 mM in physiological buffers in the absence of modifiers [3, 4]. At the concentration that we generally use (40 mM), the diameter of the micelle is estimated to be about 4–5 nm [5, 6]. The shape and size of SDS micelles are in a dynamic equilibrium and are influenced by the absolute detergent concentration as well as the relative concentrations of SDS and alpha-synuclein [7]. Despite their micellar nature and their dynamic behavior, SDS micelles are an important model for studying the structure of alpha-synuclein, since the NMR signals from the bound state of the protein are directly detectable. Importantly, SDS undergoes a phase transition at temperatures somewhat below room temperature [5], so we typically use a higher than ambient temperature (40 °C) for these NMR experiments [8]. Also, use of deuterated SDS helps to increase the NMR signal from the sample by reducing the relaxation of the signal via proton-proton spin diffusion.
2. **Lipid vesicles**: Artificial lipid vesicles (small unilamellar vesicles, SUVs, and large unilamellar vesicles, LUVs) are particularly important model systems for studying the structure of alpha-synuclein [9, 10]. After hydration of dried lipid film of a given composition, methods such as sonication and extrusion through a membrane are used to generate lipid vesicles with a specific size distribution. We typically attempt to mimic the

composition of native synaptic vesicles [11–13] and also aim for a size distribution centered around the size of native synaptic vesicles [14]. Lipid vesicles are an invaluable tool for measuring the per-residue lipid binding affinity for alpha-synuclein. In addition, different sized lipid vesicles allow for the study of the curvature sensitivity of alpha-synuclein membrane binding [15]. However, lipid vesicles are much larger in size than detergent micelles, resulting in a large increase in the rotational correlation time (slower tumbling) of the bound protein, which in turn causes NMR peaks to become very broad, making them virtually undetectable by direct observation. Hence, it is effectively impossible to study the vesicle-bound conformation of alpha-synuclein directly using solution-state NMR methods (*see* **Note 1**).

3. **Lipid bicelles**: Bicelles are fragments of lipid bilayer, typically composed of long-chain phospholipids, with their edges protected from the solvent by shorter chain phospholipids or amphipathic detergent molecules [16]. Bicelles are formed by self-assembly when lipids and detergents are mixed at specific ratios in an aqueous buffer at certain temperatures and solvent ion concentrations. Bicelles with different properties have proven highly useful in structural studies of membrane-resident and membrane-binding proteins. So-called isotropic bicelles are formed at a higher detergent concentration. Relatively small, they tumble freely and are useful as models of flat lipid membranes. At lower detergent concentration, larger bicelles are formed which can spontaneously align in strong magnetic fields. These larger bicelles are useful for NMR experiments that measure residual dipolar coupling (RDC), which report on the relative orientation of different parts of a macromolecule.
4. **Nanodiscs**: Nanodiscs are also fragments of lipid bilayer, larger than micelles and smaller than SUVs, where the edge of the bilayer disk is protected from solvent by a membrane scaffold protein (typically an amphipathic helix) that is typically derived from apolipoprotein A1 [17]. The length of the scaffold protein helix determines the circumference of the nanodisc, and hence nanodiscs are relatively homogeneous in size. Further homogeneity can be achieved by using covalently circularized nanodiscs that use an enzyme-mediated circularization of the scaffold protein [18–21]. This homogeneity allows for the characterization of subtle features in NMR membrane binding profiles of alpha-synuclein, which are apparently lost in population averages of more heterogeneous conformational ensembles populated on a distribution of lipid vesicle sizes [22]. Despite being considerably smaller than SUVs, nanodiscs thus far have also proven unsuitable for direct measurement of lipid-bound conformation using standard solution-state NMR methods [22].

1.3 Different Solution-State NMR Methods and Their Applications in the Structural Study of Alpha-Synuclein

Compared to X-ray crystallography or cryo-electron microscopy, NMR spectroscopy is inherently a less sensitive method for structure determination. Hence, the protein of interest is needed at a relatively high concentration and in a relatively large volume to obtain sufficient NMR signal. Moreover, no single type of NMR experiment is enough to build a detailed structural model of the molecule. Typically each type of NMR experiment adds different sets of information, resulting in a structural model that satisfies the majority of the accumulated experimental constraints.

Atomic nuclei with a half-integer spin are detectable by NMR and are most convenient for macromolecular spectroscopy in solution. While the predominant hydrogen isotope in nature (^{1}H) is detectable by NMR, the predominant nitrogen (^{14}N) and carbon (^{12}C) isotopes are not spin 1/2 nuclei. Hence, high-resolution protein NMR methods require the protein of interest to be labeled with nitrogen and carbon isotopes of half-integral spin (^{15}N and ^{13}C). This is achieved by expressing the recombinant protein in bacterial cells grown in a minimal media supplemented by labeled nitrogen and carbon sources.

The most frequently used protein NMR spectrum is a ^{1}H-^{15}N HSQC, or proton-nitrogen correlation spectrum, obtained using a ^{15}N-labeled protein sample. In this spectrum, the magnetization is transferred within each amide group in the protein (^{1}H to ^{15}N and back to ^{1}H for direct detection), resulting in peaks that are resolved in two dimensions (^{1}H and ^{15}N). Analysis of the spectrum at a level of individual protein residues requires that the peak assignments are known (this is accomplished using additional experiments). The peak positions and intensities in an HSQC spectrum are highly sensitive to changes in the conformation and environment of the protein, and form the basis for assays that compare different conformational states. Nearly all of the experimental NMR approaches that will be covered here rely primarily on the ^{1}H-^{15}N HSQC experiment.

2 Materials

2.1 Materials for Sample Preparation and Purification

2.1.1 Strains and Plasmids

1. **Plasmids**: We use full-length human wild-type alpha-synuclein gene cloned in pT7-7 backbone, originally a gift from Dr. Peter Lansbury's Lab, as the bacterial expression vector. The plasmid is 2892 bp in size, contains an ampicillin resistance gene, and features an IPTG-inducible promoter upstream to a 420 bp insert.
2. **Bacterial strain**: Expression of the recombinant protein production is typically induced in a BL21(DE3) strain of *E. coli* after transformation with the above plasmid.

2.1.2 Buffers, Solutions, and Nutrient Media

1. **LB-agar plates with ampicillin**: In a 500 mL bottle, add Miller's LB agar 18.5 g and water up to 500 mL. Autoclave and let it cool down to 50–60 °C (warm to touch) before adding 500 μL of ampicillin stock (50 mg/mL), mix, and dispense into sterile 10 cm petri dishes, 25 mL each (total 20 dishes). Allow to cool down and solidify. Can be stored at 4 °C for up to 2 weeks.
2. **SOC medium**: We use commercially available SOC medium for outgrowth of transformed bacteria before plating in selection media. As an alternative, LB medium can also be used.
3. **LB media**: Standard LB media (1 L) contains tryptone 10 g, NaCl 10 g, and yeast extract 5 g and pH adjusted to 7; autoclave to sterilize.
4. **M9 salts (5×)**:

   $Na_2HPO_4 \cdot 7H_2O$, 64 g.
   KH_2PO_4, 15 g.
   NaCl, 2.5 g.
   NH_4Cl, 5.0 g.
   Deionized H_2O, to 1 L.
5. **M9 minimal media for ^{15}N and ^{13}C labeling**:

   Add in sequence (*see* **Note 2**):

   Deionized water, 700 mL.

   1 M $MgCl_2$, 1 mL.

   1 M $CaCl_2$, 100 μL.

   5× M9 salts, 200 mL

   BME vitamins (100×), 10 mL.

   NH_4Cl (^{15}N), 1 g—Cambridge Isotope Laboratories # NLM-467-PK.

   D-glucose (U-$^{13}C_6$), ^{13}C enriched, 2 g—Cambridge Isotope Laboratories # CLM-1396-PK (for only ^{15}N labeling use unlabeled D-glucose (anhydrous), 4 g).

   Adjust volume to 1 L.

   Add ampicillin stock (50 mg/mL), 1 mL.

   Filter sterilize into a sterile bottle.
6. **IPTG (isopropyl-β-D-1-thiogalactopyranoside) solution**: IPTG is a competitive inhibitor of the *lac* repressor and induces transcription of the genes under control of a *lac*-inducible promoter, in this case, alpha synuclein.

   Prepare IPTG stock solution by dissolving 2 g of IPTG in 10 mL water (840 mM).
7. **Lysis buffer (50 mL, prepared fresh every time)**:

   20 mM Tris–Cl (1 mL of 1 M Tris–Cl at pH 8.0).

1 mM EDTA (1 mL of 0.5 M EDTA).
1 mM PMSF (0.5 mL of 100 mM PMSF).
1 mM DTT (if purifying cysteine mutants).
Purified water to 50 mL.

8. **Buffer A**:

   1 L: 25 mM Tris–Cl pH 8.0, 20 mM NaCl, 1 mM EDTA.
   Filter and degas before storing.

9. **Buffer B**:

   1 L: 25 mM Tris–Cl pH 80, 1 M NaCl, 1 mM EDTA.
   Filter and degas before storing.

10. **HPLC Buffer A**:

    0.1% Trifluoroacetic acid (TFA) in water.

11. **HPLC Buffer B**:

    90% Acetonitrile and 0.1% TFA in water.

12. **NMR Buffer**:

    10 mM Na_2HPO_4, 100 mM NaCl, 10% v/v D_2O, pH 6.8.

13. **Dialysis Buffer**:

    25 mM Tris–Cl pH 8.0, 20 mM NaCl, 1 mM EDTA.

14. **MTSL Stock Solution**:

    MTSL is dissolved in DMSO (300 mM).
    Aliquot in 30 μL volumes and store at −20 °C.

15. **PRE Buffer**:

    10 mM Na_2HPO_4, 100 mM NaCl, pH 6.8 (2 L).

16. **SDS Stock Solution**:

    80 mM SDS in NMR buffer.

17. **Reagents for paramagnetic relaxation enhancement experiments**:

    MTSL (*S*-(1-oxyl-2,2,5,5-tetramethyl-2,5-dihydro-1H-pyrrol-3-yl)methyl methanesulfonothioate), CAS number 81213-52-7—Toronto Research Chemicals # O875000.

    DMSO (dimethyl sulfoxide), CAS number 67-68-5—Sigma # D8418.

    DTT (1,4-Dithiothreitol), CAS number 3483-12-3.

18. **Specialized Lab equipment**:

    Bacterial culture incubator with shaker capable for 3 L flasks.

    UV-Vis spectrophotometer.

    Refrigerated centrifuge for 1 L tubes.

    Refrigerated centrifuge for 50 mL tubes.

    Ultracentrifuge for 25 mL tubes.

Lyophilizer.

Vacuum concentrator/desiccator.

Bath sonicator.

FPLC system.

HPLC system.

2.2 NMR Equipment and Other Materials

1. **Spectrometer hardware**: An NMR spectrometer with a high-field magnet and three-channel cryoprobe. We use a 600 MHz Bruker® Avance III spectrometer equipped with three-channel cryoprobe for data acquisition. For certain experiments (e.g., titration), an automated sampling system (Bruker® SampleJet) reduces the need for human intervention.
2. Glass NMR tubes (5 mm): for the SampleJet system special NMR tubes are needed and are available from Bruker®.
3. **Software**:
   (a) NMRpipe (free, available from https://www.ibbr.umd.edu/nmrpipe/install.html).
   (b) CCPNmr analysis V2 (free, available from https://www.ccpn.ac.uk/v2-software/software/analysis). Other softwares for the same purpose include NMRViewJ, Sparky, and Bruker TopSpin, and the reader is advised to consult the respective software manuals for detailed instructions and tutorials.

3 Methods

3.1 Sample Preparation

3.1.1 Site-Directed Mutagenesis of Plasmids

Standard site-dependent mutagenesis protocols are used to mutagenize the plasmid according to the project requirements. We introduce a stop codon at position 103 to make truncation mutants. Fewer number of peaks from truncation mutants result in reduced peak overlap, making triple-resonance sequence assignment process easier. In addition, for long-distance contact measurements using paramagnetic relaxation enhancement (PRE) effect, single cysteine mutation is introduced at various locations on the protein via site-directed mutagenesis.

3.1.2 Growth of Recombinant ^{15}N-Labeled and ^{13}C, ^{15}N-Labeled Alpha-Synuclein Samples

1. Transform competent *E. coli* strain BL21(DE3) with the expression plasmid, plate on LB-agar plates with ampicillin, and incubate overnight at 37 °C. It is important to use freshly transformed cells every time.
2. Next day, pick a single colony from the plate and grow in starter culture (2 mL LB broth with ampicillin 50 μg/mL).
3. Prepare M9 minimal media for isotopic labeling (1 L).

4. In the evening, inoculate 100 mL labeling media with 1 mL starter culture in a small flask and incubate at 37 °C with shaking overnight.
5. Next morning, inoculate the remaining 900 mL labeling media with 100 mL overnight culture and grow in shaker incubator with growth monitoring until $OD_{600} = 0.6$ (takes about 2 h) and then add 1 mL IPTG stock solution (final concentration 0.84 mM) to induce protein production.
6. Monitor growth every 15–30 min (more frequently in the later part) until the bacterial growth plateaus. It is important not to wait until a decline in growth starts. Usually it takes 2–3 h after induction. In our hands, the final OD_{600} is in the range of 1.2–1.5.
7. Centrifuge the culture at 10,000 × *g* for 15 min at 4 °C and discard the supernatant.
8. Scoop out the bacterial pellet in a 50 mL polypropylene tube using a spatula and store it in −20 °C or lower until ready for purification (stays good for months).

3.1.3 Purification of Alpha-Synuclein Protein

Alpha-synuclein is purified based on its physical properties without any affinity tag. However, the purification methods differ for full-length and truncated versions of the protein. The first few steps of the purification are the same for both and involve preparation of bacterial lysate by sonication and removal of insoluble and particulate material by ultracentrifugation.

1. Allow the bacterial pellet to thaw at room temperature.
2. Resuspend the bacterial pellet in 50 mL lysis buffer by vortexing and sonicate on ice for 10 min, until running watery.
3. Ultracentrifuge the sonicate for 45 min at 200,000 × *g* (40,000 rpm on a Beckman 50.2 Ti rotor) and collect the supernatant.

Purification of Full-Length Alpha-Synuclein [23]

1. Use 1 M HCl to drop the pH of the ultracentrifuge supernatant to 3.5.
2. Centrifuge at 40,000 × *g* for 15 min at 4 °C, and collect supernatant.
3. Readjust the pH of the supernatant to 7.5 using 1 M NaOH.
4. Measure the volume of the solution and slowly add ammonium sulfate, 0.291 g/mL, while stirring on ice, to achieve a 50% salt saturation.
5. Keep stirring for at least 1 h to overnight on ice.
6. Full-length alpha-synuclein will precipitate out of the solution and is centrifuged at 40,000 × *g* to pellet the protein.

7. The pellet is dissolved in 25 mL 20 mM Tris buffer (pH 8.0) and dialyzed through a 3.5 kDa membrane against water, changing the dialysis buffer twice.
8. After dialysis, the protein is flash frozen in liquid nitrogen and lyophilized to store indefinitely.
9. SDS-PAGE on a 15% gel using samples obtained at different stages of growth and purification to confirm purity.

Purification of Truncated-Length (1-102) Alpha-Synuclein

1. Add 1% w/v streptomycin sulfate to ultracentrifuge supernatant (for 50 mL add 0.5 g) and stir at 4 °C for 30 min to precipitate nucleic acids; centrifuge at 40,000 × *g* for 15 min at 4 °C and collect supernatant.
2. Add ammonium sulfate to supernatant, 0.116 g/mL (5.80 g for 50 mL), and slowly stir at 4 °C for 10 min.
   (a) Ideally no pellet is present. Go to **step 3** directly.
   (b) If a pellet is present, spin down at 40,000 × g for 15 min at 4 °C and collect supernatant. Proceed to **step 3**.
3. Add ammonium sulfate to supernatant, 0.129 g/mL (6.45 g for 50 mL), and slowly stir at 4 °C for 1 h.
4. Spin down at 40,000 × *g* for 15 min at 4 °C and decant supernatant in a new tube to store for future analysis. Proceed with the pellet.
5. Resuspend pellet in 50 mL lysis buffer.
6. Dialyze through a 3.5 kDa dialysis membrane against 1 L of dialysis buffer overnight at 4 °C.
7. Cation-exchange chromatography:
   (a) Equilibrate a CM Sepharose™ column of 20 mL column volume (CV) with 2 CV of buffer A before loading sample.
   (b) Elute sample fractions with a gradually increasing salt gradient from 0 to 40% buffer B over 4 CV. Collect 2.5 mL fractions during elution.
   (c) Run a SDS-PAGE (18% gel) with the chromatography fractions and pool relevant fractions to proceed to next step. Usually alpha-synuclein elutes between 250 and 350 mM NaCl concentration.
8. Dialyze the pooled fraction from FPLC through a 3.5 kDa membrane against 1 L of 5% acetic acid overnight at 4 °C, changing dialysis buffer once.
9. Filter through 0.22 μm membrane filter before proceeding to next step.

10. Reversed-phase HPLC using C4 column and using HPLC separation module with inline absorbance detector set at 229 nm.
    (a) Column equilibration: 90% HPLC buffer A and 10% HPLC buffer B at 4.45 mL/min for 10 min.
    (b) Sample loading: 80% sample, 10% HPLC buffer A and 10% HPLC buffer B, then equilibration with 80% HPLC buffer A and 10% HPLC buffer B for 45 min.
    (c) Elution program:

Time (min)	Flow (mL/min)	Buffer A (%)	Buffer B (%)
0	4.45	80	20
5	4.45	58	42
37	4.45	50	50
39	4.45	0	100
41	4.45	80	20
51	4.45	0	100
53	4.45	80	20

    (d) Collect and analyze all major peaks by SDS-PAGE. Alpha-synuclein elutes as a major peak at around 20 min on our system.
11. Spin for 2 h in vacuum concentrator to remove acetonitrile from HPLC buffer.
12. Dialyze in water to neutralize pH.
13. Flash freeze in liquid nitrogen and lyophilize for storage.

3.1.4 Paramagnetic Labeling of Single-Cysteine Mutants

Paramagnetic relaxation enhancement (PRE) experiments allow us to determine long-range contacts between different parts of a protein. Due to its high sensitivity, this method is able to characterize intramolecular or intermolecular contacts in transient low-population states of the protein. Residues close to a paramagnetic label (also known as spin label) experience faster relaxation of magnetic resonance and hence the signal intensity is decreased on decreasing distance with an inverse sixth power dependence. The experimental question determines the location of the paramagnetic label. For example, the broken helix state of alpha-synuclein can be captured by PRE experiments by putting a label on the first helix and observing the secondary peak intensity changes in the second helix.

Usually paramagnetic labels are incorporated via a single-cysteine mutation. We discuss here the method of incorporating a nitroxide spin label at a single-cysteine mutant of alpha-synuclein.

1. Dissolve ^{15}N alpha-synuclein at 200–400 μM concentration in 1 mL of PRE buffer.
2. Add 30 μL of 300 mM MTSL stock solution and allow the reaction to proceed for a few hours at room temperature.
3. Dialyze in 1 L of PRE buffer through a 3.5 kDa cutoff membrane with a change in buffer once, to remove all traces of the unbound spin label.
4. Add 10% D_2O by volume to prepare the sample ready for NMR experiments.

3.1.5 Preparation of NMR Sample

Dissolving Protein for NMR Sample Preparation

Lyophilized alpha-synuclein is weighed out and dissolved in NMR buffer to prepare a stock solution of 200–300 μM. After dissolving the protein thoroughly by inversion (no vortexing), the protein is filtered through a 100 kDa cutoff filter to remove any possible oligomeric species. This protein solution can be added to the additives at 1:1 volume ratio to make the final NMR sample.

SDS Micelles

SDS micelle stock solution (80 mM) is mixed with the dissolved protein sample at 1:1 volume ratio to make the final concentration of 40 mM SDS. Using deuterated SDS (SDS D25, 98%—Cambridge Isotope Laboratories # DLM-197-PK) helps to improve the S/N ratio due to slower relaxation.

Lipid Vesicles

Lipid SUVs are prepared by sonication of hydrated lipid films. The major steps are as follows:

1. **Combining lipids**: Lipids are procured as stock solutions in chloroform. Depending on the volume of 2× stock solution needed for the experiments, mix different lipid stock solutions in a borosilicate glass tube at a molar ratio required for the experiment. We use a molar ratio of DOPC:DOPE: DOPS = 60:25:25 to approximate the lipid composition of native synaptic vesicles [11–13]. Remember to handle chloroform in a chemical hood.
2. **Drying lipids**: The lipid mixture is dried in a constant flow of an inert gas (nitrogen or boron) while rolling the tube to form a thin film on the side. After the solvent evaporates, keep the film under inert gas flow for a further 20 min. After that, dry the film further in a vacuum desiccator/concentrator for 2 h. The tube can be flushed with inert gas, sealed with paraffin film, and kept at −20 °C for a few days until ready for hydration.

3. **Hydrating lipids**: The lipid film is hydrated in appropriate volume of NMR buffer and vortexed vigorously to make a cloudy suspension.
4. **Sonication**: The suspension is sonicated in a bath sonicator at room temperature for 15–30 min or until it is optically clear. This ensures that the vesicles are of small enough size to become non-scattering to visible light.
5. **Ultracentrifugation**: The SUV suspension is ultracentrifuged at 150,000 × *g* (60,000 rpm on a Sorvall S120-AT2 rotor) to pellet larger particles. The supernatant is pipetted out carefully and used within a day to prepare NMR samples (added to protein solution AT 1:1 volume ratio).

3.2 NMR Experiments

Samples are pipetted into NMR tubes using thin glass pipettes. The sample should be of sufficient volume to fill the coil of the spectrometer. For a 5 mm NMR tube, a sample volume of 500–600 μL is typical. To prevent oxidation of lipids the tubes are flushed with argon or nitrogen before capping.

3.2.1 Setting Up Spectrometer

The basic steps of setting up spectrometer for any experiment are the same. It consists of temperature equilibration, sample insertion, tuning and matching, shimming, determination of proton 90° hard pulse length, and setting up the experimental acquisition parameters before starting the experiment.

1. **Temperature**: For free-state and lipid SUV-bound-state experiments we use a temperature of 10 °C, whereas for SDS micelle-bound-state experiments the optimal temperature is much higher, 40 °C. This is because at lower temperature SDS undergoes a phase transition and the peaks broaden beyond the limit of detection [8]. It is important to adjust the temperature set point based on a calibration curve for the actual temperature, which is periodically measured as part of spectrometer maintenance. Some versions of the spectrometer operating software (TopSpin for Bruker) allow the user to set the corrected temperature within the software.
   - edte↵ (to open the temperature control panel).
   - Adjust the set temperature 5 K at a time to allow the control system to perform smoothly.
2. **Sample insertion**: Sample is inserted into the bore of the magnet once the temperature has equilibrated and is allowed to equilibrate for another ~5 min.
   - bsmsdisp↵ (to open the controller window).
   - Toggle "lift" to control the airflow that lifts the shuttle with sample.

3. **Locking**: Locking refers to setting up a feedback loop that maintains the main magnetic field so that it does not drift in time. This is important for high-resolution experiments where the signal is averaged over multiple measurements from days of experiment, and it is critical to have the magnetic field stable over time. Locking utilizes a separate NMR channel based on the deuterium nuclear frequency, which does not overlap with frequencies of other nuclei interrogated (^{1}H, ^{15}N, ^{13}C).
   - lockdisp↵ (to open up window showing lock signal).
   - Turning off lock in the controller window will show a sweeping lock signal. Adjust the lock field to center the signal and then adjust the lock phase. Finally toggle on the "lock" button on the controller window.
4. **Tuning and matching**: The probe coils are tuned and impedance-matched to maximize the energy absorbed from the sample at the resonance frequencies of each nucleus. Sample temperature and solvent conditions affect the magnetic field "seen" by the sample, and that in turn affects the resonance frequencies. For this reason it is important to tune and match the spectrometer every time the sample temperature or buffer conditions change.
   - Open a dataset.
   - For instruments with tuning and matching servo controllers, entering atma↵ will automatically tune and match all the channels for the sample. Further fine-tuning can be achieved with manual adjustment after atmm↵.
5. **Shimming**: Shimming the magnet ensures that all parts of the sample experience the same magnetic field so that there is minimal inhomogeneity of the signal from one spatial region of the sample to another. This is usually achieved by adjusting the power going into the shim coils, additional coils in the magnet that modulate the shape of the magnetic field. Most modern spectrometers allow automated shimming using pulsed field gradient imaging, which is more accurate and convenient for the user.
   - topshim gui↵ (from here one can adjust the shimming parameters and perform a z-axis shimming).
6. **Determining ^{1}H 90° hard pulse length**: Pulse lengths and delays in all pulse programs are calculated from a properly calibrated ^{1}H 90° hard pulse length. Even the ^{15}N and ^{13}C pulse lengths can be calculated from it to some degree of accuracy. To measure it, a 1D proton experiment is set up with a single pulse at the same power as the intended experiment and at a small pulse length (1 μs) that tips the magnetization vector just enough to result in an observable signal. The

acquired signal is Fourier transformed and phase corrected so that the solvent peak is upright and symmetrical. Then the pulse width is increased until tipping of the magnetization vector through a full 360° is observed, realigning it with the axis of the main magnetic field. Upon iterative fine adjustment of the 360° pulse length aiming to eliminate any observed signal, the 90° pulse width (a quarter of 360° pulse width) is recorded in the pulse program parameters. Most other parameters of the pulse program concerning pulse lengths and delays are recalculated based on the ^{1}H 90° pulse width automatically.

- Open a single-pulse 1D proton experiment.
- Check for the transmitter power attenuation value in dB (it should be the same as used in the other experiments).
- ns 1↵ (to set the number of scans to 1).
- rg 1↵ (to set receiver gain to 1).
- p1 1↵ (to set a 1 μs pulse).
- zg↵ (to run the experiment).
- ft↵ (to do a Fourier transform).
- apk↵ (for automated phase correction); one can also correct the phase manually.
- Use a P1 value (pulse length) close to four times that expected for a 90° pulse length (this is different for different spectrometers and an earlier experiment can serve as an approximate guide).
- zg↵ (to run the experiment).
- fp↵ (apply Fourier transformation and phase correction).
- Iteratively adjust P1 value to extinguish the signal (minimal and equal amounts of residual positive and negative lobes). This is the 360° pulse length.
- A quarter of this is the 90° pulse length for that transmitter power. Note down the value.

3.2.2 The ^{1}H-^{15}N HSQC Experiment

The ^{1}H-^{15}N HSQC experiment forms the basis of many other experiments for protein structural analysis and is the most frequently performed experiment for this reason. This experiment separates individual backbone amide resonances in two dimensions and is sensitive to subtle conformational changes, reflected by a movement of the corresponding peaks or a change in peak intensity.

A ^{15}N-labeled sample is needed for this experiment, although ^{13}C and ^{15}N double-labeled sample can also be used with a CN (carbon-nitrogen) decoupling pulse (explained later). Depending on the spectrometer software, there will likely be several versions of the ^{1}H-^{15}N HSQC pulse program, differing in ways the high-intensity water signal is suppressed or the way the indirect dimension (^{15}N) is acquired. We generally use a gradient-enhanced version of the experiment available in Bruker TopSpin catalog (hsqcetfpf3gpsi). After the basic setup of the spectrometer with the sample, the following steps are needed for the HSQC experiment:

1. **Setting up the spectral window**: A spectral window is defined by the spectral midpoint and spectral width. It is critical to select an appropriate spectral window for each sample so that all the resonances are encompassed within the window and are represented at their true frequencies (i.e., not folded or aliased). This is more important for the indirect (^{15}N) dimension. By convention, the spectral midpoint of the proton dimension is chosen at the water resonance (4.7 ppm), while amide nitrogen midpoint is optimized for each experiment between 112 and 122 ppm (these resonances usually fall within the range of 90–140 ppm). To optimize that, initially a wider spectral width is chosen, a lower resolution spectrum is acquired, and then the appropriate spectral midpoint and width is chosen so as to encompass all resonances (*see* **Note 3**).
2. **Choosing the number of data points**: The number of total points determines the digital resolution of the spectrum. In practice, for the proton dimension 1024 or 2048 points are chosen and for the nitrogen dimension 128–512 points are good enough for a quality spectrum.
3. **Optimizing the number of scans**: Because NMR signal is attenuated at each magnetization transfer step, for multidimensional NMR experiments a number of repeat experiments at each point ("scans" in NMR terminology) are averaged to achieve an optimal signal-to-noise (S/N) ratio. For a twofold increase in S/N, four times the number of scans are required. Since the time taken for acquisition depends linearly on the number of scans as well as the number of data points in the indirect dimension, these numbers are judiciously optimized for each experiment to fit the requirements. For the protein concentrations mentioned, usually 4–16 scans are sufficient to get a decent spectrum.
4. **Starting the experiment**: After optimization of the experimental parameters, the experiment is started. For a double-labeled sample the CN decoupling is invoked by the flag -DLABEL_CN in the zgoptions field.

3.2.3 Triple-Resonance Experiments

Triple-resonance experiments are most commonly used to determine peak assignments for new mutants or characterize secondary structure from the carbon (secondary) chemical shifts. ^{13}C and ^{15}N double-labeled samples are required for this purpose. Pairs of triple-resonance experiments provide complementary information about *i*th and ($i - 1$)th residue connectivity. Such pairs are as follows:

- HNCACB (both i and $i - 1$) and CBCACONH ($i - 1$ only), for CA and CB chemical shifts.
- HNCA (both i and $i - 1$) and HNCOCA ($i - 1$ only), for CA chemical shifts.
- HNCACO (both i and $i - 1$) and HNCO ($i - 1$ only), for CO chemical shifts.

The basic steps of setting up a triple-resonance experiment are similar to HSQC. In order to set the carbon spectral window properly, a proton-carbon 2D spectrum is acquired by setting the number of data point in nitrogen dimension to one. Additionally, depending on the pulse program, there could be a maximum limit to the number of data points in one or both indirect dimensions. The number of scans needs to be optimized depending on the sensitivity of each triple-resonance experiment (*see* **Notes 4** and **5**).

3.2.4 N-N NOESY Experiment

Nuclear Overhauser effect spectroscopy (NOESY) experiments provide information about the spatial proximity of atoms in protein residues and thereby on their secondary and/or tertiary structure. These class of experiments rely on through-space magnetization transfer (nuclear Overhauser effect resulting from dipole-dipole interaction) to spatially proximal residues. In an alpha-helix, nuclear Overhauser effect (NOE) can be identified up to four or more residues away in both C- and N-terminal directions. In contrast, in disordered or highly extended regions of a protein, NOE transfers along the polypeptide chain are weaker and do not extend as far. In the N-N NOESY experiment, a NOE transfer step is sandwiched between two $^{1}H–^{15}N$ HSQC experiments. The resulting $^{1}H–^{1}H–^{15}N$ correlation spectrum has strong peaks in the $^{1}H–^{1}H$ diagonal line, resulting from the same residue. Off-diagonal peaks result from NOEs between different residues and are informative for both resonance assignment and secondary structure analysis. Careful optimization of the NOE mixing time is critical to enable observation of cross-peaks without contamination from spin-spin diffusion effects, which give rise to cross-peaks that result from multiple spin transfers. In our hands, for alpha-synuclein bound to deuterated SDS micelles, we achieved good results with 200–300 ms mixing times.

3.2.5 Paramagnetic Relaxation Enhancement (PRE) Experiment

As mentioned previously, the PRE experiment is used to establish long-range contacts between different parts of the protein (or to identify intermolecular contact sites between two protein molecules). This is done by comparing the HSQC peak intensity between matched spin-labeled and unlabeled sample. Matched samples are produced as follows.

1. Make twice the volume of a MTSL-labeled sample and split into two equal volumes (*see* Subheading 3.1.4 above).
2. To one sample, add 0.5 M DTT (in water) to a final concentration of 5 mM to remove the MTSL label from the protein. In the other sample add same volume of water to maintain equivalent protein concentrations (*see* **Note 6**).
3. Acquire HSQC spectra of the matched samples using identical data acquisition times and parameters and plot the intensity ratios of resonances for the labeled/unlabeled samples versus residue number.
4. This plot will show a dip in the ratio around the labeling site, where NH groups are covalently restricted to be in close spatial proximity to the cysteine-attached spin label. Any other regions of the protein that approaches within ~20 Angstroms of the labeling site will also exhibit a decrease in the intensity ratio, providing the desired information on long-range contacts in the protein.

3.3 Data Analysis and Interpretation

3.3.1 Processing Raw Data from Instrument

NMR raw data consists of regularly sampled data points of the free induction decay of the NMR signal. Once acquisition of NMR data is finished, the next task is to process the raw time domain data appropriately and transform it into the frequency domain, and then perform required analyses and calculations. Instrument manufacturers have their own software to do these tasks but the fundamental steps are the same. We find it convenient to convert different spectrometer raw data formats to NMRpipe format and then perform the processing using NMRpipe. NMRpipe contains GUI-based conversion scripts that can be invoked by typing bruker or varian, as appropriate, in a C shell terminal. The GUI parses the experiment parameter files and creates a fid.com conversion script that converts the data to NMRpipe format.

A typical Bruker data conversion script may look like this:

```
bruk2pipe -in ./ser \           # Read data from Bruker ser file
-bad 0.0 -aswap -DMX -decim 2856 -dspf 20 -grpdly 67.9860382080078 \# Specifications of the Bruker data format

 -xN      1024 -yN     256 \# Total points (real + imaginary)
 -xT       512 -yT     128 \# Real points
 -xMODE   DQD -yMODE Echo-Antiecho \# Acquisition mode of receiver
 -xSW    7002.801 -ySW 1845.018 \# Spectral width
 -xOBS   700.273 -yOBS 70.966 \# Observed frequency (MHz)
 -xCAR   4.873 -yCAR   119.681 \# Carrier frequency (ppm)
 -xLAB   HN -yLAB      15N \# Nucleus name for display
-ndim  2 -aq2D         States \# Dimensions and 2D acquisition mode
-out ./test.fid -verb -ov      # Output file
```

The goal of the subsequent processing step is to transform the time domain data into frequency domain, so that the peaks under investigation can be resolved. Processing for each dimension can be generally summarized as follows:

1. **Apodization**: The acquired free induction decay signal is often truncated in order to prevent sample and probe overheating due to prolonged high-power decoupling pulses during acquisition. This truncation causes an artifact in the Fourier transformed signal in frequency domain, which is minimized by using a method called apodization. In apodization, the acquired time domain signal is multiplied by a smoothly decaying function of choice (apodization function or window function) so that the resultant signal gradually decays to zero over the actual number of data points. Commonly used apodization functions are cosine squared, cosine bell, Gaussian, and Kaiser among others, and the choice depends on the intended application [24, 25].
2. **Zero filling**: This process of adding extra trailing zeroes to the acquired signal adds to the digital resolution of the signal as well as improves the running efficiency of the fast Fourier transform (FFT) algorithm when the total data points are brought up to a power of two.
3. **Fourier transform**: The Fourier transform is a mathematical operation that converts time domain data to frequency domain data.
4. **Phase correction**: Depending on which phase of the cycle the spectrometer starts collecting data, there is a phase delay in the Fourier transformed signal. This is corrected so that the frequency domain peaks are perfectly in-phase (upright) to allow quantitative measurements to be carried out.

For a 2D, 3D, or higher dimension spectrum, the same steps above are performed on each dimension after appropriately transforming the matrix (*see* **Note 7**).

Below is a typical 2D spectrum (e.g., HSQC) processing script (*see* **Note 8**):

```
#!/bin/csh

#
# Basic 2D Phase-Sensitive Processing:
#   Cosine-Bells are used in both dimensions.
#   Use of "ZF -auto" doubles size, then rounds to power of 2.
#   Use of "FT -auto" chooses correct Transform mode.
#   Imaginaries are deleted with "-di" in each dimension.
#   Phase corrections should be inserted by hand.
```

```
nmrPipe -in test.fid \                                  # Read input file
|nmrPipe -fn SP -off 0.5 -end 1.00 -pow 1 -c 1.0 \ # Apodization (Dim 1)
| nmrPipe -fn ZF -auto                  \ # Zero fill (Dim 1)
| nmrPipe -fn FT -auto                  \ # Fourier Transform (Dim 1)
|nmrPipe -fn PS -p0 -72 -p1 0.00 -di -verb   \ # Phase correction (Dim 1)
| nmrPipe -fn TP                  \ # Transpose
|nmrPipe -fn SP -off 0.5 -end 1.00 -pow 1 -c 0.5 \ # Apodization (Dim 2)
| nmrPipe -fn ZF -auto                  \ # Zero fill (Dim 2)
| nmrPipe -fn FT -auto                  \ # Fourier Transform (Dim 2)
|nmrPipe -fn PS -p0 -90 -p1 0.00 -di -verb   \ # Phase correction (Dim 2)
 -ov -out test.ft2                              # Write output file
```

3.3.2 Spectrum Display and Measurements

In order to perform further tasks such as peak picking and assignment, measurements, and plotting, a number of spectra typically need to be displayed simultaneously. This necessitates the use of specialized data processing software like CCPNmr Analysis, NMRViewJ, Sparky, or TopSpin. The choice of the software relies on the individual. The following common steps are performed:

1. **Peak picking**: **Fully manual peak picking or** automated peak picking of the spectrum followed by manual verification/adjustment.
2. **Peak assignment**: Where preexisting resonance assignments are available, they can be transferred from a peak list or a related spectrum based on the location of the signals. Ambiguous or unassigned peaks must be assigned with the help of triple-resonance experiments that provide connectivity information (*see* **Note 5**).
3. **Peak measurements**: Peak height and volume can be used to quantify the signal for a particular resonance, and are used in comparative analyses. In addition the chemical shift values for each nucleus in the protein can be determined and exported to standard file formats (BMRB NMR-STAR, for example) for various analyses and for deposition in the BMRB databank repository.

3.3.3 Data Analysis and Visualization

To address different questions about molecular structure, the data is analyzed and visually represented in ways that capture the feature of interest. Commonly used analyses are described below:

1. **Carbon (secondary) chemical shifts**: The carbon secondary chemical shifts help in characterizing the secondary structure of the protein (Fig. 1). Secondary chemical shift is defined as the difference of observed chemical shift of a particular nucleus and the chemical shift of the nucleus if the polypeptide were a random coil. Random coil chemical shift values can be generated for any given polypeptide using temperature, pH and nearest neighbor corrections of empirically determined values [26–28].

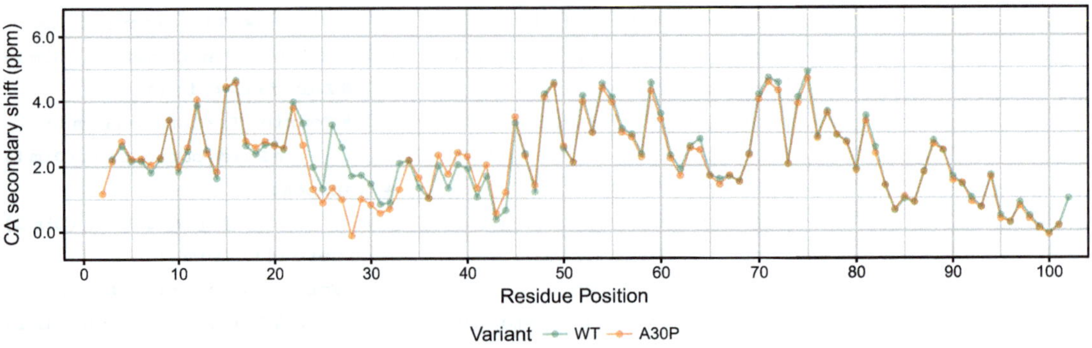

Fig. 1 Alpha-carbon (CA) secondary chemical shifts for SDS micelle-bound wild-type (WT) alpha-synuclein and its Parkinson's disease-associated mutant A30P. Introduction of proline at position 30 causes a disruption of alpha-helix, indicated by decreased CA secondary shift values around the mutation site [10]. The break between helix-1 and helix-2 of the broken-helix state results in near-zero CA secondary shifts at positions 43 and 44 [1, 2]

A Web utility to calculate random coil chemical shifts is available at https://spin.niddk.nih.gov/bax/nmrserver/Poulsen_rc_CS/

Alpha-carbon secondary shift values are generally positive in alpha-helix, negative in beta-sheet, and around zero for random coil. The difference of the alpha- and beta-carbon secondary chemical shift also follows this trend, and is independent of any potential referencing errors (*see* **Note 9**).

2. **NOEs**: Signals from NOE transfer between two residues can be seen in both forward and reverse directions, the peaks being diagonally opposed in the spectrum. Although an average of the forward and reverse NOE peak intensities are plotted in many cases, plotting them separately and observing the correlation acts as a validity check for the data (Fig. 2). Alpha helical regions are characterized by high backbone sequential NH-NH NOE intensities due to the short inter-proton distances of 2.8 Å, whereas random coil or highly extended regions exhibit minimal sequential NH-NH NOE transfer due to a larger distance and internal motion between adjacent residues [29].

3. **Paramagnetic relaxation enhancement**: The data are typically acquired as two matched spectra, with and without the paramagnetic label. After peak assignment, the peak intensity ratios (*see* **Note 10**) of labeled and unlabeled samples are plotted against the residue position to reveal residues spatially proximal to the label as a dip in the corresponding signals (*see* **Note 11**).

4. **Lipid binding experiments**: When bound to lipid vesicles (or any large object), alpha-synuclein (or any protein) experiences an increased rotational correlation time (slower tumbling rate), causing the solution-state NMR peaks to broaden to the

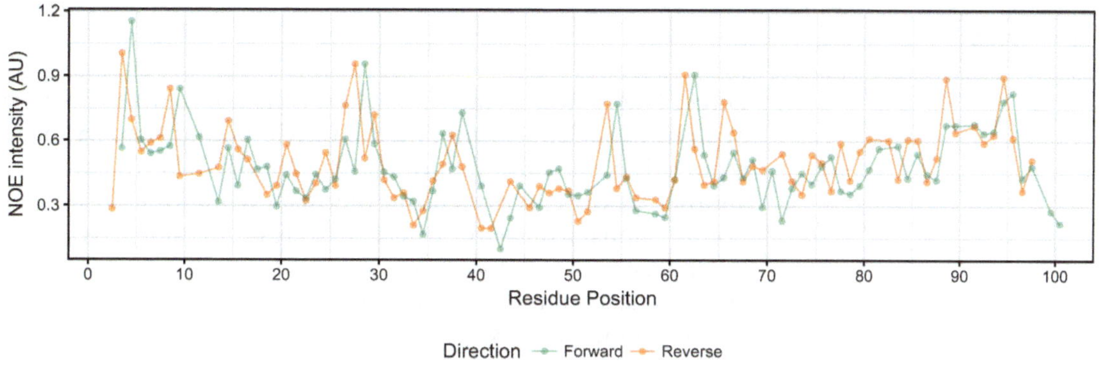

Fig. 2 Nuclear Overhauser effect (NOE) intensities for sequential NH–NH cross-peaks for SDS micelle-bound wild-type alpha-synuclein for positions 1-102. Forward and reverse NOE transfer intensities are plotted separately against residue number. Higher NOE intensities are observed within the two helices of the micelle-bound broken-helix state and lower intensity is seen for several residues in the linker region. As expected, there is good correlation of the NOE transfer intensities in the forward and reverse directions

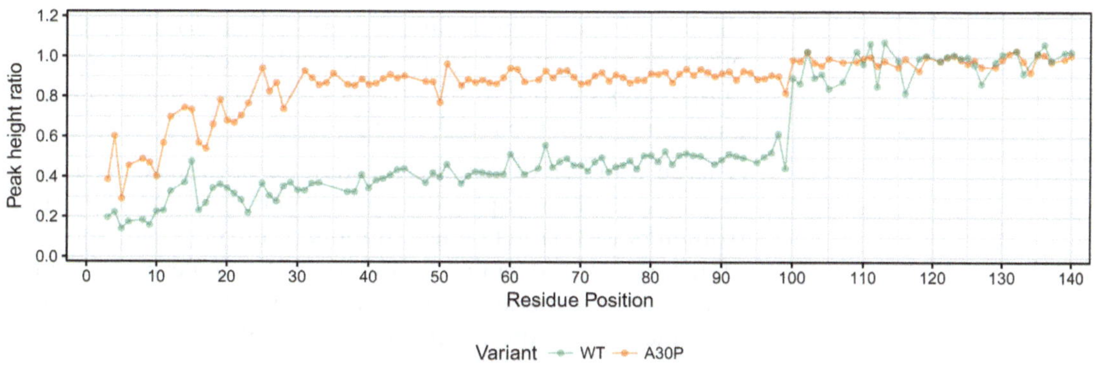

Fig. 3 Comparison of lipid vesicle binding profiles of wild-type (WT) alpha-synuclein and its Parkinson's disease-associated mutant A30P. The free protein fraction at each residue is plotted for 50 μM protein in the presence of small unilamellar vesicles (SUVs) comprised of 10 mM total phospholipids at a ratio of DOPC: DOPE:DOPS = 60:25:15. For the WT protein, the lipid-binding domain (~residues 1-102) binds somewhat cooperatively, reflecting the formation of the extended-helix state on the vesicle surface. For the A30P mutant, binding is weaker overall, but is relatively preserved N-terminal to the mutation site, whereas binding of C-terminal to the mutation site is greatly decreased. This results in a binding mode where the helix-1 region, or part of it, is bound while the helix-2 region is released from the membrane surface. Such states can potentially drive membrane-induced synuclein aggregation [9, 30–33]

extent of becoming undetectable. Hence, any observed signal from a mixture of alpha-synuclein with lipids represents only the free fraction of the protein. This fraction can be calculated at a residue-resolved level from the peak intensity ratio with and without lipids (Fig. 3). In addition, the binding curves can be calculated by performing a titration series of lipids and these curves can be used to obtain partition coefficients or apparent dissociation constants (*see* **Note 12**).

4 Notes

1. Direct NMR measurements of a highly helical soluble structure of alpha-synuclein resembling the vesicle-bound state has been reported recently using fluorinated alcohols [34]. In addition, another report describes using thermal cycling of lipid vesicle-bound alpha-synuclein, where hysteresis of alpha-synuclein secondary structure formation appears to result in a metastable alpha-helical soluble state upon dissociation from the model vesicle membranes [35].
2. Calcium can form insoluble salts with phosphate and sulfate ions. Therefore, it is important to dilute the $CaCl_2$ into a large volume prior to the addition of sulfates or phosphates.
3. In some cases a number of resonances, especially glycines, may be well separated in the ^{15}N dimension from the remainder of the spectrum. In this case, judicious selection of a narrower spectral width in the ^{15}N dimension may allow folding or aliasing of these signals into a region that does not contain other signals, preventing overlap while increasing resolution (a narrower spectral width provides a higher resolution if sampled using the same number of points).
4. A large number of parameters may be optimized to achieve maximum S/N in triple-resonance experiments. Modern spectrometers with largely linear amplifiers automatically set many of the relevant parameters and typically achieve satisfactory results. More experienced users may nevertheless want to optimize individual experiments for individual samples further.
5. A more complete description of triple-resonance experiments and their use in both obtaining resonance assignments and determining secondary structure can be found in a previous MiMB protocol from our group [36].
6. We typically use DTT to reduce the disulfide bond between the spin label and the protein. In this approach, any non-covalently associated spin label will remain bound to the protein, and will clause equivalent broadening in both samples, leading to no PRE effect. An alternative approach is to reduce the spin label itself, usually via addition of a reductant such as ascorbic acid. In this case, non-covalently associated spin label will be reduced and will result in an intensity decrease leading to a nonspecific PRE effect at the site of non-covalent binding (usually seen at aromatic side chains). This issue is specific to MTSL-type spin labels. Metal-chelating groups do not exhibit this problem, but are slightly more complicated to work with.
7. An optional linear prediction is sometimes performed one or more of the indirect dimensions before the apodization step. This method extrapolates additional points in the free

induction decay based on the trend of the acquired data points, and helps to improve the resolution in the indirect dimensions.

8. A library of processing scripts are present in the nmrDraw program, included in the NMRpipe suite. This program also allows the user to manually determine the proper phase correction values for each dimension. Since a detailed description of the different processing options is beyond the scope of this manuscript, the reader is encouraged to consult the NMRpipe tutorial to learn how to use the program.
9. By definition, chemical shift is the resonant frequency of a particular nucleus relative to a standard. A proper referencing of chemical shift values is critically important when performing analyses such as those based on secondary chemical shifts. In protein NMR spectroscopy the most commonly used standard is DSS (4,4-dimethyl-4-silapentane-1-sulfonic acid) whose strong methyl proton signal is used as the ^{1}H zero ppm reference. In practice, it is prudent to measure a DSS standard under the same temperature and buffer conditions, and indirectly reference the chemical shift values of the ^{15}N and ^{13}C nuclei based on standard ratios [37]. Of further note, the processing program NMRpipe sets the ^{1}H zero ppm reference based on the water chemical shift at user-specified temperature of the sample.
10. Paramagnetic relaxation enhancement of NMR signals of nuclei in spatial proximity to an unpaired electron results in an increased relaxation rate and a broadening of the resonance. The PRE effect is most accurately measured by directly measuring the difference in the line widths of the signals in a labeled and unlabeled sample. Because broadening leads to decreased signal intensity, the PRE effect can also be evaluated using the more easily obtained intensity ratio data.
11. A detailed protocol for PRE-based distance measurement has been described previously in this series [38].
12. NMR peaks of vesicle-bound alpha-synuclein are undetectable by traditional solution-state NMR methods due to faster relaxation decay resulting in line broadening. This phenomenon can be effectively utilized to study the exchange kinetics between NMR-visible "bright" state and NMR-invisible "dark" state using a method called dark-state excitation saturation transfer (DEST) [39]. In DEST, the transient "dark" state is selectively excited using off-resonance narrow-bandwidth pulses and the resonant signal is detected when the excited molecules exchange with the free-state molecules in solution. Details of exchange kinetics of the amyloid beta-peptide bound to a larger complex (e.g., protofibrils) have been revealed by this method [40, 41] and the method has also been applied to the study of membrane-bound alpha-synuclein [42, 43].

References

1. Bussell R, Eliezer D (2003) A structural and functional role for 11-mer repeats in alpha-synuclein and other exchangeable lipid binding proteins. J Mol Biol 329(4):763–778
2. Chandra S, Chen X, Rizo J, Jahn R, Südhof TC (2003) A broken α-helix in folded α-synuclein. J Biol Chem 278(17):15313–15318
3. Dominguez A, Fernandez A, Gonzalez N, Iglesias E, Montenegro L (1997) Determination of critical micelle concentration of some surfactants by three techniques. J Chem Educ 74(10):1227
4. Fuguet E, Ràfols C, Rosés M, Bosch E (2005) Critical micelle concentration of surfactants in aqueous buffered and unbuffered systems. Anal Chim Acta 548(1–2):95–100
5. Mazer NA, Benedek GB, Carey MC (1976) An investigation of the micellar phase of sodium dodecyl sulfate in aqueous sodium chloride solutions using quasielastic light scattering spectroscopy. J Phys Chem 80(10):1075–1085
6. Duplatre G, MR FM (1996) Size of sodium dodecyl sulfate micelles in aqueous solutions as studied by positron annihilation lifetime spectroscopy. J Phys Chem 3654 (96):16608–16612
7. Georgieva ER, Ramlall TF, Borbat PP, Freed JH, Eliezer D (2010) The lipid-binding domain of wild type and mutant alpha-synuclein: compactness and interconversion between the broken and extended helix forms. J Biol Chem 285(36):28261–28274
8. Eliezer D, Kutluay E, Bussell R, Browne G (2001) Conformational properties of alpha-synuclein in its free and lipid-associated states. J Mol Biol 307(4):1061–1073
9. Bodner CR, Dobson CM, Bax A (2009) Multiple tight phospholipid-binding modes of α-Synuclein revealed by solution NMR spectroscopy. J Mol Biol 390(4):775–790
10. Bussell R, Eliezer D (2004) Effects of Parkinson's disease-linked mutations on the structure of lipid-associated alpha-synuclein. Biochemistry 43(16):4810–4818
11. Breckenridge WC, Morgan IG, Zanetta JP, Vincendon G (1973) Adult rat brain synaptic vesicles II. Lipid composition. Biochim Biophys Acta Gen Subj 320(3):681–686
12. Deutsch JW, Kelly RB (1981) Lipids of synaptic vesicles: relevance to the mechanism of membrane fusion. Biochemistry 20 (2):378–385
13. Takamori S, Holt M, Stenius K, Lemke EA, Grønborg M, Riedel D et al (2006) Molecular anatomy of a trafficking organelle. Cell 127 (4):831–846
14. Barenholz Y, Gibbes D, Litman BJ, Goll J, Thompson TE, Carlson RD (1977) A simple method for the preparation of homogeneous phospholipid vesicles. Biochemistry 16 (12):2806–2810
15. Middleton ER, Rhoades E (2010) Effects of curvature and composition on α-synuclein binding to lipid vesicles. Biophys J 99 (7):2279–2288
16. Dürr UHN, Soong R, Ramamoorthy A (2013) When detergent meets bilayer: birth and coming of age of lipid bicelles. Prog Nucl Magn Reson Spectrosc 69:1–22
17. Bayburt TH, Grinkova YV, Sligar SG (2002) Self-assembly of discoidal phospholipid bilayer nanoparticles with membrane scaffold proteins. Nano Lett 2(8):853–856
18. Yusuf Y, Massiot J, Chang Y-T, Wu P-H, Yeh V, Kuo P-C et al (2018) Optimization of the production of covalently circularized nanodiscs and their characterization in physiological conditions. Langmuir 34(11):3525–3532
19. Nasr ML, Baptista D, Strauss M, Sun ZYJ, Grigoriu S, Huser S et al (2016) Covalently circularized nanodiscs for studying membrane proteins and viral entry. Nat Methods 14 (1):49–52
20. Nasr ML, Wagner G (2018) Covalently circularized nanodiscs; challenges and applications. Curr Opin Struct Biol 51:129–134
21. Popp M, Antos JM, Ploegh HL (2009) Current protocols in protein science. Curr Protoc Protein Sci Chapter 15:Unit 15.3
22. Viennet T, Wördehoff MM, Uluca B, Poojari C, Hoyer W, Etzkorn M et al (2017) A structural and kinetic link between membrane association and amyloid fibril formation of α-Synuclein. BioRxiv 1:1–31
23. Coelho-Cerqueira E, Carmo-Gonçalves P, Sá Pinheiro A, Cortines J, Follmer C (2013) α-Synuclein as an intrinsically disordered monomer—Fact or artefact? FEBS J 280 (19):4915–4927
24. Kaiser J, Schafer R (1980) On the use of the I0-sinh window for spectrum analysis. IEEE Trans Acoust 28(1):105–107
25. Cavanagh J, Fairbrother WJ, Palmer AG, Rance M, Skelton NJ (2007) Experimental aspects of NMR spectroscopy. In: Protein NMR spectroscopy. Elsevier, New York, pp 114–270
26. Kjaergaard M, Brander S, Poulsen FM (2011) Random coil chemical shift for intrinsically

disordered proteins: effects of temperature and pH. J Biomol NMR 49(2):139–149
27. Kjaergaard M, Poulsen FM (2011) Sequence correction of random coil chemical shifts: correlation between neighbor correction factors and changes in the Ramachandran distribution. J Biomol NMR 50(2):157–165
28. Schwarzinger S, Kroon GJ, Foss TR, Chung J, Wright PE, Dyson HJ (2001) Sequence-dependent correction of random coil NMR chemical shifts. J Am Chem Soc 123 (13):2970–2978
29. Bax A (1989) Two-dimensional NMR and protein structure. Annu Rev Biochem 58:223–256
30. Georgieva ER, Ramlall TF, Borbat PP, Freed JH, Eliezer D (2008) Membrane-bound alpha-synuclein forms an extended helix: long-distance pulsed ESR measurements using vesicles, bicelles, and rodlike micelles. J Am Chem Soc 130(39):12856–12857
31. Anderson VL, Ramlall TF, Rospigliosi CC, Webb WW, Eliezer D (2010) Identification of a helical intermediate in trifluoroethanol-induced alpha-synuclein aggregation. Proc Natl Acad Sci U S A 107(44):18850–18855
32. Dikiy I, Eliezer D (2012) Folding and misfolding of alpha-synuclein on membranes. Biochim Biophys Acta 1818(4):1013–1018
33. Dikiy I, Fauvet B, Jovičić A, Mahul-Mellier AL, Desobry C, El-Turk F et al (2016) Semisynthetic and in vitro phosphorylation of alpha-Synuclein at Y39 promotes functional partly helical membrane-bound states resembling those induced by PD mutations. ACS Chem Biol 11(9):2428–2437
34. Sung Y-H, Eliezer D (2018) Structure and dynamics of the extended-helix state of alpha-synuclein: intrinsic lability of the linker region. Protein Sci 22(6):1–50
35. Rovere M, Sanderson JB, Fonseca-Ornelas L, Patel DS, Bartels T (2018) Refolding of helical soluble α-synuclein through transient interaction with lipid interfaces. FEBS Lett 592:1464
36. Eliezer D (2006) Characterizing residual structure in disordered protein states using nuclear magnetic resonance. In: Bai Y, Nussinov R (eds) Protein folding protocols. Humana Press, Totowa, pp 49–68
37. Wishart DS, Bigam CG, Yao J, Abildgaard F, Dyson HJ, Oldfield E et al (1995) 1H, 13C and 15N chemical shift referencing in biomolecular NMR. J Biomol NMR 6(2):135–140
38. Eliezer D (2012) Distance information for disordered proteins from NMR and ESR measurements using paramagnetic spin labels. Methods Mol Biol 895(1):127–138
39. Anthis NJ, Clore GM (2015) Visualizing transient dark states by NMR spectroscopy. Q Rev Biophys 48(1):35–116
40. Fawzi NL, Ying J, Torchia DA, Clore GM (2010) Kinetics of amyloid β monomer-to-oligomer exchange by NMR relaxation. J Am Chem Soc 132(29):9948–9951
41. Fawzi NL, Ying J, Ghirlando R, Torchia DA, Clore GM (2011) Atomic-resolution dynamics on the surface of amyloid-β protofibrils probed by solution NMR. Nature 480 (7376):268–272
42. Fusco G, Pape T, Stephens AD, Mahou P, Costa AR, Kaminski CF et al (2016) Structural basis of synaptic vesicle assembly promoted by α-synuclein. Nat Commun 7:1–11
43. Fusco G, De Simone A, Gopinath T, Vostrikov V, Vendruscolo M, Dobson CM et al (2014) Direct observation of the three regions in α-synuclein that determine its membrane-bound behaviour. Nat Commun 5 (May):1–8

Chapter 14

Measuring Mitochondrial Dysfunction Caused by Soluble α-Synuclein Oligomers

Eric S. Luth and Irina G. Stavrovskaya

Abstract

Accumulation of misfolded αSyn and mitochondrial dysfunction are central features of Parkinson's disease. Growing evidence points to a relationship between these two phenomena as oligomeric α-synuclein (αSyn) can interact with mitochondria and impair their function. Standardization of methods to prepare αSyn oligomers and isolate functional mitochondria will facilitate efforts to expand upon early findings. Here we present detailed protocols for preparing soluble αSyn oligomers; for isolating functional mitochondria from mouse tissue; and for simultaneously measuring several aspects of mitochondrial physiology. These protocols will benefit future studies aimed at characterizing the mitotoxicity of αSyn species isolated from the brains of synucleinopathy patients as well as efforts to identify small molecules and genetic or environmental alterations that prevent αSyn-induced mitochondrial dysfunction.

Key words α-Synuclein, Oligomer, Mitochondria, Mitochondrial dysfunction, Calcium retention, Parkinson's disease

1 Introduction

The accumulation of misfolded α-synuclein (αSyn) is a key step in the pathogenesis of Parkinson's disease (PD). Dominantly inherited point mutations in the SNCA gene and multiplications of the wild-type SNCA locus increase the abundance of monomeric, unfolded αSyn which, unlike native helical multimers [1–6], aggregates into insoluble amyloid fibrils by way of soluble oligomeric intermediates [7–11]. These soluble oligomers are believed to be primarily responsible for αSyn-induced neuronal degeneration [12–16]. Consistent with the mitochondrial dysfunction that is also characteristic of PD brains and animal models, mitochondria themselves are subcellular targets of toxic αSyn species [17–26].

We previously reported that complex I-related mitochondrial physiology is disrupted upon direct exposure to soluble αSyn oligomers, but not monomeric or fibrillar αSyn species, in the context of high Ca^{2+} [27]. The changes in mitochondrial physiology

Tim Bartels (ed.), *Alpha-Synuclein: Methods and Protocols*, Methods in Molecular Biology, vol. 1948,
https://doi.org/10.1007/978-1-4939-9124-2_14,

observed in our reductionist system (i.e., direct application of purified αSyn protein to isolated mitochondria) mirror the mitochondrial impairments caused by overexpression of αSyn in cells and in vivo, namely altered Ca^{2+} retention time, accelerated mitochondrial depolarization and swelling, and enhanced NAD(P)H oxidation [20–26]. Here we present detailed protocols for preparing soluble, mitotoxic αSyn oligomers; for isolating intact, functional mitochondria from mouse liver and brain tissue; and for simultaneously measuring the aforementioned aspects of mitochondrial physiology. Our method of preparing αSyn oligomers is based on a protocol to prepare preformed αSyn fibrils that, when applied to tissue, cells, or organelles, recapitulate features of PD [28–31]. Due to their stability, reliability, and high reproducibility we used isolated liver mitochondria as an initial model system for testing αSyn toxicity. Importantly, αSyn similarly compromised the function of mitochondria isolated from brain tissue.

There are several applications of these protocols that will provide valuable insight into PD pathogenesis and potential therapeutics. Efforts to develop drugs that prevent the aggregation of αSyn into toxic oligomers or stabilize aggregation-resistant native forms would benefit from knowledge of the specific αSyn species found in diseased brains that are most damaging to mitochondria. Various αSyn species could be isolated from the brains of synucleinopathy patients and compared with in vitro-prepared αSyn oligomers in these established assays of mitochondrial function. Complementary studies could use these protocols to examine how genetic or environmental variability affects mitochondrial sensitivity to αSyn oligomers and identify drugs that may protect against oligomer-induced mitochondrial dysfunction.

2 Materials

Unless stated otherwise, all solutions are prepared using analytical grade reagents purchased from Millipore-Sigma and ultrapure water (sensitivity of 18 MΩ cm at 25 °C). All reagents are prepared and stored at room temperature (if not otherwise stated) and waste materials are discarded following disposal regulations.

2.1 Preparation of Oligomeric αSyn

1. Lyophilized recombinant αSyn (*see* **Note 1**).
2. PBS: 10 mM phosphate, 2.7 mM KCl, 137 mM NaCl, pH 7.4.
3. 0.22 μm syringe filters.
4. Spectrophotometer, Uvikon 943 (Bio-Tek, USA), or equivalent capable of measuring absorbance at 280 nm.
5. Eppendorf LoBind protein or genomic microcentrifuge tubes (Eppendorf, Germany).

6. Liquid nitrogen and appropriate storage containers.
7. −80 °C freezer.
8. Parafilm M laboratory film.
9. 37 °C incubator.
10. Nutating mixer.
11. Probe-type ultrasonicator (Sonic Dismembrator model or similar) with microtip.
12. Benchtop refrigerated ultracentrifuge capable of 100,000 × *g*.

2.2 Isolation of Mouse Liver Mitochondria

1. 10 cm Petri dish.
2. 50 mL Plastic or glass beaker.
3. Straight, sharp-pointed dissection scissors.
4. Three 50 mL disposable conical-bottom tubes.
5. Potter-Elvehjem glass Teflon homogenizers 40 and 15 mL volume.
6. Dounce glass-glass homogenizers 40 and 15 mL volume.
7. 40 mL Round-bottom centrifuge tubes.
8. 30 mL Round-bottom ultracentrifuge tubes.
9. Plastic disposable transfer pipets.
10. Fixed-angle JA-20 centrifuge rotor (Beckman, USA) and appropriate refrigerated high speed centrifuge (for example Sorvall RC 5C plus) with a maximum speed of 20,000 rpm, both pre-chilled to 4 ° C.
11. Spectrophotometer, Uvikon 943 (Bio-Tek, USA), or equivalent capable of measuring absorbance at 540 nm.
12. Fatty acid-free bovine serum albumin (BSA).
13. 1 M HEPES-KOH: Dissolve 119.15 g of HEPES in 400 mL of distilled water and adjust pH to 7.4 using KOH. Bring the solution to 500 mL with water and store at RT.
14. 0.5 M EGTA-KOH: Dissolve 19.02 g of ethylene glycol-bis (2-aminoethylether)-N,N,N′,N′-tetraacetic acid (EGTA) in 80 mL of distilled water and adjust pH to 7.4 using KOH (*see* **Note 2**). Bring the solution to 100 mL with water and store at room temperature (RT).
15. Isolation buffer for liver #1 ($IB1_L$) (*see* **Note 3**): 240 mM D-sucrose, 10 mM HEPES, pH 7.4, 1 mM ethylene glycol-bis (2-aminoethylether)-*N*,*N*,*N*′,*N*′-tetraacetic acid (EGTA) stock with pH 7.4 adjusted by KOH and 0.5% fatty acid-free bovine serum albumin (BSA). Dissolve 82.152 g of D-sucrose in 900 mL of distilled water (*see* **Note 4**). Add 10 mL of 1 M HEPES-KOH and 20 mL of 0.5 M EGTA-KOH. Adjust with KOH pH to 7.4. Bring the solution to 1 L with water. Sterilize the solution using sterile vacuum filter units with a pore size of

0.22 μm (*see* **Notes 5** and **6**). Add 5 g of fatty acid-free bovine serum albumin (BSA, Sigma-Aldrich), do not shake it, and leave it overnight at 4 ° C to dissolve completely. Sterile $IB1_L$ solution can be stored at 4 ° C for several months (*see* **Note 7**).

16. Isolation buffer for liver #2 ($IB2_L$): 240 mM D-sucrose, 10 mM HEPES-KOH pH 7.4. Dissolve 82.152 g of D-sucrose in 900 mL of distilled water. Add 10 mL of 1 M HEPES-KOH. Adjust with KOH pH to 7.4 at 4 ° C. Bring the solution to 1 L with water. Sterilize the solution using sterile vacuum filter units with a pore size of 0.22 μm. Sterile $IB2_L$ solution can be stored at 4 ° C for several months.
17. 0.5 M EDTA-KOH: Dissolve 19.02 g of ethylenediaminetetraacetic acid (EDTA) disodium salt dihydrate in 80 mL of distilled water and adjust pH to 7.4 using KOH (*see* **Note 2**). Bring the solution to 100 mL with water and store at RT. Use this stock to prepare 10 mM solution.
18. Incubation buffer for liver mitochondria ($IB3_L$): 240 mM D-sucrose, 10 mM HEPES-KOH pH 7.4, 3 μM EDTA-KOH pH 7.4. Take 50 mL of $IB2_L$ and add 15 μL of 10 mM EDTA-KOH.
19. Bicinchoninic acid kit to determine mitochondrial protein concentration and 2.0 mg/mL bovine serum albumin (BSA) as a standard solution (*see* **Note 8**).

2.3 Isolation of Mouse Non-synaptosomal Brain Mitochondria

1. **Items 1–14** and **19** from Subheading 2.2.
2. Beckman Coulter swinging rotor SW 28, swinging rotor buckets, and appropriate Beckman Coulter Optima XL90 floor ultracentrifuge with a maximum speed of 99,000 rpm (both Beckman, USA), both pre-chilled to 4 °C.
3. Isolation buffer for brain #1 ($IB1_B$): 320 mM D-sucrose, 10 mM HEPES-KOH, pH 7.4, 0.5 mM EGTA-KOH, pH 7.4. Dissolve 109.54 g of D-sucrose in 900 mL of distilled water. Add 20 mL of 0.5 M HEPES-KOH and 10 mL of 0.5 M EGTA-KOH. Adjust with KOH pH to 7.4. Bring the solution to 1 L with water. Sterilize the solution using sterile vacuum filter units with a pore size of 0.22 μm. Sterile $IB1_B$ solution can be stored at 4 ° C for several months.
4. Isolation buffer for brain # 2 ($IB2_B$): 320 mM D-sucrose, 10 mM HEPES-KOH, pH 7.4, 1 mM EGTA-KOH, and 0.5 mg/mL BSA. Take 50 mL of $IB1_B$ and add 25 mg of BSA; do not stir it, and leave overnight at 4 ° C.
5. Isolation buffer for brain # 3 ($IB3_B$): 320 mM D-sucrose, 10 mM HEPES-KOH, pH 7.4, 0.02 mM EGTA-KOH. Dissolve 10.95 g of D-sucrose in 90 mL of distilled water. Add 2 mL of 0.5 M HEPES-KOH and 4 μL of 0.5 M EGTA-KOH.

Adjust with KOH pH to 7.4. Bring the solution to 100 mL with water. Sterilize the solution using sterile vacuum filter units with a pore size of 0.22 μm. Sterile $IB3_B$ solution can be stored at 4 °C for several months.

6. 1 M KH_2PO_4: Dissolve 6.80 g of KH_2PO_4 in 40 mL of distilled water. Adjust pH with KOH to 7.4 and bring the solution to 50 mL with water. Sterilize the solution using sterile vacuum filter units with a pore size of 0.22 μm. Sterile solution is stable for several months at RT.
7. 0.1 M ATP: Dissolve 551.14 mg of adenosine 5′-triphosphate disodium salt hydrate (ATP) powder in 8 mL of distilled water. Adjust pH to 7.4 using KOH. Bring the solution to 10 mL with water, aliquot in 500 μL samples, and store at −20 °C.
8. 1 M $MgCl_2$: Dissolve 50.8 g of magnesium chloride hexahydrate ($MgCl_2$) in 200 mL of distilled water. Bring the solution to 250 mL with water and store at RT. Use this solution for further dilutions.
9. Incubation buffer for non-synaptosomal brain mitochondria ($IB4_B$): 100 mM D-sucrose, 65 mM KCl, 10 mM HEPES-KOH, pH 7.4, 1 mM KH_2PO_4, 150 μM of ATP, 150 μM $MgCl_2$, 3 μM EDTA-KOH. Prepare basic incubation—dissolve 34.2 g of D-sucrose and 4.8 g KCl in 900 mL of distilled water. Add 20 mL of 0.5 M HEPES-KOH. Adjust with KOH pH to 7.4. Bring the solution to 1 L with water. Sterilize the solution using sterile vacuum filter units with a pore size of 0.22 μm. Sterile $IB4_B$ solution can be stored at 4 °C for several months. Before experiment take 50 mL of basic incubation buffer and add 50 μL 1 M KH_2PO_4, 7.5 μL 1 mM ATP, 7.5 μL 1 mM $MgCl_2$, and 15 μL of 10 mM EDTA-KOH. Discard solution after experiment.
10. 30% (w/v) Ficoll 400: Add 7 mL of distilled water to 3 g Ficoll 400 in 50 mL VWR conical-bottom disposable tube and leave it overnight to dissolve completely (*see* **Note 9**).

2.4 Measurements of Mitochondrial Function

1. 0.5 M Succinate-KOH: Dissolve 2.95 g of succinic acid in 40 mL of distilled water. Adjust pH with KOH to 7.4 at RT and bring the solution to 50 mL with water. Aliquot the solution in 0.5 mL aliquots and store at −20 ° C. Do not freeze and thaw.
2. 0.5 M Glutamate-KOH: Dissolve 3.68 g of L-glutamic acid in 40 mL of distilled water. Adjust pH with KOH to 7.4 at RT and bring the solution to 50 mL with water. Aliquot the solution in 0.5 mL aliquots and store at −20 °C. Do not freeze and thaw.
3. 0.5 M Malate: Dissolve 3.35 g of L-(−)-malic acid in 40 mL of distilled water. Adjust pH with KOH to 7.4 at RT and bring the solution to 50 mL with water. Aliquot the solution in 0.5 mL aliquots and store at −20 ° C. Do not freeze and thaw.

4. 1 M Rotenone dissolved in ethanol: Aliquot the solution in 0.5 mL aliquots and store at −20 ° C. Use to prepare 1 mM solution.
5. 1 M Calcium chloride solution for addition to mitochondria.
6. 0.42 mM Calcium-Green 5N (Thermo Fisher, USA): Dissolve 500 μg of Calcium-Green 5N hexapotassium salt in 500 μL of distilled water. Stock solution must be protected from light and can be stored up to 6 months at −20 °C. Final concentration of Calcium-Green 5N in incubation buffer is 125 nM.
7. 10 mM Tetramethylrhodamine methyl ester (TMRM) (Thermo Fisher, USA): Dissolve 25 mg of lyophilized powder in 5 mL of DMSO. To prepare a stock solution at 100 μM, add 10 μL of the 10 mM solution into 990 μL of DMSO. All TMRM stock solutions must be protected from light and can be stored up to 6 months at −20 °C. Final concentration of TMRM in incubation buffer is 60 nM.
8. Multichannel dye cuvette-based fluorimeter (C&L Instruments, Inc., USA).

3 Methods

3.1 Preparation of Oligomeric αSyn (See Note 10)

1. Dissolve lyophilized αSyn in PBS that has been freshly filtered through a 0.22 μm syringe filter (*see* **Note 11**). Allow the sample to equilibrate for 5 min and gently mix.
2. Measure absorption of the αSyn solution at 280 nm using a UV-VIS spectrophotometer and determine the protein concentration using Beer's law and a theoretical molar extinction coefficient at 280 nm of 0.412 $(mg/mL)^{-1}$ cm^{-1}.
3. Dilute αSyn solution to 2 mg/mL with additional PBS (*see* **Notes 12** and **13**).
4. Prepare 500 μL aliquots of 2 mg/mL monomeric αSyn in 1.5 mL Eppendorf LoBind microcentrifuge tubes (*see* **Notes 14** and **15**).
5. Flash freeze 2 mg/mL aliquots by submerging them in liquid nitrogen for approximately 1 min. Store tubes at −80 °C for later use.
6. To induce aggregation of αSyn, place 2 mg/mL αSyn aliquots on a nutating mixer at 37 °C for 5 days (*see* **Notes 16** and **17**).
7. Sonicate the fibrils with a mircotip probe using ten brief strokes (*see* **Notes 18** and **19**). Allow sample to cool for 10 s. Repeat this sonication/cooling cycle four more times for a total of 50 strokes (*see* **Note 20**).
8. To enrich for soluble mitotoxic species, spin sonicated αSyn samples for 30 min at 100,000 × *g* at 4 °C and collect the supernatant (Fig. 1, *see* **Notes 21** and **22**).

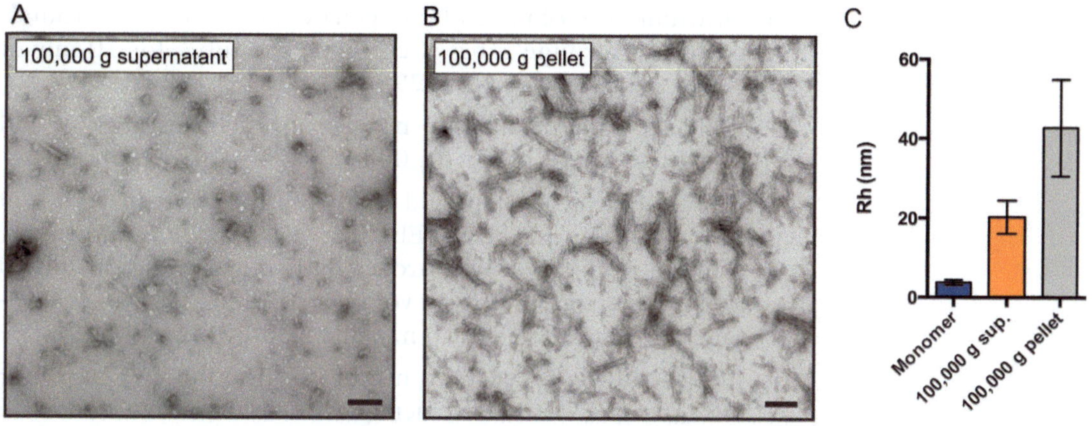

Fig. 1 100,000 *g* soluble fraction of sonicated αSyn fibrils contains oligomers. (**a**, **b**) Representative electron micrographs of sonicated αSyn fibrils fractionated into a 100,000 *g* supernatant (**a**) and pellet (**b**). *Scale bars = 100 nm.* (**c**) Comparison of the hydrodynamic radii (Rh) of αSyn monomer and the 100,000 *g* supernatant and pellet of sonicated αSyn fibrils as measured by dynamic light scattering. Error bars represent SEM of 3, 5, and 3 replicates for fibrils, 100,000 *g* supernatant, and 100,000 *g* pellet, respectively. *Reproduced from Luth* et al., *2014, with permission from ASBMB*

3.2 Mouse Liver Mitochondria Isolation

All steps should be performed on ice and/or in a cold room at 0–4 °C as described previously [27, 32] (*see* **Notes 23** and **24**). All solutions and tubes should be kept on ice during the entire procedure. Alternative mitochondrial isolation approaches from liver and/or other tissue are also expected to provide suitable starting material, although the user should be aware of the characteristics of any given preparation with regard, for example, to purity. If using other preparations, proceed to Procedures Subheading 3.4.

1. Sacrifice the mouse by cervical dislocation and rapidly remove liver from the animal. Place it immediately in a pre-weighed 50 mL plastic or glass beaker containing ice-cold 20–30 mL of the isolation buffer 1 for liver ($IB1_L$) (*see* **Notes 25** and **26**).
2. Weigh the liver in a beaker.
3. Mince the tissue in small pieces with scissors and change $IB1_L$ several times to remove blood contamination.
4. Pour the suspension into a pre-chilled 40 mL Potter-Elvehjem glass Teflon homogenizer and add $IB1_L$ to a total volume of 40 mL.
5. Homogenize the suspension using 7–8 up-and-down strokes with the pestle rotating by hand or in case of use of electrical homogenizer—at 500–600 rpm (*see* **Notes 27** and **28**). Suspension should look homogeneous. Prepare 10% w/v homogenate.

6. Distribute the obtained homogenate into two 40 mL round-bottom centrifuge tubes. Centrifuge the tubes for 10 min at 800 × *g* at 4 °C (*see* **Note 29**).
7. Pour the supernatant into new 40 mL centrifuge tubes and centrifuge for 10 min at 11,000 × *g* at 4 ° C (*see* **Note 30**).
8. Discard the supernatant and resuspend the pellet in 15 mL of $IB1_L$ in the 15 mL Potter-Elvehjem glass Teflon homogenizer using 4–5 up-and-down strokes by hand. Suspension should look homogeneous. Adjust volume to 40 mL, mix, and centrifuge the suspension for 10 min at 11,000 × *g* at 4 °C.
9. Resuspend the pellet in 15 mL of isolation buffer #2 for liver ($IB2_L$) in the Potter-Elvehjem glass Teflon homogenizer with 4–5 up-and-down strokes by hand. Suspension should look homogeneous. Adjust volume to 40 mL, mix, and centrifuge the suspension for 10 min at 11,000 × *g* at 4 °C.
10. Resuspend final pellet in 0.3–0.5 mL of $IB2_L$ and store on ice (*see* **Note 31**).
11. Add 10 μL of mitochondrial suspension into 990 μL of $IB2_L$ to measure protein concentration by BCA-based method using a spectrophotometer.

3.3 Mouse Non-synaptosomal Brain Mitochondria Isolation by Discontinuous Ficoll Gradient

1. All isolation steps have to be done on ice, and all operations have to be carried out at 0–4 °C as described previously [27, 32].
2. Sacrifice the mouse by cervical dislocation, decapitate, and open the skull rapidly. Remove the brain from the animal (*see* **Notes 25** and **26**) and place it immediately in a pre-weighed 50 mL plastic or glass beaker containing ice-cold 20–30 mL of the isolation buffer 1 for brain ($IB1_B$).
3. Weigh the brain tissue.
4. Transfer the brain on an ice-chilled Petri dish containing 10–20 mL of $IB1_B$ for dissection.
5. Remove the posterior region containing the pons and cerebellum, and divide the remainder into two hemispheres. The two cortical hemispheres are freed from the hippocampus and the striatum.
6. Transfer cortical hemispheres in a 50 mL plastic disposable beaker containing 20–30 mL of $IB1_B$, and mince the tissue with scissors.
7. Pour the suspension in a pre-chilled 40 mL Dounce glass-glass homogenizer and add $IB1_B$ to a total volume of 40 mL.
8. Homogenize the suspension using a loose-fitting pestle with 4–6 up-and-down strokes followed by 8 strokes with a tight-

fitting pestle (*see* **Notes 27** and **28**). Suspension should look homogeneous. Prepare 10% w/v homogenate.

9. Transfer the obtained homogenate into 40 mL round-bottom centrifuge tubes. Centrifuge the tubes for 3 min at 1300 × *g* at 4 °C.
10. Collect the supernatant and centrifuge for 3 min at 1300 × *g* at 4 °C.
11. Collect the supernatant and centrifuge for 8 min at 1700 × *g* at 4 °C.
12. Collect the pellet and resuspend it in 8 mL of $IB1_B$ in the 15 mL Dounce glass-glass homogenizer using a loose-fitting pestle with 4 up-and-down strokes followed by 8 strokes with a tight-fitting pestle and 4–5 up-and-down strokes by hand. Suspension should look homogeneous. Adjust volume to 10 mL and mix.
13. Place 7 mL of 10% (w/v) Ficoll 400 solution into each of two 30 mL ultracentrifuge tubes. Add 7 mL of 7.5% (w/v) Ficoll 400 solution dropwise on top of the 10% Ficoll 400 solution using plastic disposable transfer pipet. Add 5 mL of homogenate on top of the 7.5% Ficoll 400 solution. Carefully place the centrifuge tube into swinging rotor buckets and balance them with $IB1_B$. Centrifuge the tubes for 20 min at 99,000 × *g* at 4 °C.
14. Remove all layers (*see* **Note 32**) consecutively from the top down with disposable transfer plastic pipette. Collect non-synaptosomal mitochondrial fraction at the bottom of the tubes.
15. Resuspend the pellet in 15 mL of $IB1_B$ in the 15 mL Dounce glass-glass homogenizer using a loose-fitting pestle and 4 up-and-down strokes followed by 8 strokes with a tight-fitting pestle and 4–5 up-and-down strokes by hand. Centrifuge the suspension for 8 min at 12,000 × *g* at 4 °C.
16. Resuspend the pellet in 10 mL of $IB2_B$. Centrifuge the suspension for 8 min at 12,000 × *g* at 4 °C.
17. Resuspend the pellet in 0.1–0.2 mL of $IB3_B$ in 1.5 mL Eppendorf tube and store it on ice (*see* **Notes 31** and **33**).
18. Add 10 μL of mitochondrial suspension into 990 μL of $IB1_B$ to measure protein concentration by BCA-based method using a spectrophotometer.

3.4 Measurement of Mitochondrial Parameters

Measuring mitochondrial Ca^{2+} retention time (CRT) is an efficient parameter to assess αSyn toxicity. CRT is determined as the amount of Ca^{2+} sequestered by mitochondria without incurring structural damage and is expressed in nmol/mg protein. Ca^{2+} concentration in the mitochondrial suspension is measured as Calcium-Green 5N fluorescence signal. A higher signal corresponds to an increase of Ca^{2+} concentration in the buffer. Addition of Ca^{2+} into the incubation buffer induces an increase of Calcium-Green 5N fluorescence followed by a decrease due to the uptake of Ca^{2+} by mitochondria and gradual increase of Calcium-Green 5N signal due to Ca^{2+} release from mitochondria. CRT assay may be performed simultaneously with the assessment of other mitochondrial parameters such as membrane potential ($\Delta\Psi_m$) using TMRM, redox state of pyridine nucleotides by monitoring NAD(P)H autofluorescence, and Ca^{2+}-induced swelling of mitochondrial suspension by observing changes in absorbance [27] (Fig. 2).

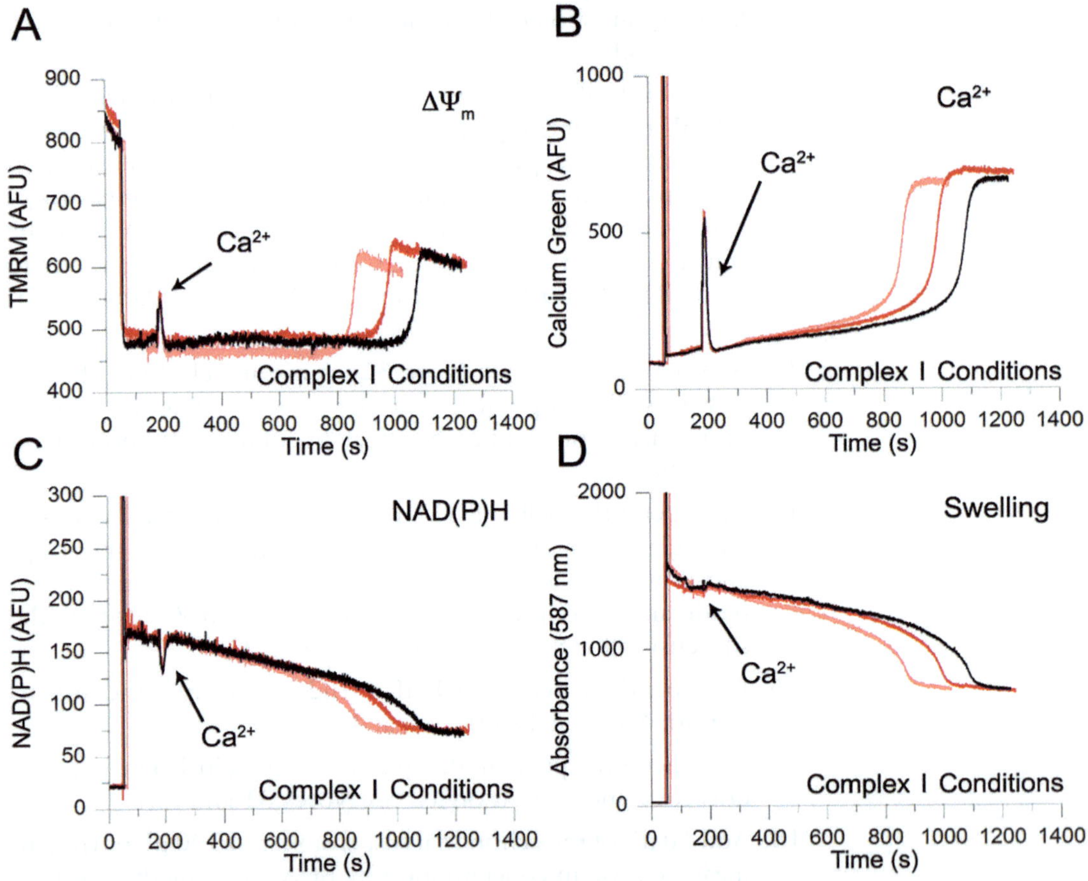

Fig. 2 Measurement of liver mitochondrion parameters. (**a–d**) Traces showing basic parameters of isolated liver mitochondria ($\Delta\Psi_m$) (**a**), extramitochondrial Ca^{2+}fluorescence (**b**), NAD(P)H autofluorescence (**c**), and mitochondrial swelling (**d**) simultaneously measured by a multichannel fluorimeter and recorded in the presence of a single 20 μM Ca^{2+} addition, 5 mM glutamate, and 5 mM malate as substrates, and either PBS vehicle (black traces) or 1 μM (red traces) or 2 μM (pink traces) oligomeric αSyn. Spikes result from addition of mitochondria; arrows are used to indicate the time of Ca^{2+} addition. *Reproduced from Luth* et al., *2014, with permission from ASBMB*

1. Use a cuvette-based multichannel dye fluorimeter with the following excitation and emission wavelengths:
   (a) 482/535 nm for Calcium-Green 5N.
   (b) 548/574 nm for TMRM.
   (c) 350/450 nm for NAD(P)H oxidation.
   (d) 587 nm excitation and emission for mitochondrial absorbance

For liver mitochondria:

2. Incubate 5 mM glutamate/malate or 5 mM succinate plus 1 μM rotenone in 2 mL of $IB3_L$ in the stirring cuvette.
3. Add liver mitochondria at 0.25 mg/mL and wait for 1 min for mitochondria to equilibrate.
4. Add αSyn or vehicle at desired concentration and wait for 1 min for αSyn and mitochondria to mix.
5. Add a single Ca^{2+} addition of 20–40 nmol/mg protein.
6. Collect measurements until Calcium-Green 5N fluorescence reaches a plateau at its maximum value (approximately 20 min).
7. Record time from Ca^{2+} addition to plateau as CRT.
8. For measurements of other mitochondrial parameters ($\Delta\Psi_m$, redox state of pyridine nucleotides, or mitochondrial swelling), adjust fluorescence excitation/emission wavelengths as described in **step 1** and repeat **steps 2–6** (*see* **Note 34**).

For brain mitochondria:

9. Incubate 5 mM glutamate/malate or 5 mM succinate plus 1 μM rotenone in 2 mL of $IB4_B$.
10. Add brain mitochondria at 0.10 mg/mL and wait for 1 min for mitochondria to equilibrate.
11. Add αSyn or vehicle at desired concentration and wait for 1 min for αSyn and mitochondria to mix.
12. Add aliquots of Ca^{2+}, each addition 200 nmol/mg protein, every 2 min until the Calcium-Green-5N fluorescence does not return to baseline following the spike induced by Ca^{2+} addition (Fig. 3).
13. Record the total amount of added Ca^{2+} as the Ca^{2+} retention capacity (CRC) (*see* **Note 35**).

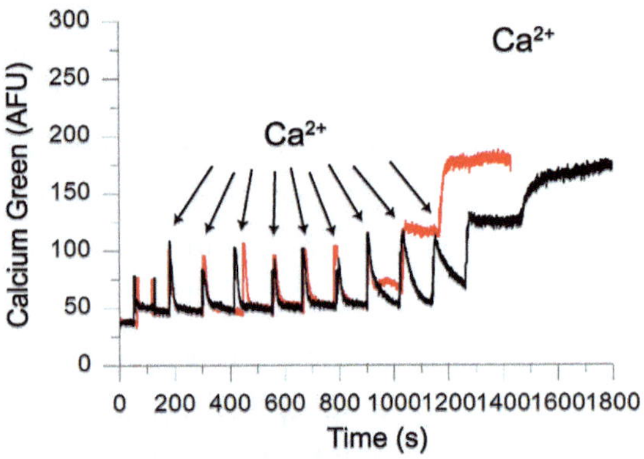

Fig. 3 Measurement of CRC of brain mitochondria. Traces of the Ca-Green 5N fluorescence of vehicle-treated (black) and 2 μM sonicated αSyn-treated brain mitochondria (red) respiring using glutamate and malate. *Reproduced from Luth* et al., *2014, with permission from ASBMB*

4 Notes

1. This protocol assumes an ability to purify recombinant monomeric αSyn from *E. coli*. For this procedure and associated reagents, please see Mulvihill et al., this issue. Alternatively, recombinant αSyn can be purchased from commercial sources (for example, we have successfully used αSyn purchased from rPeptide) but this may be cost prohibitive due to the amount used here.
2. EGTA and EDTA solutions are very acidic and dissolve poorly in water. Add KOH pellets until reaching a pH ~7.0, and then continue to adjust the pH with a 5 M KOH solution.
3. The buffer composition for mitochondria isolation and incubation may vary. For example, examination of the literature will show differences in relative concentrations of sucrose and/or mannitol, in the choice of buffering agent and counterion, the inclusion/exclusion of divalent cation chelators, osmolarity, and ionic strength. The buffers described here have been used by our group for respiration and calcium overload experiments.
4. Use deionized water (Milli-Q, "Millipore") for mitochondrial buffer preparation.
5. Airtight glass containers or Thermo Scientific Nalgene disposable bottles are strongly recommended for storage of mitochondrial buffers.
6. Sterilize all buffers by filtering through 0.22 μm membrane (we recommend using Thermo Scientific Nalgene disposable filter units that come with bottles). The buffers are stable for at least 3 months if kept sterile.

7. Do not freeze isolation and incubation buffers.
8. Alternative protein concentration approaches can be used, but linear range and potential interference must be examined.
9. Do not stir Ficoll 400 with a pipet or plastic stick because it is very sticky.
10. We have previously described two ways of generating αSyn oligomers that have similar effects on the function of isolated mitochondria [27]. An alternative method to the one presented here involves isolating short-lived, "on-pathway" oligomers formed during the lag phase of aggregation (as monitored by thioflavin T fluorescence). The duration of the lag phase, and therefore the time window for isolating mitotoxic oligomers, can be extremely variable and depends on, among other factors, temperature, pH, protein concentration and purity, buffer composition, speed and type of mixing, and presence of small amounts of oligomeric aggregation seeds. To improve experimental reproducibility, we recommend preparing αSyn oligomers via sonication of αSyn fibrils as described here.
11. Buffer choice depends on desired experimental conditions and may influence aggregation kinetics. We have achieved similar effects on mitochondria with oligomers prepared using 10 mM ammonium acetate. Though both ammonium and acetate may affect mitochondrial function at high concentrations, the final concentration of ammonium acetate experienced by mitochondria in our assay (~140 μM) had no effect when compared to addition of an equal volume of mitochondrial isolation buffer.
12. If the concentration is less than 2 mg/mL, the solution can be added to an additional tube of lyophilized αSyn after which the concentration should again be determined and adjusted to 2 mg/mL with PBS.
13. Using size-exclusion chromatography, we have observed that newly dissolved recombinant αSyn exists in a monomeric form. It is possible that trace amounts of higher order species may be present at this stage. To enrich for monomers, the sample can be filtered through a 0.22 μm syringe filter; however, because fibril formation is desired, this is not necessary.
14. It is important to add sufficient volume to each tube such that the action of the nutator (*see* Subheading 3.1, **step 6**) allows for effective mixing of the solution. The αSyn solution will eventually be diluted 1:1 just prior to sonication in this tube. A volume of 500 μL affords enough room for additional PBS and the sonicator tip.
15. The use of LoBind tubes is critical to prevent protein loss; aggregated αSyn can stick to the sides of standard microcentrifuge tubes.

16. We recommend sealing tubes with Parafilm to prevent sample evaporation and flicking tubes to coat the interior surface with the αSyn solution to ensure even mixing while nutating.
17. After 5 days, flick the tube to observe insoluble fibrils that should be visible to the naked eye. If no fibrils are visible when held up to light, continue incubating at 37 degrees.
18. We used a Sonic Dismembrator Model 300 (Fisher Scientific, USA) at 50% power.
19. Begin the sonication with the microtip submerged at the bottom of the tube. Each stroke consists of a downward movement of the tube to bring the sonicator microtip almost to the surface of the liquid followed by an upward movement to return to the original tube position and should last approximately 1 s. Removing and reinserting the sonicator microtip from the solution will create unwanted bubbles.
20. Alternatively, a programmable Ultrasonic Liquid Processor (Misonix, USA) equipped with a microtip (settings: amplitude of 20, sonication for 1 s on and 1 s off for a total of 60 s) can be used to produce αSyn oligomers with comparable bioactivity on mitochondria as those produced via our standard manual sonication technique.
21. If an ultracentrifuge is unavailable, centrifugation using a standard benchtop centrifuge for 30 min at 20,000 × *g* will also enrich for soluble oligomers.
22. After centrifugation at 100,000 × *g*, large (60–100 nm long) fibril fragments will appear in the pellet, while the supernatant will be composed largely of smaller, more rounded oligomers (Fig. 1, Reproduced from Luth et al., 2014, with permission from ASBMB).
23. Quality and quantity of isolated mitochondria are very important because contamination of the mitochondrial preparation with cytosolic structures has the potential to interfere with or mask the effect of αSyn.
24. Omit protease inhibitors in the buffers because they interfere with Complex I activity.
25. Animals have to be sacrificed without CO_2 or isoflurane, because they interfere with mitochondrial respiratory complex activity.
26. Liver or brain tissue must be removed and placed in ice-cold buffer as quickly as possible.
27. During tissue homogenization steps, do not draw a vacuum and avoid bubble creation.
28. Avoid very fast pestle rotation during homogenization.
29. Do not use detergents to clean centrifuge tubes.

30. Carefully remove fat from centrifuge tube walls with a delicate wipe after the second centrifugation.
31. We recommend a concentration of at least 60–80 mg/mL of mitochondria.
32. After centrifugation at 99,000 × *g* for 20 min at 4 °C the layers (from the top to bottom) are comprised of a clear layer (isolation buffer), a white myelin-containing layer, a light yellow synaptosomal mitochondrion-containing layer, and non-synaptosomal mitochondrion pellet at the bottom of the tube.
33. Resuspend the final non-synaptosomal brain mitochondrion pellet using Eppendorf (or any other brand) micropestle that matches the conical contour of Eppendorf 1.5 mL microcentrifuge tubes.
34. The pattern of Ca^{2+}-induced changes in TMRM fluorescence, NAD(P)H autofluorescence, and light scattering usually coincides with the pattern of CaGreen-5N fluorescence signal.
35. Measurement of other parameters (TMRM fluorescence, NAD(P)H autofluorescence, and light scattering) using brain mitochondria may be less informative.

References

1. Bartels T, Choi JG, Selkoe DJ (2011) α-Synuclein occurs physiologically as a helically folded tetramer that resists aggregation. Nature 477:107–110
2. Westphal CH, Chandra SS (2013) Monomeric synucleins generate membrane curvature. J Biol Chem 288:1829–1840
3. Luth ES, Bartels T, Dettmer U et al (2015) Purification of α-synuclein from human brain reveals an instability of endogenous multimers as the protein approaches purity. Biochemistry 54:279–292
4. Gould N, Mor DE, Lightfoot R et al (2014) Evidence of native α-synuclein conformers in the human brain. J Biol Chem 289:7929–7934
5. Wang L, Das U, Scott DA et al (2014) α-Synuclein multimers cluster synaptic vesicles and attenuate recycling. Curr Biol 24:2319–2326
6. Iljina M, Tosatto L, Choi ML et al (2016) Arachidonic acid mediates the formation of abundant alpha-helical multimers of alpha-synuclein. Sci Rep 6:33928
7. Chartier-Harlin M-C, Kachergus J, Roumier C et al (2004) Alpha-synuclein locus duplication as a cause of familial Parkinson's disease. Lancet (London, England) 364:1167–1169
8. Singleton AB, Farrer M, Johnson J et al (2003) Alpha-Synuclein locus triplication causes Parkinson's disease. Science 302:841–841
9. Kara E, Lewis PA, Ling H et al (2013) α-Synuclein mutations cluster around a putative protein loop. Neurosci Lett 546:67–70
10. Spillantini MG, Schmidt ML, Lee VM et al (1997) Alpha-synuclein in Lewy bodies. Nature 388:839–840
11. Dettmer U, Newman AJ, Soldner F et al (2015) Parkinson-causing α-synuclein missense mutations shift native tetramers to monomers as a mechanism for disease initiation. Nat Commun 6:7314
12. Danzer KM, Haasen D, Karow AR et al (2007) Different species of alpha-synuclein oligomers induce calcium influx and seeding. J Neurosci 27:9220–9232
13. Winner B, Jappelli R, Maji SK et al (2011) In vivo demonstration that alpha-synuclein oligomers are toxic. Proc Natl Acad Sci U S A 108:4194–4199
14. Karpinar DP, Balija MBG, Kügler S et al (2009) Pre-fibrillar alpha-synuclein variants with impaired beta-structure increase neurotoxicity in Parkinson's disease models. EMBO J 28:3256–3268

15. Volles MJ, Lansbury PT (2003) Zeroing in on the pathogenic form of alpha-synuclein and its mechanism of neurotoxicity in Parkinson's disease. Biochemistry 42:7871–7878
16. Fusco G, Chen SW, Williamson PTF et al (2017) Structural basis of membrane disruption and cellular toxicity by α-synuclein oligomers. Science 358:1440–1443
17. Schapira AH, Cooper JM, Dexter D et al (1989) Mitochondrial complex I deficiency in Parkinson's disease. Lancet (London, England) 1:1269
18. Parker WD, Boyson SJ, Parks JK (1989) Abnormalities of the electron transport chain in idiopathic Parkinson's disease. Ann Neurol 26:719–723
19. Keeney PM, Xie J, Capaldi RA et al (2006) Parkinson's disease brain mitochondrial complex I has oxidatively damaged subunits and is functionally impaired and misassembled. J Neurosci 26:5256–5264
20. Devi L, Raghavendran V, Prabhu BM et al (2008) Mitochondrial import and accumulation of alpha-synuclein impair complex I in human dopaminergic neuronal cultures and Parkinson disease brain. J Biol Chem 283:9089–9100
21. Martin LJ, Pan Y, Price AC et al (2006) Parkinson's disease alpha-synuclein transgenic mice develop neuronal mitochondrial degeneration and cell death. J Neurosci 26:41–50
22. Stichel CC, Zhu X-R, Bader V et al (2007) Mono- and double-mutant mouse models of Parkinson's disease display severe mitochondrial damage. Hum Mol Genet 16:2377–2393
23. Hsu LJ, Sagara Y, Arroyo A et al (2000) Alpha-synuclein promotes mitochondrial deficit and oxidative stress. Am J Pathol 157:401–410
24. Büttner S, Habernig L, Broeskamp F et al (2013) Endonuclease G mediates α-synuclein cytotoxicity during Parkinson's disease. EMBO J 32:3041–3054
25. Parihar MS, Parihar A, Fujita M et al (2009) Alpha-synuclein overexpression and aggregation exacerbates impairment of mitochondrial functions by augmenting oxidative stress in human neuroblastoma cells. Int J Biochem Cell Biol 41:2015–2024
26. Shavali S, Brown-Borg HM, Ebadi M et al (2008) Mitochondrial localization of alpha-synuclein protein in alpha-synuclein overexpressing cells. Neurosci Lett 439:125–128
27. Luth ES, Stavrovskaya IG, Bartels T et al (2014) Soluble, prefibrillar α-synuclein oligomers promote complex I-dependent, Ca2+-induced mitochondrial dysfunction. J Biol Chem 289:21490–21507
28. Luk KC, Song C, O'Brien P et al (2009) Exogenous alpha-synuclein fibrils seed the formation of Lewy body-like intracellular inclusions in cultured cells. Proc Natl Acad Sci U S A 106:20051–20056
29. Volpicelli-Daley LA, Luk KC, Patel TP et al (2011) Exogenous α-synuclein fibrils induce Lewy body pathology leading to synaptic dysfunction and neuron death. Neuron 72:57–71
30. Luk KC, Kehm V, Carroll J et al (2012) Pathological α-synuclein transmission initiates Parkinson-like neurodegeneration in nontransgenic mice. Science 338:949–953
31. Polinski NK, Volpicelli-Daley LA, Sortwell CE et al (2018) Best practices for generating and using alpha-Synuclein pre-formed fibrils to model Parkinson's disease in rodents. J Parkinson's Dis 68:1–20
32. Stavrovskaya IG, Baranov SV, Guo X et al (2010) Reactive gamma-ketoaldehydes formed via the isoprostane pathway disrupt mitochondrial respiration and calcium homeostasis. Free Radic Biol Med 49:567–579

Chapter 15

Monitoring α-Synuclein Proteotoxicity in *Drosophila* Models

Aron Szabo and George K. Tofaris

Abstract

Parkinson's disease is the second most common neurodegenerative disease without cure. It is characterized by α-synuclein accumulation and aggregation in dopaminergic and other types of neurons. Because α-synuclein accumulation leads to a toxic gain of function, its ectopic expression in *Drosophila* has been a useful in vivo model for testing modifiers of its toxicity. This chapter describes four assays: the rapid iterative negative geotaxis, rough eye phenotype, quantification of dopaminergic neuronal loss, and measurements of circadian effects.

Key words Drosophila, Behavior, In vivo models, Morphology analysis, Neurotoxicity

1 Introduction

Parkinson's disease (PD) is the second most common neurodegenerative disorder, affecting 1% of people over the age of 60. Clinically, it is characterized primarily by a movement disorder causing resting tremor, bradykinesia, rigidity, postural instability, and diverse non-motor symptoms including dementia, which in community-based studies was reported in up to 80% of patients with a long disease duration. This latter finding indicates that PD is a diffuse neurodegenerative disorder. Similarly, detailed neuropathological studies have shown that intraneuronal inclusions called Lewy bodies, which are one of the cardinal histological features of the disease, are detected in numerous cortical areas and often correlate with the extent of cognitive decline [1]. Despite this diffuse evolution, the presentation to health services is commonly due to the loss of a critical number of dopaminergic neurons in the substantia nigra, but in patients with diffuse Lewy bodies disease dementia may be the presenting feature. α-Synuclein is also the major component of glial cytoplasmic inclusions that define multiple system atrophy (MSA), a movement disorder characterized by cerebellar ataxia, parkinsonism, and autonomic dysfunction. In MSA, unlike Lewy body disease, α-synuclein inclusions are found

Tim Bartels (ed.), *Alpha-Synuclein: Methods and Protocols*, Methods in Molecular Biology, vol. 1948,
https://doi.org/10.1007/978-1-4939-9124-2_15,

in glial cells. The identification of additional causative genes of familial PD that clinically mimic sporadic disease such as *LRRK2, VPS35, Pink1*, and *Parkin* has suggested new mechanisms of pathogenicity [2]. A major limitation in translating this basic understanding into therapies is the lack of animal models that faithfully recapitulate the human condition. In this respect, the use of simple model organisms to interrogate in vivo the function of genes against biochemical and cellular readouts has facilitated the mapping of disease-causing genes with a fly ortholog into pathways (e.g., mitophagy and endosomal-lysosomal trafficking) and suggested novel therapeutic strategies in PD.

Drosophila melanogaster is a tractable organism for modeling epistatic interactions in the context of pathological phenotypes. Some of the advantages of utilizing *Drosophila* for genetic studies include its short life span of 45–60 days, relatively low cost, and evolutionary conservation of most genes. About 75% of human genes have a *Drosophila* ortholog [3]. The development of the Gal4-UAS system [4] has transformed the tractability of this organism for gene interaction studies. This combines the yeast transcriptional activator Gal4 with UAS (upstream activation sequence), an enhancer to which Gal4 specifically binds to activate gene transcription. By fusing Gal4 to a tissue-specific promoter, it is possible to drive the expression or knockdown of genes of interest in specific cell populations [5].

Drosophila melanogaster has been extensively used to model pathogenic processes, including neurodegenerative diseases. It has been especially informative in the elucidation of the function of the recessive PD genes *Pink1* and *Parkin* in mitochondrial homeostasis. The earliest evidence that Pink1 and Parkin function in the same pathway came from phenotypic observations in *Pink1* and *Parkin* knockout flies which both exhibited swollen mitochondria with broken cristae, muscle dysfunction, dopaminergic neuron degeneration, and reproductive defects [6, 7]. Parkin overexpression could rescue the Pink1 mutant phenotypes whereas Pink1 overexpression did not rescue the Parkin mutant ones [7–9] suggesting a linear cascade with Pink1 upstream of Parkin. This has been delineated mechanistically in mammalian cells where Pink1 activation phosphorylates ubiquitin and Parkin, activating Parkin-mediated ubiquitination and mitochondrial autophagy [10]. Additional insights came from the study of the LRRK2 ortholog in flies, which provided early in vivo evidence for a role of this kinase in endosomal trafficking, lysosomal function [11, 12], and synaptic autophagy [13].

Flies do not have the *α-synuclein* (SNCA) gene and therefore are not a suitable system to investigate its function. However, in sporadic PD, neurodegeneration is thought to arise from accumulation and misfolding of α-synuclein into oligomeric fibrillar species, which acquire a toxic gain of function, partly by impairing membrane fusion events [14]. In this context, ectopic expression of α-synuclein in different tissues using cell-type-specific drivers has been shown to recapitulate cardinal features of Parkinson's pathology: Feany and colleagues first expressed human α-synuclein in the nervous system of flies using the elav/Gal4 system [15]. Transgenic flies showed age-dependent, progressive degeneration of the dorsomedial (PPM1/2) cluster of dopaminergic neurons. The neurodegeneration was associated with an age-dependent decline in the climbing ability of transgenic flies using a negative geostatic assay. In addition, transgenic flies were shown to have neuronal α-synuclein-immunoreactive cytoplasmic aggregates, which exhibited a fibrillar ultrastructure under electron microscopy. The fly eye consists of 760 unit eyes, called ommatidia, which contain photoreceptors and bristles. Transgenic expression of α-synuclein in the fly eye using GMR/Gal4 causes an architectural distortion, known as rough eye phenotype that is associated with retinal degeneration with vacuolization [15]. The rough eye phenotype induced by α-synuclein toxicity is reversed by specific modifiers that were also shown to directly regulate α-synuclein degradation in mammalian cells [16, 17]. The observation that modifiers such as the E3 ligase NEDD4 are protective in fly as well as rat models of α-synuclein proteotoxicity [17] supports the notion that simplified in vivo assays in flies may be especially helpful in the study of proteostasis, where key enzymes are conversed across species.

Sleep timing is partly governed by the circadian clock, which can be readily assayed in *Drosophila* models [18]. Sleep disturbance is well documented in Parkinson's patients and there is emerging evidence that circadian defects occur in PD [19]. Recently, circadian defects were observed in fly models expressing α-synuclein [20] suggesting that this may be another simple behavioral readout for in vivo quantification of α-synuclein toxicity.

Here, we describe four assays of α-synuclein proteotoxicity in *Drosophila*: (1) rapid iterative negative geotaxis (RING) as a readout of motor activity, (2) rough eye phenotype, (3) tyrosine hydroxylase (TH) immunostaining of the adult brain to track the loss of specific dopaminergic neuronal subpopulations, and (4) measurement of locomotor activity in response to circadian rhythms as a potentially non-motor defect.

2 Materials

The following *Drosophila* lines are required: *UAS-α-synuclein* wild-type (8146) and *A53T* mutant (8148), available from the Bloomington *Drosophila* Stock Center. Driver lines such as *GMR-Gal4* for eye-specific expression, *ddc-Gal4* or *TH-Gal4* for expression in the dopaminergic neurons, *elav-Gal4* for pan-neuronal expression, and triple mutant *TP-α-synuclein* flies [21] are available from a number of laboratories. Transgenic flies expressing α-synuclein and either RNAi or cDNA construct for the modifier of interest can be generated to assay its knockdown or overexpression against α-synuclein toxicity. Flies are typically aged for 30 days post-eclosion.

2.1 Equipment

1. Dissecting stereomicroscope.
2. Sylgard-coated Petri dishes for dissection.
3. Dumont No. 5 forceps.
4. Insect pins.
5. Microscope slides.
6. No. 1 and No. 2 coverslips.
7. Imaging fluorescent microscope.
8. Light- and temperature-controlled incubators.
9. DAM2 Drosophila Activity Monitor (TriKinetics).
10. Glass tubes and tube caps for the monitors (TriKinetics).
11. PSIU9 Power Supply Interface Unit (TriKinetics).
12. DC Power Supply (TriKinetics).
13. Data collection computer.
14. FaasX software for data analysis.
15. Digital camera and timer.
16. Custom-made RING apparatus (*see* Fig. 1) and 50 mL Falcon tubes.
17. JEOL JSM 6390 or similar scanning electron microscope.
18. Denton DV-503 vacuum evaporator.
19. SEM stubs.

2.2 Reagents

1. Fly food—Recipes can be found at the Bloomington Drosophila Stock Center website (https://bdsc.indiana.edu/information/recipes/index.html).
2. Phosphate-buffered saline (PBS).
3. Triton X-100.
4. Schneider's Drosophila medium.
5. 4% Paraformaldehyde solution (PFA) in PBS.
6. Normal goat serum (NGS).
7. Primary antibody: Monoclonal mouse anti-tyrosine hydroxylase (TH) (Millipore, MAB318, 1:500).
8. Fluorescently labeled secondary antibody (e.g., Alexa 488-coupled goat anti-mouse IgG (1:1000)).
9. SlowFade Diamond mounting medium.
10. 100% ethanol.
11. Milli-Q-grade water.

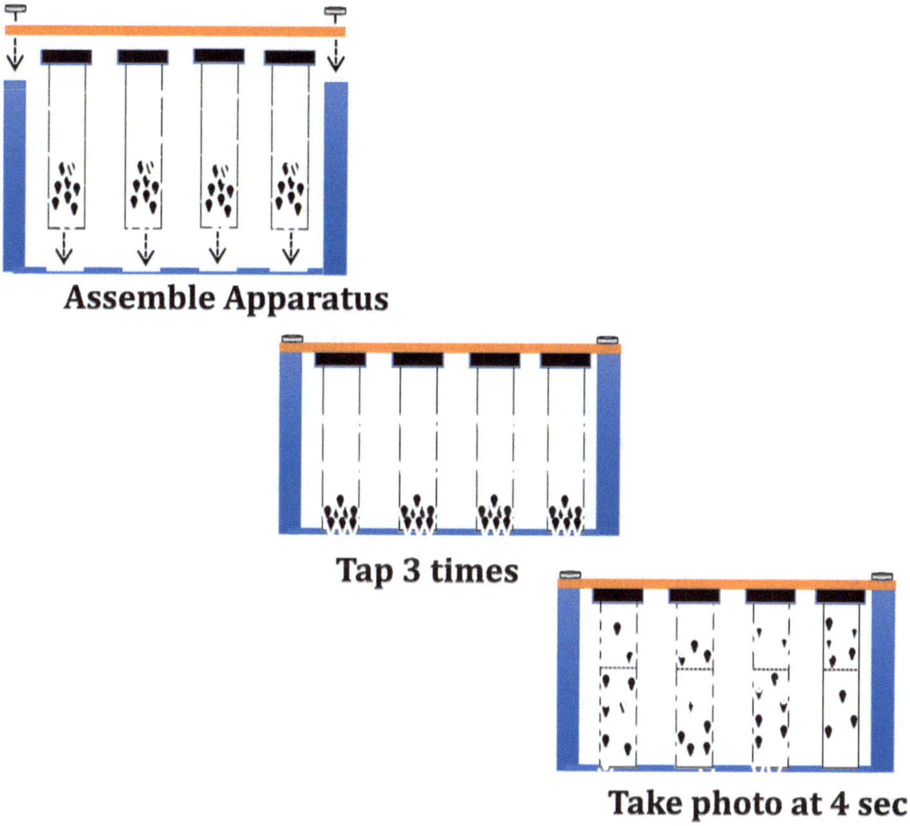

Fig. 1 Schematic illustration of the custom-made equipment for RING. This is comprised of an open-faced box with a removable lid that can be assembled in the laboratory to hold the geotaxis vials of choice in place. A digital camera is placed approximately 30 cm in front of the RING apparatus containing flies. Gentle tapping of the apparatus causes a startle-induced fall of the flies to the bottom of each vial. Images are taken at 4 s after tapping and the percentage of flies above a predefined height (in our case 6 cm) is estimated

3 Methods

3.1 Locomotion Assay

Locomotor function can be assayed using the rapid iterative negative geotaxis (RING) protocol, which is an adaptation from Gargano et al. [22].

1. Transfer 10–20 male flies in a clean vial without anesthetizing them with CO_2 as flies can take up to 24 h to fully recover from anesthesia.
2. Use 5 vials per line in each experiment (total 50 flies per line) and mark each vial with a line at the 6 cm height (*see* **Note 1**).
3. Allow 15–20 min for the flies to acclimate to the environment.
4. For simultaneous testing of multiple vials, transfer the vials into a custom-made RING apparatus (Fig. 1).

5. Place a digital camera in front of the apparatus (approximately 30 cm) so that the lens is at mid-height of the vials, focus and zoom the camera to capture the image of the whole vial, and set a timer to 4.0 s.
6. Tap the apparatus three times sharply ensuring that the tapping is hard enough for all the flies to startle and drop to the bottom of each vial.
7. With the third tap simultaneously start the timer and take the picture at 4 s. Repeat at least five times.
8. Quantify the percentage of flies above the 6 cm mark. In our experience this is a more accurate measurement of the locomotion defect in the α-synuclein model compared to the average distance traveled.
9. Repeat the experiment from at least two independently derived transgenic lines (*see* **Note 2**).

3.2 Rough Eye Phenotype

Although the rough eye phenotype is visible by light microscopy, for more accurate quantification of the effect of modifiers we used scanning electron microscopy (*see* **Notes 3–6**).

1. Fix age-matched, male flies in 70% ethanol.
2. Dehydrate the flies in 100% ethanol. The flies are then critically point dried and mounted on SEM stubs.
3. Sputter coat the samples with 15 nm gold nanoparticles using a Denton DV-503 vacuum evaporator or similar equipment and image on a JEOL JSM 6390 or similar scanning electron microscope.
4. Unless a complete rescue is observed, quantitate the eye phenotype, by randomly selecting 20 flies per genotype. Score the images blindly using objective features of severity. Examine the eyes for abnormal bristle orientation, ommatidial fusion or pitting, and disorganization of the ommatidial array.
5. Score the phenotype by giving one point if any of the above features is present, 2 points if it affects more than 5% of the eye, and 4 points if it affects more than 25% of the eye.

3.3 Quantification of the Loss of PPM1/2 Cluster Dopaminergic Neurons in the Brain

1. Obtain adult flies expressing *UAS-α-synuclein A53T* with *ddc-Gal4* or *elav-Gal4* and age them together with the single-transgene controls to 25 days post-eclosion.
2. Anesthetize flies under CO_2, rinse them in 100% ethanol for 20 s, and place them in ice-cold PBS transferring as little ethanol as possible (*see* **Note 7**).
3. Pin one fly submerged in ice-cold Schneider's *Drosophila* medium onto a Sylgard-coated dish with two insect pins.

4. Incise between the eyes on the top of the head using the forceps. Extend the opening and gently peel off the eyes one followed by the other.
5. Remove other appendages such as the proboscis and clean the brain from tracheae and floating membranes (*see* **Note 8**).
6. Place brains in 1 mL ice-cold PFA in a microcentrifuge tube and incubate rocking for 1 h at 25 °C (*see* **Note 9**).
7. Let the brains settle to the bottom of the tube, aspirate PFA, and rinse quickly twice with PBS + 0.05% triton X-100 (PBST).
8. Wash thrice for 20 min nutating in 1 mL PBST at room temperature (RT).
9. Block with 1 mL 5% NGS in PBST at RT.
10. Add anti-TH primary antibody at 1:500 in 5% NGS, and incubate at 4 °C for 48 h.
11. Wash as for point 8.
12. Add secondary antibody in 5% NGS, and incubate at 4 °C for 48 h in the dark.
13. Wash as for point 8 in the dark.
14. Prepare a microscope slide by placing a drop of SlowFade (40 μL) in the center and create a bridge by positioning broken No. 2 coverslips on the sides of the drop.
15. With a pipette and a cutoff P200 tip, transfer the brains from the tube into the drop of SlowFade.
16. Align the brains so that the posterior side faces upwards.
17. Cover them with a No. 1 coverslip and seal with nail polish.
18. Proceed to imaging on an epifluorescent or confocal fluorescent microscope.
19. Identify TH^+ dopaminergic neurons of the PPM1/2 cluster as shown in Fig. 2 and count them.

3.4 Analysis of Circadian Rhythms in Locomotor Activity

1. Place anesthetized flies individually into glass tubes containing 5% sucrose and 2% Bacto agar designed for DAM2 monitors and close them with a cotton plug (*see* **Note 10**).
2. Fill the 32 channels of DAM2 with tubes representing one genotype.
3. Place the DAM2-s into the incubators and connect them to Power Supply Interface Unit through the phone cables.
4. Entrain flies for 3 days to 12-h light (L):12-h dark (D) cycles and then leave them in constant darkness (DD) for 10 days at a constant temperature of 25 °C.
5. Extract the locomotor activity data from the recordings only for the DD period of the experiment using the DAMSystem3

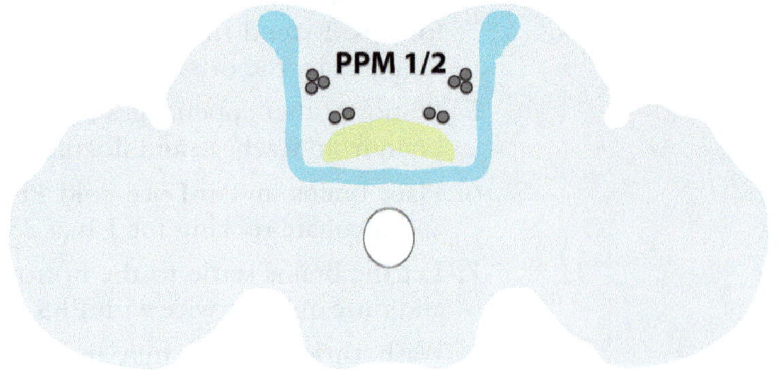

Fig. 2 Localization of PPM 1/2 neurons in the adult *Drosophila* brain. Schematic diagram of the adult fly brain (grey) viewed from the posterior side. The neuropil structures of the mushroom body (blue) and the fan-shaped body (yellow) are indicated as an aid for orientation. Cell bodies of the TH+ PPM 1/2 cluster are shown in the dorsomedial region of the brain. Note that their relative position can vary between brains

Data Acquisition Software and DAMFileScan. Use bins of 5 min.

6. Apply the FaasX software (MacIntosh only, http://neuro-psi.cnrs.fr/spip.php?article298&lang=en) to analyze locomotor activity data. Actogram allows the visualization of activity patterns throughout the days. Cycle_p calculates the exact period length in DD. Follow the instructions in the startup kit (*see* **Note 11**).

4 Notes

1. We did not observe any desensitization of the flies on the geotaxis assay up to ten consecutive trials spaced 1 min apart.
2. It is critical *not* to reuse the testing vials because new flies placed into used vials do not climb to the same extent as in fresh vials.
3. We found that the α-synuclein-induced eye damage is observed more consistently when flies were crossed and larvae kept at 29 °C. Because the Gal4/UAS system is heat sensitive we observed that a more consistent phenotype at 29 °C that correlated with a higher level of α-synuclein expression. Under these conditions the rough eye phenotype was detected from day 1.
4. Expression of the A53T mutant α-synuclein consistently caused a more severe phenotype than wild-type α-synuclein, which can be mild depending on the driver. The mutant construct is more suitable for screening assays.

5. Homozygous GMR-Gal4 in flies per se can cause a rough eye phenotype and therefore it is important that the driver is heterozygous.
6. We noticed that the level of expression of α-synuclein varies between different heterozygous GMR-Gal4 lines. It is therefore necessary to test and identify the best line that achieves the highest expression of α-synuclein at 29 °C without toxicity when the driver alone is present.
7. The ethanol wash before brain dissection is critical as too little wash will not soften the cuticle while too much will affect the efficiency of dissection by lysing the fat body.
8. Leaving tracheae on the brain will make it float, and thus reduces evenness of staining later.
9. Conditions for fixation with PFA depend on the antibody used for staining—Shorter times or fixation on ice can increase success of staining.
10. Ideally, flies are raised and aged on a light-dark schedule for circadian experiments to facilitate later synchronization to LD.
11. Besides circadian rhythms, sleep can also be analyzed in LD by applying a bin length of 1 min instead of 5 min and using the Matlab script insomniac (Lesley Ashmore, University of Pennsylvania).

Acknowledgements

GKT is funded by a Wellcome Trust Intermediate Clinical Fellowship (097479/Z/11/Z), the Wellcome Beit Prize (097479/Z/11/A), and the NIHR Oxford Biomedical Research Centre.

References

1. Tofaris GK, Spillantini MG (2005) Alpha-synuclein dysfunction in Lewy body diseases. Mov Disord 20(Suppl 12):S37–S44
2. Tofaris GK (2012) Lysosome-dependent pathways as a unifying theme in Parkinson's disease. Mov Disord 27(11):1364–1369
3. Reiter LT, Potocki L, Chien S, Gribskov M, Bier E (2001) A systematic analysis of human disease-associated gene sequences in Drosophila melanogaster. Genome Res 11 (6):1114–1125
4. Brand AH, Perrimon N (1993) Targeted gene expression as a means of altering cell fates and generating dominant phenotypes. Development 118(2):401–415
5. Dietzl G, Chen D, Schnorrer F, Su KC, Barinova Y, Fellner M, Gasser B, Kinsey K, Oppel S, Scheiblauer S, Couto A, Marra V, Keleman K, Dickson BJ (2007) A genome-wide transgenic RNAi library for conditional gene inactivation in Drosophila. Nature 448 (7150):151–156
6. Greene JC, Whitworth AJ, Kuo I, Andrews LA, Feany MB, Pallanck LJ (2003) Mitochondrial pathology and apoptotic muscle degeneration in Drosophila parkin mutants. Proc Natl Acad Sci U S A 100(7):4078–4083
7. Yang Y, Gehrke S, Imai Y, Huang Z, Ouyang Y, Wang JW, Yang L, Beal MF, Vogel H, Lu B (2006) Mitochondrial pathology and muscle

and dopaminergic neuron degeneration caused by inactivation of Drosophila Pink1 is rescued by Parkin. Proc Natl Acad Sci U S A 103 (28):10793–10798

8. Clark IE, Dodson MW, Jiang C, Cao JH, Huh JR, Seol JH, Yoo SJ, Hay BA, Guo M (2006) Drosophila pink1 is required for mitochondrial function and interacts genetically with parkin. Nature 441(7097):1162–1166
9. Park J, Lee SB, Lee S, Kim Y, Song S, Kim S, Bae E, Kim J, Shong M, Kim JM, Chung J (2006) Mitochondrial dysfunction in Drosophila PINK1 mutants is complemented by parkin. Nature 441(7097):1157–1161
10. McWilliams TG, Muqit MM (2017) PINK1 and Parkin: emerging themes in mitochondrial homeostasis. Curr Opin Cell Biol 45:83–91
11. Dodson MW, Zhang T, Jiang C, Chen S, Guo M (2012) Roles of the Drosophila LRRK2 homolog in Rab7-dependent Hum Mol lysosomal positioning. Hum Mol Genet 21 (6):1350–1363
12. MacLeod DA, Rhinn H, Kuwahara T, Zolin A, Di Paolo G, McCabe BD, Marder KS, Honig LS, Clark LN, Small SA, Abeliovich A (2013) RAB7L1 interacts with LRRK2 to modify intraneuronal protein sorting and Parkinson's disease risk. Neuron 77(3):425–439
13. Soukup SF, Kuenen S, Vanhauwaert R, Manetsberger J, Hernández-Díaz S, Swerts J, Schoovaerts N, Vilain S, Gounko NV, Vints K, Geens A, De Strooper B, Verstreken P (2016) A LRRK2-dependent endophilin A phosphoswitch is critical for macroautophagy at presynaptic terminals. Neuron 92(4):829–844
14. Tofaris GK, Goedert M, Spillantini MG (2017) The transcellular propagation and intracellular trafficking of α-synuclein. Cold Spring Harb Perspect Med 7(9). pii: a024380). https://doi.org/10.1101/cshperspect.a024380
15. Feany MB, Bender WW (2000) A Drosophila model of Parkinson's disease. Nature 404 (6776):394–398
16. Davies SE, Hallett PJ, Moens T, Smith G, Mangano E, Kim HT, Goldberg AL, Liu JL, Isacson O, Tofaris GK (2014) Enhanced ubiquitin-dependent degradation by Nedd4 protects against α-synuclein accumulation and toxicity in animal models of Parkinson's disease. Neurobiol Dis 64:79–87
17. Alexopoulou Z, Lang J, Perrett RM, Elschami M, Hurry ME, Kim HT, Mazaraki D, Szabo A, Kessler BM, Goldberg AL, Ansorge O, Fulga TA, Tofaris GK (2016) Deubiquitinase Usp8 regulates α-synuclein clearance and modifies its toxicity in Lewy body disease. Proc Natl Acad Sci U S A 113 (32):E4688–E4697
18. Szabó A, Papin C, Zorn D, Ponien P, Weber F, Raabe T, Rouyer F (2013) The CK2 kinase stabilizes CLOCK and represses its activity in the Drosophila circadian oscillator. PLoS Biol 11(8):e1001645
19. Breen DP, Vuono R, Nawarathna U, Fisher K, Shneerson JM, Reddy AB, Barker RA (2014) Sleep and circadian rhythm regulation in early Parkinson disease. JAMA Neurol 71 (5):589–595
20. Gajula Balija MB, Griesinger C, Herzig A, Zweckstetter M, Jäckle H (2011) Pre-fibrillar α-synuclein mutants cause Parkinson's disease-like non-motor symptoms in Drosophila. PLoS One 6(9):e24701
21. Karpinar DP, Balija MB, Kügler S, Opazo F, Rezaei-Ghaleh N, Wender N, Kim HY, Taschenberger G, Falkenburger BH, Heise H, Kumar A, Riedel D, Fichtner L, Voigt A, Braus GH, Giller K, Becker S, Herzig A, Baldus M, Jäckle H, Eimer S, Schulz JB, Griesinger C, Zweckstetter M (2009) Pre-fibrillar alpha-synuclein variants with impaired beta-structure increase neurotoxicity in Parkinson's disease models. EMBO J 28(20):3256–3268
22. Gargano JW, Martin I, Bhandari P, Grotewiel MS (2005) Rapid iterative negative geotaxis (RING): a new method for assessing age-related locomotor decline in Drosophila. Exp Gerontol 40(5):386–395

Chapter 16

Detection of α-Synuclein in Biological Samples Using Mass Spectrometry

Ann Brinkmalm, Annika Öhrfelt, Payel Bhattacharjee, and Henrik Zetterberg

Abstract

Here we describe a method using mass spectrometry to characterize and quantify immuno-enriched α-synuclein forms from biochemically fractionated brain tissue.

Key words Parkinson's disease, α-Synuclein, Mass spectrometry, CSF, Brain

1 Introduction

The presynaptic neuronal protein α-synuclein can misfold and form seeds that may aggregate further into intracellular inclusions that are called Lewy bodies. These inclusions are characteristic of Parkinson's disease (PD) and DLB [1], but are also frequently seen in AD [2]. In PD and in other synucleinopathies, α-synuclein concentrations in the CSF are typically lower than in healthy controls [3, 4], while in AD and CJD, its concentrations are increased and correlate with T-tau, indicating that α-synuclein might also be a nonspecific marker of neurodegeneration [4–8]. Increased CSF levels of α-synuclein have also been reported in DLB, where a competition might exist between the aggregation of α-synuclein into Lewy bodies and its release from degenerating synapses, complicating the data interpretation [9]. In agreement with this hypothesis, a recently published, multiple reaction monitoring mass spectrometry assay revealed significantly increased CSF concentrations of α-, β-, and γ-synuclein in AD and CJD, but not in the "classical" synucleinopathies like PD [10]. Currently available assays for α-synuclein measure the total amounts of the protein and not the Lewy body-specific isoforms. The availability of sensitive and specific assays for these pathogenic isoforms would resolve the issue of having the biomarker results influenced or confounded

Tim Bartels (ed.), *Alpha-Synuclein: Methods and Protocols*, Methods in Molecular Biology, vol. 1948,
https://doi.org/10.1007/978-1-4939-9124-2_16, © Springer Science+Business Media, LLC, part of Springer Nature 2019

by the release of native α-synuclein from degenerating synapses. However, there are some preliminary reports of increased concentrations of α-synuclein oligomers in the CSF of PD patients [11, 12]. Recent studies described sensitive assays that detect the amplified biochemical signal of α-synuclein seeds that may be Lewy body derived in the CSF from PD patients but not in healthy controls [13, 14], opening a promising avenue for using CSF α-synuclein as a biomarker. Since α-synuclein is highly expressed in red blood cells, blood contamination during CSF collection might limit its diagnostic value [15, 16]. For the very same reason, blood tests for α-synuclein pathology in the brain may lack the specificity needed for it to be an informative clinical biomarker. Nevertheless, as peripheral Lewy body pathology, such as in the salivary gland and the gut, has been reported in PD [17], blood or salivary tests for α-synuclein seeds might be worth exploring in the future as a biomarker of PD and other dementias associated with Lewy bodies, such as DLB and AD. Altogether, additional fluid biomarkers are needed that are specific for pathological inclusions of α-synuclein. To that end, we need methods to characterize α-synuclein in biological samples. The method of choice to gain maximal information is mass spectrometry (MS).

We here describe an MS-based method to characterize and quantify immuno-enriched α-synuclein forms from biochemically fractionated brain tissue (Fig. 1).

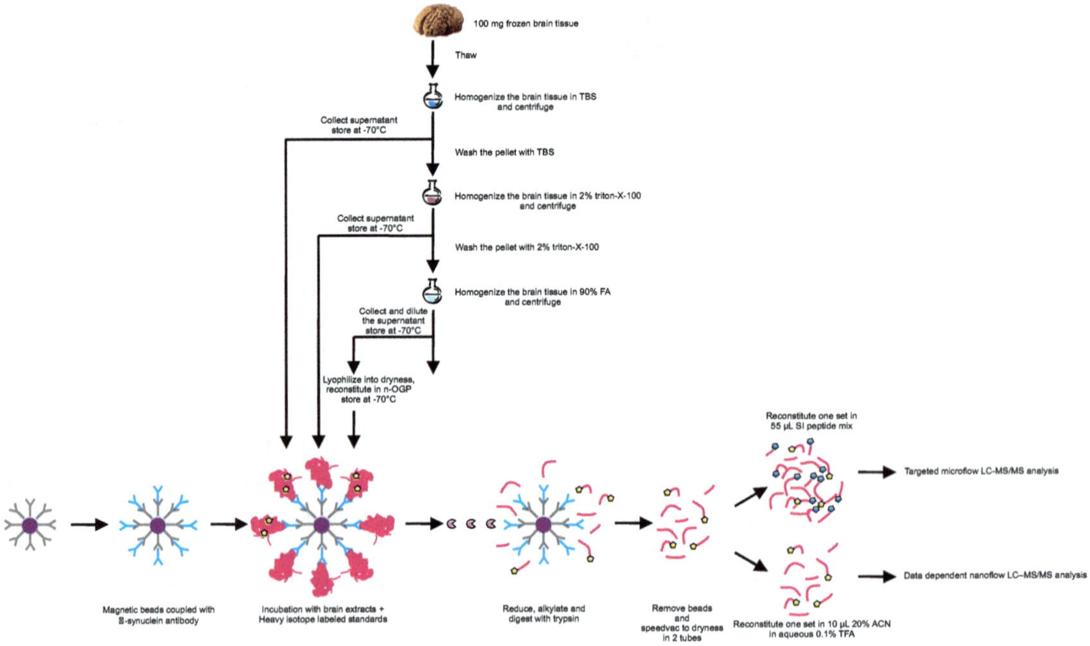

Fig. 1 Schematic of experimental method workflow

2 Materials

All solutions are prepared using ultrapure water (prepared by purifying deionized water, to attain a sensitivity of 18 MΩ-cm at +25 °C) and analytical grade reagents. Prepare and store all reagents at +2–8 °C (unless indicated otherwise). Follow all waste disposal regulations when disposing waste materials.

2.1 Tissue Sample Preparation Components

1. Snap-frozen blocks of brain tissue were stored at −70 °C pending analyses. Working on dry ice the blocks are cut in approximately 100 mg brain tissue. The tissues are transferred into pre-weighed polypropylene tubes and weighed. The tissue weight for each sample is then calculated.
2. Tris-buffered saline (TBS): 137 mM NaCl, 0.1 M Tris–HCl, pH 7.6. Weigh 1.6 g NaCl and 485 mg Tris base and transfer to a beaker. Add 150 mL water, mix, and then adjust pH with HCl. Transfer to a graduated cylinder and make up to 200 mL of water. Store at +2–8 °C.
3. TBS containing protease inhibitor: The protease inhibitor stock solution (25×) is prepared by solving two tablets of complete protease inhibitor (Roche, Ref no: 11697498001) in 4 mL water and can be used immediately or aliquoted and stored at −20 °C for 3 months. Working solution is prepared immediately before use. Add 4 mL of the protease inhibitor stock solution (25×) to 96 mL of TBS.
4. 2% triton X-100 containing protease inhibitor: Dissolve 1 mL 100% triton X-100 in 49 mL TBS containing protease inhibitors with magnetic stirring at +4 °C. the solution is prepared immediately before use.
5. 90% Formic acid (FA) aqueous solution: Inside a fume hood, add 45 mL of 100% FA to a graduated cylinder and make up to 50 mL with water. Mix thoroughly and transfer to an airtight bottle. Keep at +2–8 °C until further use.
6. Aqueous solution of 72.5 mM NaCl: Add 42.5 g NaCl to 0.5 L water to prepare 1.45 M NaCl stock solution, and then the stock solution is further diluted (1:20 v/v) with water to prepare working solution of 72.5 mM NaCl during brain tissue homogenate preparation.
7. 100 mM Na-phosphate (pH 7.4): Weigh 2.68 g $Na_2HPO_4 \times 2\ H_2O$ and 6.8 g $NaH_2PO_4 \times H_2O$ and transfer to a beaker. Add 100 mL water and stir until dissolved. Adjust pH to 7.4. Transfer the solution to 250 mL graduated cylinder and adjust the volume to 200 mL.
8. 0.3% *n*-Octyl β-D-glucopyranoside (*n*-OGP) solution: Dissolve 0.6 g *n*-OGP in 200 mL 100 mM Na-phosphate (pH 7.4), and

mix well on a roller at +4 °C for 1–2 h. the solution is stored at +2–8 °C until further use.

9. Protein DC assay reagent (Bio-Rad Laboratories).

2.2 Immunoprecipitation Components

1. Monoclonal anti-α-synuclein Clone 42 antibody solution: Monoclonal anti-α-synuclein Clone 42 (BD Transduction Laboratories, Material number: 610787) (250 mg/L) is stored undiluted at −20 °C pending analyses.
2. Magnetic beads: Magnetic Dynabeads™ Sheep anti-mouse IgG (Invitrogen).
3. Phosphate-buffered saline (PBS) stock solution A: Add 42.5 g NaCl to 0.5 L water.
4. PBS stock solution B: Weigh 6.7 g $Na_2HPO_4 \times 2H_2O$ and 17 g $NaH_2PO_4 \times H_2O$ and transfer to a beaker. Add 450 mL water and stir until dissolved. Adjust pH to 7.4. Transfer the solution to 500 mL graduated cylinder and adjust the volume to 500 mL.
5. PBS working solution: Mix 100 mL stock solution A with 100 mL stock solution B and 800 mL water in a glass bottle and store up to 3 weeks in +2–8 °C.
6. PBS containing 2.5% Tween-20: To 1 mL 10% Tween-20 add 3 mL PBS.
7. PBS containing 0.025% Tween-20: Mix 1 mL 2.5% Tween-20 and 99 mL PBS in a glass bottle and store up to 3 weeks at +2–8 °C.
8. 20 mM Dimethyl pimelimidate dihydrochloride (DMP; Sigma–Aldrich) in 0.2 M triethanolamine (TEA; Sigma–Aldrich): Dissolve 25 mg DMP (MW: 259.18 g/mol) in 5 mL 0.2 M TEA (pH 8.2) and mix thoroughly. The solution is freshly prepared.
9. 50 mM Tris–HCl (pH 7.5): Weigh 1.21 g Tris base and transfer to a beaker. Add 150 mL water, mix, and then adjust pH with HCl. Transfer to a graduated cylinder and make up to 200 mL of water. Store at +2–8 °C.
10. Roti-block solution: Dilute 1 mL Roti®-block (10×) (Carl Roth) with 9 mL water.
11. 50 mM Ammonium hydrogen carbonate (NH_4HCO_3) (pH ~8.0): Weigh 0.395 g NH_4HCO_3 and transfer to a 100 mL graduated cylinder. Add 80 mL water, mix, and make up to 100 mL. Store at +2–8 °C.

2.3 On-Bead Digestion Components

1. 10 mM Dithiothreitol (DTT): Weigh 1.5 mg DTT in a polypropylene tube, add 1 mL 50 mM NH_4HCO_3, and vortex.

2. 10 mM Iodoacetamide (IAM): Weigh 1.85 mg IAM in a polypropylene tube, add 1 mL 50 mM NH_4HCO_3, and vortex.
3. 0.01% Hydrochloric acid aqueous (0.01% HCl) solution: Inside a fume hood, add 27 μL 37% HCl to a graduated cylinder and make up to 100 mL with water. Mix thoroughly and transfer to an airtight bottle. Keep at +2–8 °C until further use.
4. Trypsin stock solution: Dissolve 20 μg sequencing-grade modified trypsin (Promega) in 200 μL 0.01% HCl. Aliquot 10 μL (0.1 g/L) stock solution and store at −20 °C until use.
5. Trypsin solution: Thaw 10 μL (0.1 g/L) trypsin stock solution, add 190 μL 50 mM NH_4HCO_3, and vortex.
6. 10% FA aqueous solution: Inside a fume hood, add 5 mL of 100% FA to a graduated cylinder and make up to 50 mL with water. Mix thoroughly and transfer to an airtight bottle. Keep at +2–8 °C until further use.

2.4 Stable Isotope-Labeled Standards

1. 15 N α-synuclein working solution (10 fmol/μL): Dissolve 1 mg recombinant α-synuclein (see chapter on expression of labeled protein by Eliezer et al., this issue) uniformly labeled with 15 N α-synuclein (14,624 Da) in 1 mL water (resulting in 1 g/L stock solution) and mix thoroughly. Prepare working solution by adding 14.6 μL stock solution (1 g/L) to 1 mL water (resulting in 1 pmol/μL 15 N α-synuclein solution). Dilute this solution 100 times in water to prepare the final working solution of 15 N α-synuclein (10 fmol/μL).
2. Stable isotope-labeled standard peptides (SI peptides): α-Synuclein peptides were supplied by Thermo Fisher Scientific with at least one stable isotopically labeled amino acid. The purity of the peptides was >95%. Dissolve SI peptides in 1 mL water, aliquot, and store stock solutions at −70 °C. Prepare SI peptide mix by adding each peptide stock solution to 50 mM NH_4HCO_3 and diluting to a concentration of 10 fmol/μL per peptide.

2.5 Consumables, Solutions, and Buffers for LC-MS/MS

1. Loading buffer: Add 1 mL 100% FA, 40 mL acetonitrile (ACN), and 1960 mL water to a glass bottle, and sonicate for 10 min.
2. Buffer A: Add 1 mL 100% FA and 1 L water to a glass bottle, and sonicate for 10 min.
3. Buffer B: Add 840 mL ACN, 160 mL water, and 1 mL 100% FA to a glass bottle, and sonicate for 10 min.
4. 20% ACN in aqueous 0.1% TFA: Add 20 mL ACN, 10 mL 1% TFA, and 70 mL water to a glass bottle. The solution can be stored at +2–8 °C for 1 month.

5. LC vials: Polypropylene LC vials.
6. Nano LC-separation column: Acclaim PepMap RSLC C18, 2 μm, 100 Å, 75 μm i.d. × 15 cm, nanoViper (Thermo Fisher Scientific).
7. Nano LC-trapping column: Acclaim PepMap 100 C18, length 500 mm; i.d. 75 μm, particle size 2 μm) (Thermo Fisher Scientific).
8. Micro LC-separation column: Hypersil Gold-C18 column (length 100 mm, i.d. 2.1 mm, particle size 1.9 μm) (Thermo Fisher Scientific).

3 Methods

3.1 Biochemical Fractionation of Brain Tissue

During the entire process below the samples should be kept on ice unless otherwise specified.

1. Thaw the samples of approximately 100 mg brain tissue.
2. Homogenize the brain tissue in 1 mL ice-cold TBS containing protease inhibitor.
3. Centrifuge the homogenate at 31,000 *g* for 1 h at +4 °C.
4. Collect the supernatant (soluble fraction) and make aliquots to store at −70 °C.
5. Wash the pellet two times with TBS containing protease inhibitor. After each wash the samples are centrifuged at 31,000 *g* for 10 min at +4 °C and supernatants are discarded.
6. Add 0.8 mL 2% triton-X-100 containing protease inhibitor to the pellet.
7. Homogenize and sonicate for 30 s keeping the samples in an ice bath.
8. Centrifuge the homogenate at 31,000 *g* for 1 h at +4 °C.
9. Collect the supernatant (detergent-soluble fraction) and make aliquots to store at −70 °C.
10. Wash the pellet two times with 2% triton X-100 containing protease inhibitor. After each wash the samples are centrifuged at 31,000 *g* for 10 min at +4 °C and supernatants are discarded.
11. Add 0.8 mL 90% FA to the pellet.
12. Homogenize and sonicate for 30 s keeping the samples in an ice bath.
13. Centrifuge the homogenate at 31,000 *g* for 1 h at +4 °C.
14. Collect the supernatant (detergent-insoluble fraction).
15. Dilute the detergent-insoluble fraction immediately (1:20, v/v) with aqueous 72.5 mM NaCl in 50 mL conical tubes.

Remove the cap of the tubes and wrap with parafilm. Make small pores on the parafilm layer and transfer the tubes to −70 °C freezer immediately.

16. Lyophilize the frozen samples into complete dryness (*see* **Notes 1** and **2**).
17. Reconstitute the dried sample in 0.5 mL 0.3% *n*-OGP solution in 100 mM Na-phosphate (pH 7.4). Store at −70 °C.
18. The total protein concentration in each sample is determined by using Protein DC assay reagent according to the protocol from the manufacturer.

3.2 Immunoprecipitation

The methodology described here is compatible with a KingFisher® magnetic particle processor (Thermo Fisher Scientific).

1. Vortex vial with magnetic beads carefully and transfer 50 μL per sample to a new polypropylene vial.
2. Use a magnetic particle concentrator to pellet the beads for 3 min and wash with PBS (repeat three times).
3. Add 4 μL of the mouse monoclonal anti-α-synuclein antibody Clone 42 (250 mg/L).
4. Incubate for 1 h on a rocking platform at room temperature (~20–25 °C).
5. Remove the remaining unbound antibody by washing three times with PBS using the magnetic particle concentrator.
6. Cross-link the antibodies to the beads by resuspending the washed beads in 20 mM DMP in 0.2 M TEA.
7. Incubate for 30 min on a rocking platform at room temperature.
8. Stop the reaction by placing the vial in the magnetic particle concentrator, removing the supernatant, and resuspending in 50 mM Tris–HCl.
9. Incubate for 15 min on a rocking platform at room temperature.
10. Wash the beads two times with PBS using the magnetic particle concentrator.
11. Resuspend in Roti-block solution.
12. Incubate for 1 h on a rocking platform at room temperature.
13. Wash the beads two times with PBS using the magnetic particle concentrator and resuspend in PBS.
14. Add 25 μg total protein (soluble and detergent-soluble brain fractions) or 10 μg total protein (detergent-non-soluble brain fractions), 50 μL 15 N-labeled α-synuclein, 10 μL 2.5% Tween-20, 50 μL antibody-coupled magnetic beads, and PBS to a total volume of 1000 μL.

15. Incubate overnight on a rocking platform at +4 °C.
16. Transfer incubated magnetic beads to the KingFisher® magnetic particle processor, tube 1.
17. Conduct the following three wash steps (tubes 2–4) for 10 s in 1 mL of each washing buffer: (tube 2) 0.025% Tween-20 in PBS, (tube 3) PBS, and (tube 4) 50 mM NH_4HCO_3, pH ~8.0.
18. Release the washed magnetic beads in 100 μL 50 mM NH_4HCO_3 (tube 5).
19. Transfer the eluted fractions to 0.5 mL polypropylene tubes.

3.3 On-Bead Digestion

1. Reduce disulfate bridges by adding 10 μL 10 mM DTT to each sample, vortex, and centrifuge for 10 s at 2000 *g*.
2. Incubate for 30 min at +60 °C and cool down to room temperature for 15 min.
3. Perform alkylation by adding 10 μL 10 mM IAM to each sample, vortex, and centrifuge for 10 s at 2000 *g*.
4. Incubate for 30 min at room temperature in the dark.
5. Add 10 μL trypsin solution and incubate at +37 °C overnight with shaking at 400 rpm.
6. Quench the reaction by adding 10 μL 10% FA aqueous solution.
7. Centrifuge at 16,910 *g* for 10 min at +4 °C.
8. Collect the supernatant in a 0.5 mL polypropylene tube and divide the supernatant into two tubes.
9. SpeedVac to dryness.

3.4 Data-Dependent Nanoflow LC-MS/MS Analysis and Database Search

The methodology described here is compatible with a Q Exactive™ Hybrid Quadrupole-Orbitrap™ mass spectrometer coupled to a Dionex Ultimate 3000 RSLC nano system.

1. Reconstitute one set of dried samples in 10 μL 20% ACN in aqueous 0.1% TFA (*see* **Note 3**).
2. Mix thoroughly for 1 h, transfer 7 μL of each samples to LC vials, and place the vials in the autosampler of the LC system.
3. Program the LC to inject 5 μL sample, and load the peptides on the trapping column with loading buffer at 5 μL/min. After 2 min of washing the peptides are loaded on the separation column with 3% buffer B at 150 nL/min and subsequently eluted with a 50-min linear gradient of 5–40% buffer B.
4. Set the mass spectrometer to acquire positive ions in the data-dependent mode, with a scan cycle consisting of single microscans of one full scan mass spectrum followed by up to ten MS^2 scans. Settings as follows: Spray voltage (+1.7 kV); capillary temperature (+250 °C); resolution (70,000 MS, 17500

dd-MS^2); AGC target (1e6 MS, 5e4 dd-MS^2); maximum injection time (250 MS MS, 60 MS dd-MS^2); and m/z isolation window (m/z 300–1200 MS, m/z 2.0 dd-MS^2). MS/MS scans are acquired with a normalized collision energy (NCE) setting of 28.0 for the ten most intense precursor ions with an intensity greater than 1.7e4, charge state 2–4, and a dynamic exclusion for 50 s.

5. Search acquired spectra with the following search parameters: Database (Swiss-Prot), taxonomy (*Homo sapiens*), enzyme (semi-trypsin), variable modifications (acetyl [N-term] and oxidation [M]), static modification (carbamidomethyl [C]), mass values (monoisotopic), peptide mass tolerance (±15 ppm), fragment mass tolerance (±20 mmu), and maximum two missed cleavages.

3.5 Assay Development for Targeted Microflow LC-MS/MS Analysis

The methodology described here is compatible with a Q Exactive™ Hybrid Quadrupole-Orbitrap™ mass spectrometer coupled to a Dionex Ultimate 3000 system (Thermo Fisher Scientific).

1. Use nano-LC-MS/MS data to select peptides to be quantified.
2. Let synthesize stable-isotope-labeled (SI) peptides.
3. Dissolve SI peptides in LC-MS-grade water, aliquot, and store stock solutions at −70 °C.
4. To one vial of stock solution add aqueous 20% ACN resulting in 1 μM peptide concentration.
5. To select and optimize collision energy values maximizing fragment ion production infuse 1 μM IS peptide solution into the mass spectrometer with a syringe at a flow rate of 5 μL/min. After directly monitoring full mass scans and determining the most intense charge state set the mass spectrometer to acquire positive ions in the data-dependent mode, with a scan cycle consisting of one full scan mass spectrum followed by one MS^2 scan for 2 min. Settings as follows: Spray voltage (+1.7 kV); capillary temperature (+250 °C); resolution (17500); AGC target (3e6 MS, 1e5 dd-MS^2); maximum injection time (100 MS); and m/z isolation window (m/z 200–1800 MS, m/z 3.0 dd-MS^2). MS/MS scans are acquired with an inclusion list of different normalized collision energy settings from 15 to 35 in random order.
6. Thaw aliquots of the IS stock solutions, mix, and dilute in 50 mM NH_4HCO_3 to a concentration of 1 fmol/μL.
7. Thaw aliquots of the IS stock solutions, mix, and dilute in 50 mM NH_4HCO_3 to a concentration of.
8. Inject IS peptide mix several times and program the LC so as to elute the peptides off the column using several different gradients.

9. Optimize the LC gradient so that the retention times of the standard peptides are as far dispersed as possible.
10. Select PRM transitions.

3.6 Targeted Microflow LC-MS/MS Analysis

1. Redissolve the second set of dried samples in 55 μL SI peptide mix, with shaking for 1 h. Transfer the samples to LC vials and place the vials in the LC autosampler.
2. Inject 50 μL of each sample, which is loaded directly onto the analysis column with 0.1% aqueous FA at 300 μL/min.
3. Program the LC so that after 2 min of loading, the peptides are eluted off the column using the following linear gradient steps: 0 min 0%B; 4 min 17%B; 16 min 35%B; 17.5 min 100%B; and 20 min 0%B.
4. Set the mass spectrometer to acquire scheduled triplets of PRM scans at defined NCE allowing parallel detection of corresponding brain-derived α-synuclein peptide, IS peptides, and 15 N-labeled peptide derived from the 15 N-labeled recombinant α-synuclein. Mass spectrometry settings as follows:
5. Import acquired raw files into Pinpoint 1.3.0. Settings as follows: MS accuracy (±10 ppm centered at 0), MS/MS accuracy (±15 ppm), isolation mode (MS/MS with an isolation width of 3.0 μm), chromatographic peak window size (±2.0 min), peak area (four points of smoothing), and scheduling window size (±0.5 min).
6. Inspect fragment ion peaks manually for accuracy of integration and absence of interferences from other peptides than the peptide of interest.
7. Calculate the relative amounts of the α-synuclein peptide T by taking the ratio between the sums of fragment ion peak areas of the brain-derived peptides against the sum of fragment ion peak areas of the respective IS peptide. Since the purity of the IS peptides is not 100%, normalizing with respect to the amount IS peptide present in the samples gives a rough estimate about the relative amount of the endogenous peptides of α-synuclein in the brain tissue homogenates.

4 Notes

1. Evaporation of α-synuclein to dryness by vacuum centrifugation should be avoided since a major part of the protein is not recovered after reconstitution.
2. A major challenge in studying the membrane-bound or LB-associated proteins using mass spectrometry is the use of

strong ionic and nonionic detergents like SDS and Triton X-100; chaotropic agents like urea, thiourea, or guanidine; and high concentration of organic acids like FA. These reagents are essential for solubilizing these types of proteins; however, they interfere with the IP, enzymatic digestion, and mass spectral analysis by suppressing ionization and affecting chromatographic separation. In this study 2% Triton X-100 and 90% FA were used for solubilization of membrane-bound and LB-enriched detergent-insoluble fractions from brain, respectively. The organic acid was removed by the process of lyophilization followed by reconstitution of the protein in the presence of 0.3% n-OGP in 100 mM phosphate buffer (pH 7.4).

3. Choice of reconstitution buffer will influence which forms (modifications and parts of sequence) of α-synuclein will be possible to detect with LC-MS.

Acknowledgments

Work in the authors' laboratories is supported by the Michael J. Fox Foundation (MJFF) for Parkinson's Research, the Swedish Research Council, the European Research Council, the Knut and Alice Wallenberg Foundation, the Torsten Söderberg Foundation, the Eivind and Elsa K:Son Sylvans Foundation, the Märtha and Gustaf Ågrens Foundation, the Olav Thon Foundation, Swedish State Support for Clinical Research (ALFGBG), Alzheimerfonden, Stiftelsen för Gamla Tjänarinnor, and Frimurarestiftelsen.

References

1. Mollenhauer B, El-Agnaf OM, Marcus K, Trenkwalder C, Schlossmacher MG (2010) Quantification of alpha-synuclein in cerebrospinal fluid as a biomarker candidate: review of the literature and considerations for future studies. Biomark Med 4(5):683–699
2. Schneider JA, Arvanitakis Z, Leurgans SE, Bennett DA (2009) The neuropathology of probable Alzheimer disease and mild cognitive impairment. Ann Neurol 66(2):200–208
3. Hall S, Ohrfelt A, Constantinescu R, Andreasson U, Surova Y, Bostrom F et al (2012) Accuracy of a panel of 5 cerebrospinal fluid biomarkers in the differential diagnosis of patients with dementia and/or parkinsonian disorders. Arch Neurol 69(11):1445–1452
4. Mollenhauer B, Locascio JJ, Schulz-Schaeffer-W, Sixel-Doring F, Trenkwalder C, Schlossmacher MG (2011) Alpha-Synuclein and tau concentrations in cerebrospinal fluid of patients presenting with parkinsonism: a cohort study. Lancet Neurol 10(3):230–240
5. Ohrfelt A, Grognet P, Andreasen N, Wallin A, Vanmechelen E, Blennow K et al (2009) Cerebrospinal fluid alpha-synuclein in neurodegenerative disorders-a marker of synapse loss? Neurosci Lett 450(3):332–335
6. Slaets S, Vanmechelen E, Le Bastard N, Decraemer H, Vandijck M, Martin JJ et al (2014) Increased CSF alpha-synuclein levels in Alzheimer's disease: correlation with tau levels. Alzheimers Dement 10(5 Suppl): S290–S298
7. Tateno F, Sakakibara R, Kawai T, Kishi M, Murano T (2012) Alpha-synuclein in the cerebrospinal fluid differentiates synucleinopathies (Parkinson Disease, dementia with Lewy bodies, multiple system atrophy) from Alzheimer disease. Alzheimer Dis Assoc Disord 26 (3):213–216

8. Wennstrom M, Surova Y, Hall S, Nilsson C, Minthon L, Bostrom F et al (2013) Low CSF levels of both alpha-synuclein and the alpha-synuclein cleaving enzyme neurosin in patients with synucleinopathy. PLoS One 8(1):e53250
9. Kapaki E, Paraskevas GP, Emmanouilidou E, Vekrellis K (2013) The diagnostic value of CSF alpha-synuclein in the differential diagnosis of dementia with Lewy bodies vs. normal subjects and patients with Alzheimer's disease. PLoS One 8(11):e81654
10. Oeckl P, Metzger F, Nagl M, von Arnim CA, Halbgebauer S, Steinacker P et al (2016) Alpha-, beta-, and gamma-synuclein quantification in cerebrospinal fluid by multiple reaction monitoring reveals increased concentrations in Alzheimer's and Creutzfeldt-Jakob disease but no alteration in synucleinopathies. Mol Cell Proteomics 15 (10):3126–3138
11. Hansson O, Hall S, Ohrfelt A, Zetterberg H, Blennow K, Minthon L et al (2014) Levels of cerebrospinal fluid alpha-synuclein oligomers are increased in Parkinson's disease with dementia and dementia with Lewy bodies compared to Alzheimer's disease. Alzheimers Res Ther 6(3):25
12. Tokuda T, Qureshi MM, Ardah MT, Varghese S, Shehab SA, Kasai T et al (2010) Detection of elevated levels of alpha-synuclein oligomers in CSF from patients with Parkinson disease. Neurology 75(20):1766–1772
13. Fairfoul G, McGuire LI, Pal S, Ironside JW, Neumann J, Christie S et al (2016) Alpha-synuclein RT-QuIC in the CSF of patients with alpha-synucleinopathies. Ann Clin Transl Neurol 3(10):812–818
14. Shahnawaz M, Tokuda T, Waragai M, Mendez N, Ishii R, Trenkwalder C et al (2017) Development of a biochemical diagnosis of Parkinson disease by detection of alpha-synuclein misfolded aggregates in cerebrospinal fluid. JAMA Neurol 74(2):163–172
15. Barbour R, Kling K, Anderson JP, Banducci K, Cole T, Diep L et al (2008) Red blood cells are the major source of alpha-synuclein in blood. Neurodegener Dis 5(2):55–59
16. Hong Z, Shi M, Chung KA, Quinn JF, Peskind ER, Galasko D et al (2010) DJ-1 and alpha-synuclein in human cerebrospinal fluid as biomarkers of Parkinson's disease. Brain 133 (Pt 3):713–726
17. Uchihara T, Giasson BI (2016) Propagation of alpha-synuclein pathology: hypotheses, discoveries, and yet unresolved questions from experimental and human brain studies. Acta Neuropathol 131(1):49–73

Chapter 17

Single-Molecule FRET Detection of Early-Stage Conformations in α-Synuclein Aggregation

Mahdi Muhammad Moosa, Josephine C. Ferreon, and Allan Chris M. Ferreon

Abstract

Misfolding and aggregation of α-synuclein are linked to many neurodegenerative disorders, including Parkinson's and Alzheimer's disease. Despite intense research efforts, detailed structural characterization of early conformational transitions that initiate and drive α-synuclein aggregation remains elusive often due to the low sensitivity and ensemble averaging of commonly used techniques. Single-molecule Förster resonance energy transfer (smFRET) provides unique advantages in detecting minor conformations that initiate protein pathologic aggregation. In this chapter, we describe an smFRET-based method for characterizing early conformational conversions that are responsible for α-synuclein self-assembly and aggregation.

Key words Facilitated self-assembly, Intrinsically disordered protein, Neurodegenerative disease, Protein aggregation, Single-molecule spectroscopy

1 Introduction

Dysfunction of α-synuclein is linked to several neurodegenerative disorders. Collectively known as synucleinopathies, these diseases are characterized by the presence of histopathological aggregates that are rich in α-synuclein [1]. Consequently, the mechanism of α-synuclein aggregation has been of intense research focus for quite some time [2]. Nonetheless, the structural-conformational details of the early events that initiate the protein's pathologic transformation remain elusive. The early aggregation intermediates account for a minor fractional population and are in dynamic equilibrium, which make them difficult to detect and characterize using traditional ensemble measurements. Single-molecule Förster resonance energy transfer (smFRET) experiments circumvent these limitations by allowing direct detection of minor subpopulation(s) in a mixture of species.

Tim Bartels (ed.), *Alpha-Synuclein: Methods and Protocols*, Methods in Molecular Biology, vol. 1948,
https://doi.org/10.1007/978-1-4939-9124-2_17,

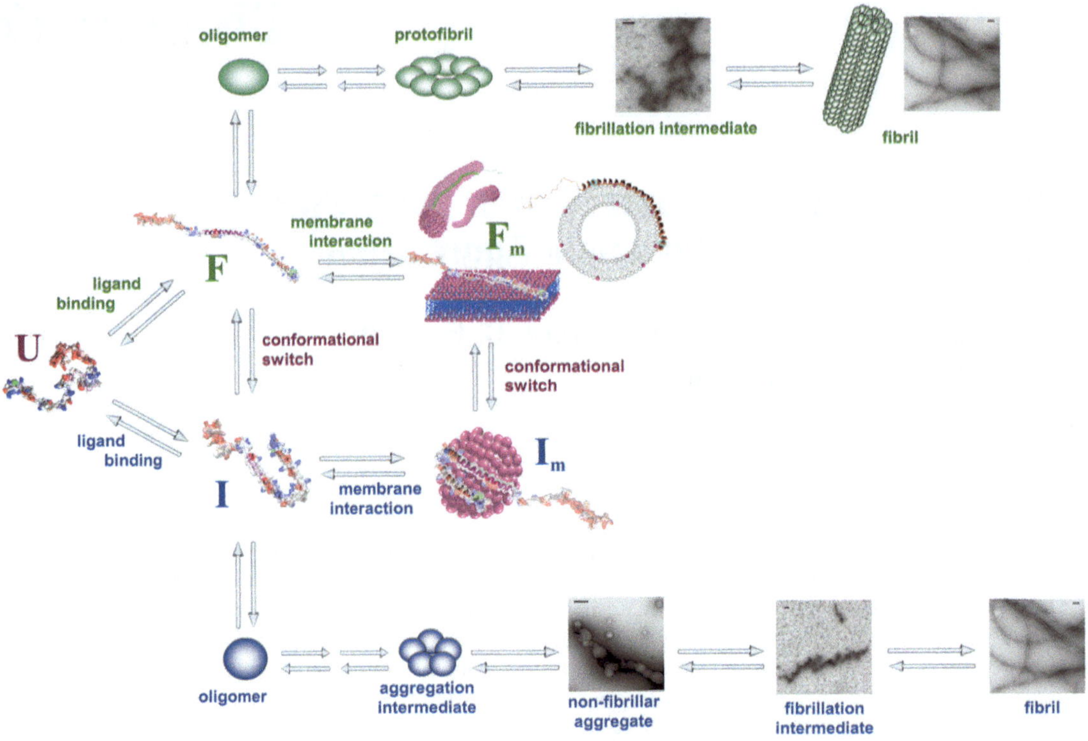

Fig. 1 α-Synuclein folding, misfolding, and aggregation conformational landscape. The intrinsic multistate character of α-synuclein influences both its function and dysfunction. The disordered protein undergoes ligand binding-coupled folding into alternate helical conformations [3], and can be induced to aggregate by physicochemical perturbations [4]. As with most aggregation-prone proteins, the early steps in α-synuclein aggregation are generally reversible, whereas the protein's higher order supramolecular assemblies become less reversible in later steps. Detailed molecular/structural characterization of this complex folding-misfolding landscape requires multiple biochemical approaches combined with different biophysical detection strategies. Adapted from [3, 4]

α-Synuclein aggregation is a multistep process, characterized by conformational switches that are modulated by multifaceted partner interactions (Fig. 1). In addition to its disordered state, monomeric α-synuclein has been shown to adopt alternate helical conformations (extended and broken helical) [3, 5]. Specific α-synuclein conformations are likely to exhibit unique misfolding/aggregation propensities and follow different assembly pathways. Here, we describe a protocol to directly observe early conformational changes associated with misfolding, self-assembly, and aggregation of α-synuclein.

2 Materials

2.1 Instrumentation

Diffusion-based confocal microscopy experiments typically involve either continuous or pulsed laser illumination of sub-femtoliter observation volume and detection of fluorescence from freely

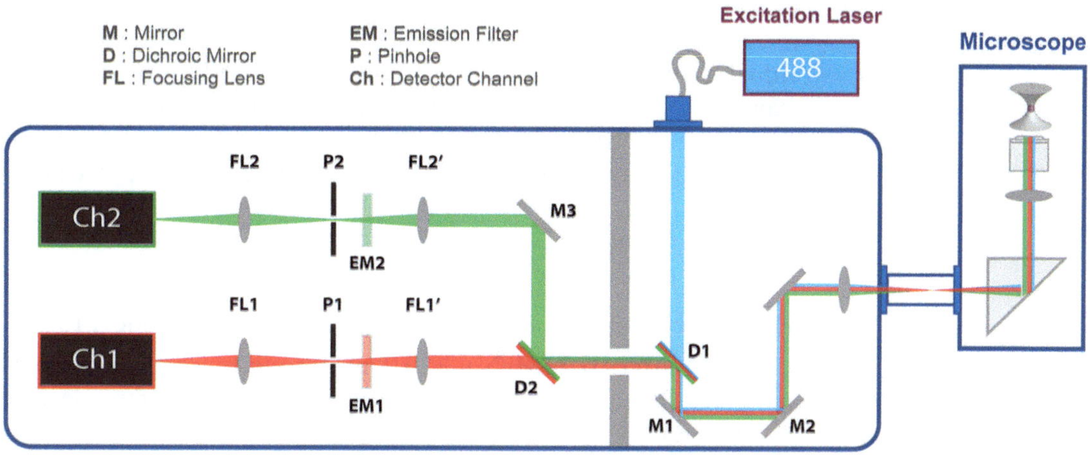

Fig. 2 Schematic of a single-molecule confocal laser microscopy system

diffusing molecules. Early diffusion-based experiments were mostly performed using custom-built confocal microscopy setups [3, 6–8]. Currently, similar microscopy systems can be commercially purchased from different vendors (e.g., Becker & Hickl GmbH, ISS Inc., PicoQuant).

The data presented below for the single-molecule method described in this chapter were collected using a customized ISS Alba microscopy system (ISS Inc., Champaign, IL) [9, 10], with the corresponding smFRET module depicted in Fig. 2. Briefly, the fluorescence excitation-detection system is coupled to an Olympus IX81 microscope that is equipped with an Olympus 60X/1.2 NA water objective lens. The excitation laser beam (488-nm laser, ISS Inc.; laser power ~115 μW) is focused on the confocal detection volume (~0.25 fL). Freely diffusing fluorophores are excited as they pass through the observation volume. Emitted fluorescence is detected through 70 μm pinholes using SPCM-ARQH-16 avalanche photodiode detectors (Excelitas, Waltham, MA).

For FRET measurements, fluorescence emission is split into donor and acceptor fluorescence by a 605 nm long-pass beam splitter dichroic. Donor and acceptor signals are further filtered using 535/50 nm and 641/75 nm band-pass emission filters, respectively. The 488 nm excitation laser is mounted on a customized laser launcher for intensity control and delivered to the Alba system via a single-mode polarization-maintained fiber (QiOptiq, Rhyl, UK).

2.2 Chemicals and Reagents

For single-molecule fluorescence applications, use chemicals and reagents that are of spectroscopy grade. Diffusion-based single-molecule methods are ultrasensitive and detect minor fluorescently active species that are typically present as impurities in non-spectroscopy-grade reagents. Such impurities significantly increase background and reduce signal-to-noise ratio. The presence of fluorescent impurities and/or chemical degradation products in individual buffers and reagents should be checked prior to actual experiments. If necessary, purification can be performed to remove unwarranted backgrounds (e.g., by recrystallization and/or treatment with activated carbon followed by mixed-bed ion-exchange resin).

A list of materials required for the protocol presented in this chapter is provided below:

1. Low-binding plasticwares (i.e., pipette tips, Eppendorf tubes).
2. αβγ buffer: 0.2 M NaCl, 10 mM sodium acetate, 10 mM NaH_2PO_4, 10 mM glycine (*see* **Note 1** for buffer selection details).
3. Thioflavin T (Th T) solution: Prepare 500 μM Th T stock solution in αβγ buffer.
4. α-Synuclein: Express and purify α-synuclein using preferred protein expression system (*see* **Note 2** for additional discussions).
5. Dye-labeled α-synuclein: Label α-synuclein at positions that are sensitive to protein conformational changes using available labeling chemistries (*see* **Note 3** for details on labeling positions, dye selection, and labeling chemistries).
6. Ensemble fluorescence cuvette cleaning solution: 5% Hellmanex™ III (Sigma, Cat # Z805939).
7. Cuvette for smFRET detection: Nunc™ Lab-Tek™ II Chamber Slide™ (Thermo Scientific; Cat # 154534)
8. 12 N and 1.2 N HCl stocks.
9. pH meter equipped with microelectrode.
10. Alignment solutions (10 nM Alexa Fluor 488 and 10 nM Alexa Fluor 594; Molecular Probes, Eugene, OR).
11. Graphing software (e.g., Origin; OriginLab, Northampton, MA).

3 Methods

3.1 Optimization of α-Synuclein Aggregation Kinetics for Single-Molecule Experiments

Confocal single-molecule techniques are powerful in detecting minor subpopulations. However, kinetic transitions should happen within a certain time window to be tractable by these techniques. On the one hand, the data collection window should be long enough (typically ≥5 min) to collect statistically significant number of single-molecule events. On the other, the process of interest should be completed within a reasonable timeframe.

For an efficient workflow, solution conditions that result in suitable protein aggregation timeframe tractable for single-molecule measurements can be determined using ensemble techniques where parallel measurements can be easily performed. The ideal final experimental "Goldilocks" condition can then be utilized for smFRET experiments.

1. Prepare α-synuclein aggregation mixtures in the desired buffer conditions using low-binding plasticwares. We performed our experiments in αβγ buffer using 20 μM α-synuclein, prepared by diluting the protein stock (895 μM) in αβγ buffer (pH 7.5). Protein solution pHs were adjusted by adding negligible volumes of HCl titrants to the buffered samples (*see* **Note 4**).
2. Wash cuvettes properly. Residual aggregate seeds from prior runs, if present, can alter proceeding aggregation kinetics experiments. We ensured complete seed removal by washing the cuvette with deionized distilled water (ddH_2O), followed by 5-min incubation with cuvette-cleaning solution and a final ddH_2O wash.
3. Monitor α-synuclein aggregation using ensemble techniques that provide a global picture of the protein's pathologic conversion. We monitored α-synuclein self-assembly and aggregation by detecting changes in Th T fluorescence (Fig. 3). Th T exhibits a pronounced enhancement of fluorescence emission upon binding to amyloid fibrils [11], and has been used extensively to study α-synuclein aggregation [12–17]. We recorded

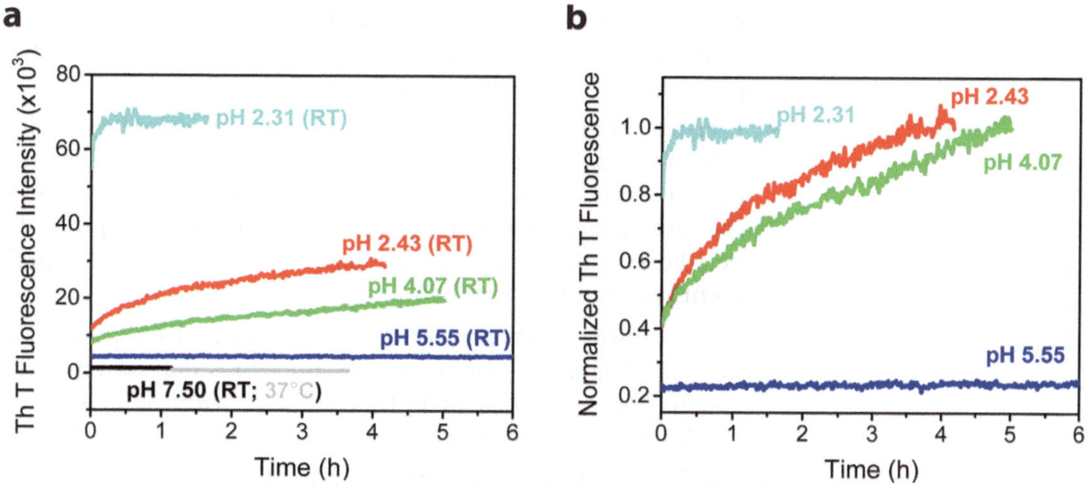

Fig. 3 Solution pH modulates α-synuclein aggregation kinetics. α-Synuclein aggregation (from the extended helical conformation [3–5]) was monitored by the increase in thioflavin T (Th T) fluorescence (λ_{Ex}: 450 nm and λ_{Em}: 482 nm). Sample conditions: 20 μM α-synuclein, 1 mM SDS, 20 μM Th T in αβγ buffer at specified pH conditions and room temperature (21.5 °C). Raw and normalized Th T fluorescence data are shown in (**a**) and (**b**), respectively. *See* **Note 6** for a brief summary on the normalization procedure

Th T fluorescence (λ_{Ex}: 450 nm and λ_{Em}: 482 nm) for up to 6 h in situ (i.e., the aggregation mix was loaded in cuvette and fluorescence signal was recorded continuously; *see* **Notes 5** and **6**).

4. Characterize the effects of physicochemical perturbations (i.e., pH, temperature, salt, molecular crowder) on the protein's aggregation (*see* **Note** 7). These perturbations can be used to modulate protein aggregation kinetics to accommodate the technical requirements for single-molecule experiments. Here, we utilized pH to tune α-synuclein aggregation (Fig. 3). Consistent with previous observations [12], our data show that acidic pH conditions increase the rate of α-synuclein aggregation significantly.

3.2 Single-Molecule Detection of Early α-Synuclein Aggregation

1. Align instrument optics to ensure high signal-to-noise ratio by maximizing signals from standard solutions. Standards can be dilute dye solutions (10 nM Alexa Fluor 488 and 10 nM Alexa Fluor 594) and/or biomolecules labeled with the donor and acceptor dyes.
2. Buffers should be checked for minimal fluorescence background by recording fluorescence in both donor and acceptor channels for buffer-only sample. The average fluorescence signal of the buffer-only sample should be subtracted from the actual protein sample fluorescence as background correction.
3. Based on the ensemble data, select solution conditions that are appropriate for smFRET experiments (*see* **Note 8**). We performed our single-molecule aggregation experiments at pH conditions between 2.43 and 4.07. α-Synuclein aggregation timescales are suitable for smFRET experiments within this pH range at our selected conditions.
4. Prepare samples for single-molecule experiments using low-binding plasticwares and transfer to the cuvette. The sample should contain ~50–100 pM labeled protein in addition to the same concentration of unlabeled α-synuclein (20 μM unlabeled α-synuclein in our experiments) used in ensemble experiments (*see* Subheading 3.1).
5. Record fluorescence signals from the donor and acceptor channels. Dual-labeled monomeric α-synuclein is detected as coincident photon bursts in both channels, whereas aggregates exhibit wavelike intensity traces (Fig. 4; *see* **Note 9**).
6. Plot FRET efficiency histograms and fit Gaussian function(s) to determine individual peak position(s) using preferred graphing software (e.g., Origin). Details on the smFRET data analysis are discussed in **Notes 9** and **10**.

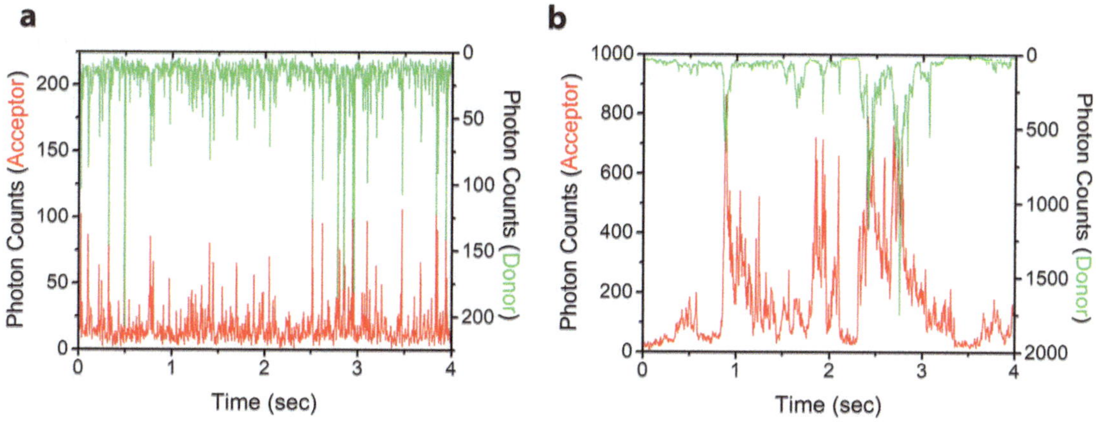

Fig. 4 smFRET trajectories: Non-aggregating versus aggregating species. α-Synuclein aggregates display wavelike trajectories in their fluorescence photon count traces. Intensity traces of monomeric and aggregating α-synuclein are shown in (**a**) and (**b**), respectively. *Top trace*: donor channel; *bottom trace*: acceptor channel

4 Notes

1. Buffer selection is dependent on the experimental technique (s) to be employed as well as limitations imposed by the system to be studied. For example, buffering agents that have high absorbance at low wavelengths in the UV should be avoided if far-UV CD experiments are planned [18, 19]. Phosphates and inorganic primary amines or carboxylic acids should be avoided if metal-induced α-synuclein aggregation is to be studied [20]. In our ensemble experiments, we varied the pH of the aggregation mix to tune α-synuclein aggregation kinetics. Therefore, our buffer components were selected for broad pH buffering range and minimal temperature effects on pH.
2. Protein samples for aggregation experiments should be of high purity. Minor impurities alter aggregation kinetics and, therefore, complicate data interpretation. Additionally, the sample must be free from any fluorescently active impurities that can increase background (and reduce signal-to-noise ratio) in single-molecule experiments.
3. FRET experiments rely on the distance dependence of energy transfer between the donor and acceptor dyes. The relationship between inter-dye distance (r) and the energy transfer efficiency (E_{FRET}) is given by

$$E_{\mathrm{FRET}} = \frac{1}{1 + (r/R_0)^6} \quad (1)$$

where R_0 is the Förster distance of the chosen dye pair at a given experimental condition. The commonly used dye pairs typically provide effective distance information in the 30–70 Å range ($R_0 \approx 50$ Å) [21]; labeling positions are ideally selected so that the inter-dye distance is within the range (and the corresponding E_{FRET} between 0.9 and 0.1). Previously, we have shown that fluorescent probes conjugated to residue positions 7 and 84 report on different α-synuclein conformations [3, 5, 22–24], with the positions spanning α-synuclein's two major helical segments. The two alternate helical conformations are detected as low and high E_{FRET} states, whereas the disordered state exhibits intermediate E_{FRET} values [3, 5]. Other labeling positions have also been successfully employed to directly visualize different α-synuclein conformations [25–27].

Once the labeling positions are decided, appropriate FRET dye pairs should be chosen (e.g., Alexa Fluor 488 and 594). Selected dyes should have suitable photophysical properties (high photostability, high quantum yield, and low triplet-state population and blinking). Additionally, a significant spectral overlap should be present between the emission spectra of the donor dye and the excitation spectra of the acceptor dye. Details on general dye selection are discussed elsewhere [28, 29].

We utilized Cys-maleimide labeling scheme where the thiol-reactive moieties react with solvent-exposed Cys residues of the dual-Cys α-synuclein mutant. Different site-specific labeling strategies and their pros/cons are discussed elsewhere [21, 28].

4. α-Synuclein stock solutions were maintained at near-physiological pH conditions (pH 7.5), where the protein exhibits minimal aggregation propensity. The absence of aggregation seeds can be ensured through brief alkaline treatment of α-synuclein solutions (i.e., by adjusting α-synuclein stock pH to between 9.5 and 11 briefly (~30 s, with gentle mixing), then re-adjusting pH back to 7.5). Additionally, the stock protein solution can also be syringe-filtered.

   For the data presented in Fig. 3, the pH of the aggregation mix was changed to the desired pH conditions and the kinetics of aggregation was recorded. We measured solution pH at the end (vs. at the start) of each aggregation experiment to minimize the dead time of the kinetics experiments.

5. In addition to Th T fluorescence, other biophysical techniques that report on changes in α-synuclein conformations and/or assembly states (e.g., far-UV CD spectroscopy, FTIR) can be employed.

6. The normalization was performed by using either the starting or the ending fluorescence signals as reference (Fig. 3b; pH 5.55 data was referenced to the pH 4.07 endpoint signal).

Normalization allows a direct comparison between the early events in protein supramolecular assembly formation.

7. Different physicochemical perturbations that are known to alter α-synuclein aggregation include changes in pH, temperature, and salt and molecular crowder concentrations [12, 30, 31].
8. At physiological pH conditions, we did not observe any detectable α-synuclein aggregation (20 μM protein) in ~2.5 h (Fig. 3a). Such slow processes are difficult to monitor continuously in single-molecule experiments because of the need to keep the ultrasensitive detection system running for the long data collection window (>6 h for physiologic condition(s) presented in Fig. 3). Many physicochemical perturbations are known to affect α-synuclein aggregation and can be used to modulate the protein's aggregation kinetics and render it compatible for smFRET experiments (*see* **Note** 7 and Subheading 3.1).
9. Correct for donor and acceptor backgrounds as measured from buffer-only sample. Also, correct for the leakage of donor emission into the acceptor channel and for direct excitation of acceptor, and calculate E_{FRET} for each binning window (e.g., 500 μs). E_{FRET} is estimated from the corrected donor (I_{D}) and acceptor (I_{A}) fluorescence intensities as

$$E_{\mathrm{FRET}} = \frac{I_{\mathrm{A}}}{I_{\mathrm{A}} + \gamma I_{\mathrm{D}}} \tag{2}$$

where γ is a correction factor that depends on the donor and acceptor quantum yields, and donor channel and acceptor channel detection efficiencies. Details on γ estimation are discussed elsewhere [3]; for most applications, it is generally acceptable to assume γ to be approximately 1.

Prepare FRET efficiency histograms from calculated E_{FRET} values and fit to Gaussian function(s) to determine peak position(s). E_{FRET} values of the peaks can be converted to inter-dye distances using Eq. (1). These estimated distances provide dimensional information of the labeled protein population(s). Representative smFRET histograms of monomeric α-synuclein along with fitted peaks are presented in Fig. 5.

Unlike monomer burst profiles (Fig. 4a), aggregating α-synuclein species exhibit wavelike donor and accepter channel fluorescence trajectories (Fig. 4b), characterized by higher photon counts. This is primarily due to prolonged residence of aggregates within the sub-femtoliter observation volume as a result of the reduced diffusion rate of higher molecular weight assemblies. Our observation on α-synuclein aggregation is similar to previous observations on aggregation-prone rhodanese mutant by Hillger et al. [32].

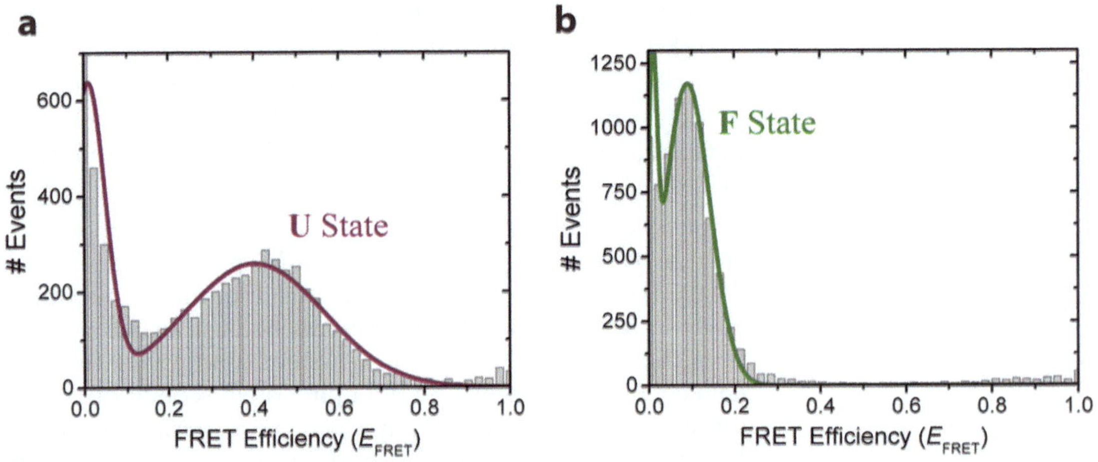

Fig. 5 E_{FRET} as a reporter for α-synuclein conformation. FRET efficiency histograms of disordered (**a**) and extended helical (**b**) α-synuclein (αβγ buffer, pH 7.5) showing intermediate- and low-E_{FRET} conformational states, respectively. The extended helical conformation was induced by α-synuclein's ligand interaction [3, 5]

A 2.5-h time trace of α-synuclein aggregation from the extended helical conformation is presented in Fig. 6. Analysis of such heterogeneous temporally evolving species cannot be performed globally as shown (for homogeneous samples) in Fig. 5. Instead, a segment-by-segment analysis of the time trace should be performed. Detailed analysis of individual zoomed-in snapshots of the time trace can reveal the presence of multiple conformational subpopulations (as indicated by the presence of multiple distinct nonzero E_{FRET} peaks) that are undergoing time-dependent reorganization (Fig. 6). A low concentration of labeled α-synuclein (~100 pM dual-labeled +20 μM unlabeled wild-type protein) ensures single-fluorescence species, especially during the early phases of protein aggregation. Importantly, the ratio of labeled to unlabeled α-synuclein can be tuned to observe different timescales of protein aggregation.

10. Several studies employed different single-molecule resolution methods to characterize α-synuclein aggregation. These studies mostly focused on later phases of the protein's aggregation. For example, Klenerman and coworkers report pathologic conformational conversion within relatively stable (i.e., irreversible) oligomers using FRET [33, 34]. Total internal fluorescence spectroscopy (TIRF) methods were also utilized to directly characterize structural features of individual α-synuclein fibrils (i.e., late irreversible supramolecular assemblies) [35]. The protocol presented in this chapter, however, is focused on characterizing α-synuclein conformations and their transitions within the reversible early protein assemblies in situ. We hypothesize that these reversible and low fractional population molecular species are the initial molecular switches that are responsible for α-synuclein dysfunction.

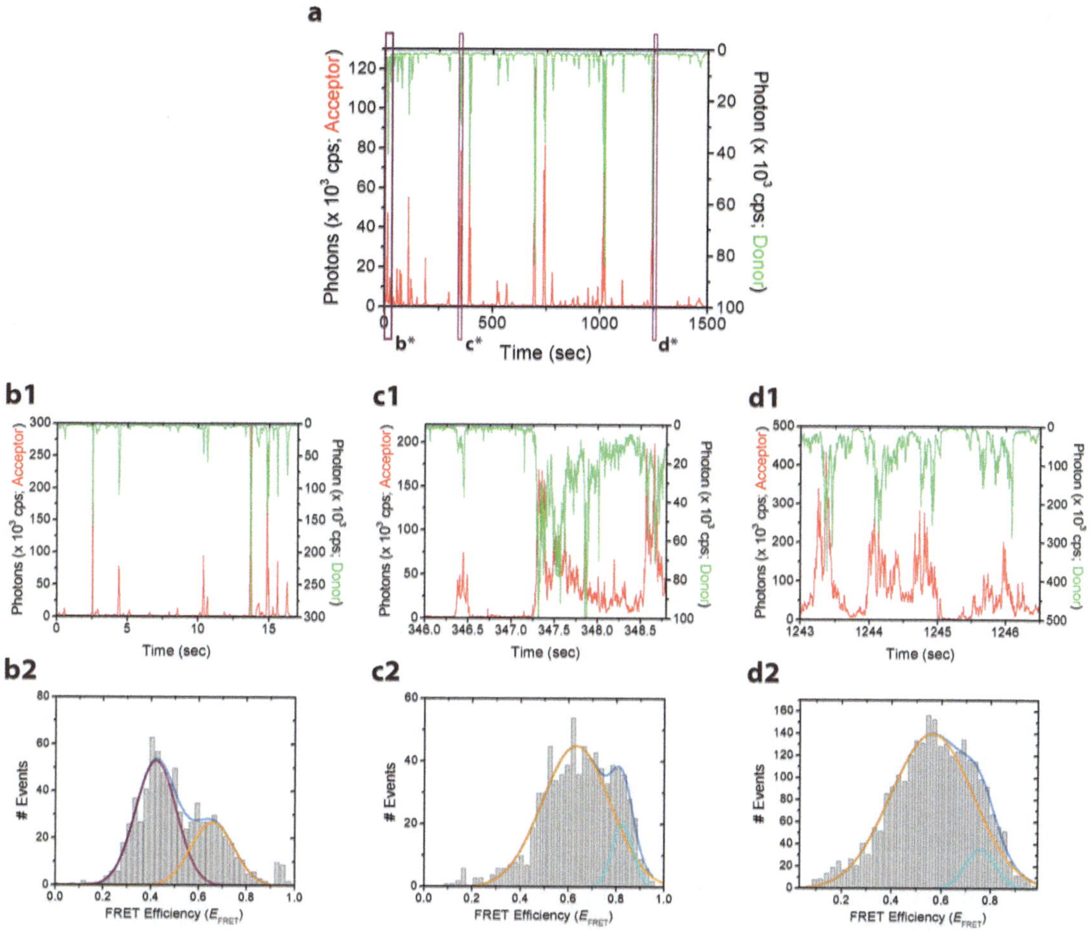

Fig. 6 smFRET detection of conformational transitions leading to α-synuclein fibrillation. A 2.5-h time trace of extended helical α-synuclein during aggregation (αβγ buffer, 1 mM SDS, pH 3.7 at room temperature; **a**). Temporal evolution of conformational subpopulations can be visualized by zooming in on individual segments of the time trace as shown in the bottom panels (**b–d**)

Acknowledgments

This work was supported by laboratory startup funds from the Baylor College of Medicine (A.C.M.F. and J.C.F.). The authors declare no competing financial interests.

References

1. Savica R, Boeve BF, Logroscino G (2016) Epidemiology of alpha-synucleinopathies: from Parkinson disease to dementia with Lewy bodies. Handb Clin Neurol 138:153–158
2. Ghosh D, Mehra S, Sahay S, Singh PK, Maji SK (2017) Alpha-synuclein aggregation and its modulation. Int J Biol Macromol 100:37–54
3. Ferreon AC, Gambin Y, Lemke EA, Deniz AA (2009) Interplay of α-synuclein binding and conformational switching probed by single-molecule fluorescence. Proc Natl Acad Sci U S A 106(14):5645–5650
4. Ferreon AC, Deniz AA (2007) α-Synuclein multistate folding thermodynamics:

implications for protein misfolding and aggregation. Biochemistry 46(15):4499–4509

5. Moosa MM, Ferreon AC, Deniz AA (2015) Forced folding of a disordered protein accesses an alternative folding landscape. ChemPhysChem 16(1):90–94
6. Deniz AA, Laurence TA, Beligere GS, Dahan M, Martin AB, Chemla DS, Dawson PE, Schultz PG, Weiss S (2000) Single-molecule protein folding: diffusion fluorescence resonance energy transfer studies of the denaturation of chymotrypsin inhibitor 2. Proc Natl Acad Sci U S A 97(10):5179–5184
7. Michalet X, Weiss S (2002) Single-molecule spectroscopy and microscopy. C R Phys 3 (5):619–644
8. Trexler AJ, Rhoades E (2009) Alpha-synuclein binds large unilamellar vesicles as an extended helix. Biochemistry 48(11):2304–2306
9. Sun Y, Coskun U, Tsoi PS, Ferreon JC, Ferreon AC, Barbieri B, Liao S-CJ (2017) Quantitative multi-parameter analysis of single molecule dynamics by PIE FastFLIM microscopy. In: Periasamy A, So PTC, König K, Xie XS (eds) SPIE BiOS. SPIE, San Francisco, p 10
10. Tsoi PS, Choi KJ, Leonard PG, Sizovs A, Moosa MM, MacKenzie KR, Ferreon JC, Ferreon ACM (2017) The N-terminal domain of ALS-linked TDP-43 assembles without misfolding. Angew Chem Int Ed Engl 56 (41):12590–12593
11. Biancalana M, Koide S (2010) Molecular mechanism of Thioflavin-T binding to amyloid fibrils. Biochim Biophys Acta 1804 (7):1405–1412
12. Uversky VN, Li J, Fink AL (2001) Evidence for a partially folded intermediate in α-synuclein fibril formation. J Biol Chem 276 (14):10737–10744
13. Hoyer W, Antony T, Cherny D, Heim G, Jovin TM, Subramaniam V (2002) Dependence of alpha-synuclein aggregate morphology on solution conditions. J Mol Biol 322 (2):383–393
14. Krumova P, Meulmeester E, Garrido M, Tirard M, Hsiao HH, Bossis G, Urlaub H, Zweckstetter M, Kugler S, Melchior F, Bahr M, Weishaupt JH (2011) Sumoylation inhibits alpha-synuclein aggregation and toxicity. J Cell Biol 194(1):49–60
15. Bartels T, Choi JG, Selkoe DJ (2011) Alpha-Synuclein occurs physiologically as a helically folded tetramer that resists aggregation. Nature 477(7362):107–110
16. Buell AK, Galvagnion C, Gaspar R, Sparr E, Vendruscolo M, Knowles TP, Linse S, Dobson CM (2014) Solution conditions determine the relative importance of nucleation and growth processes in alpha-synuclein aggregation. Proc Natl Acad Sci U S A 111(21):7671–7676
17. Galvagnion C, Buell AK, Meisl G, Michaels TC, Vendruscolo M, Knowles TP, Dobson CM (2015) Lipid vesicles trigger alpha-synuclein aggregation by stimulating primary nucleation. Nat Chem Biol 11(3):229–234
18. Kelly SM, Jess TJ, Price NC (2005) How to study proteins by circular dichroism. Biochim Biophys Acta 1751(2):119–139
19. Greenfield NJ (2006) Using circular dichroism spectra to estimate protein secondary structure. Nat Protoc 1(6):2876–2890
20. Blanchard JS (1984) Buffers for enzymes. Methods Enzymol 104:404–414
21. Ferreon AC, Moran CR, Gambin Y, Deniz AA (2010) Single-molecule fluorescence studies of intrinsically disordered proteins. Methods Enzymol 472:179–204
22. Ferreon AC, Moran CR, Ferreon JC, Deniz AA (2010) Alteration of the α-synuclein folding landscape by a mutation related to Parkinson's disease. Angew Chem Int Ed Engl 49 (20):3469–3472
23. Gambin Y, VanDelinder V, Ferreon AC, Lemke EA, Groisman A, Deniz AA (2011) Visualizing a one-way protein encounter complex by ultrafast single-molecule mixing. Nat Methods 8 (3):239–241
24. Ferreon ACM, Moosa MM, Gambin Y, Deniz AA (2012) Counteracting chemical chaperone effects on the single-molecule α-synuclein structural landscape. Proc Natl Acad Sci U S A 109(44):17826–17831
25. Veldhuis G, Segers-Nolten I, Ferlemann E, Subramaniam V (2009) Single-molecule FRET reveals structural heterogeneity of SDS-bound alpha-synuclein. Chembiochem 10(3):436–439
26. Trexler AJ, Rhoades E (2010) Single molecule characterization of alpha-synuclein in aggregation-prone states. Biophys J 99 (9):3048–3055
27. Banerjee PR, Moosa MM, Deniz AA (2016) Two-dimensional crowding uncovers a hidden conformation of alpha-Synuclein. Angew Chem Int Ed Engl 55(41):12789–12792
28. Kapanidis AN, Weiss S (2002) Fluorescent probes and bioconjugation chemistries for single-molecule fluorescence analysis of biomolecules. J Chem Phys 117 (24):10953–10964
29. Schuler B, Müller-Späth S, Soranno A, Nettels D (2012) Application of confocal single-molecule FRET to intrinsically disordered proteins. In: Uversky VN, Dunker AK (eds)

Intrinsically disordered protein analysis: Vol 2, Methods and experimental tools. Springer, New York, pp 21–45

30. Uversky VN, Cooper E M, Bower KS, Li J, Fink AL (2002) Accelerated alpha-synuclein fibrillation in crowded milieu. FEBS Lett 515 (1–3):99–103
31. Roeters SJ, Iyer A, Pletikapic G, Kogan V, Subramaniam V, Woutersen S (2017) Evidence for Intramolecular antiparallel beta-sheet structure in alpha-synuclein fibrils from a combination of two-dimensional infrared spectroscopy and atomic force microscopy. Sci Rep 7:41051
32. Hillger F, Nettels D, Dorsch S, Schuler B (2007) Detection and analysis of protein aggregation with confocal single molecule fluorescence spectroscopy. J Fluoresc 17 (6):759–765
33. Cremades N, Cohen SI, Deas E, Abramov AY, Chen AY, Orte A, Sandal M, Clarke RW, Dunne P, Aprile FA, Bertoncini CW, Wood NW, Knowles TP, Dobson CM, Klenerman D (2012) Direct observation of the interconversion of normal and toxic forms of alpha-synuclein. Cell 149(5):1048–1059
34. Tosatto L, Horrocks MH, Dear AJ, Knowles TP, Dalla Serra M, Cremades N, Dobson CM, Klenerman D (2015) Single-molecule FRET studies on alpha-synuclein oligomerization of Parkinson's disease genetically related mutants. Sci Rep 5:16696
35. Varela JA, Rodrigues M, De S, Flagmeier P, Gandhi S, Dobson CM, Klenerman D, Lee SF (2018) Optical structural analysis of individual alpha-synuclein oligomers. Angew Chem Int Ed Engl 57(18):4886–4890

Chapter 18

In Situ Peroxidase Labeling and Mass Spectrometry of Alpha-Synuclein in Rat Cortical Neurons

Ricardo Sanz, Patrick Ovando-Roche, Namrata D. Udeshi, Steven A. Carr, Chee Yeun Chung, and Vikram Khurana

Abstract

In this chapter, we describe a novel ascorbate peroxidase (APEX)-based labeling method that in combination with mass spectrometry identifies proteins in the immediate vicinity of αSyn in living rat cortical neurons. To isolate these interactions, we transduced primary cortical neurons with a lentivirus encoding APEX2 tagged to the C-terminus of alpha-synuclein (αSyn) and under the control of a synapsin promoter. Neural protein lysates were then incubated with streptavidin magnetic beads, washed, eluted from the beads, and digested overnight. The desalted peptides were then labeled with iTRAQ (4-plex) reagents and analyzed by nanoflow liquid chromatography-tandem mass spectrometry (LC-MS/MS). Collected data were analyzed using Spectrum Mill software, ultimately shedding light on αSyn physiological function and abnormal behavior during pathology.

Key words Synucleinopathies, Parkinson's disease, Alpha-synuclein, APEX, APEX2, Mass spectrometry, Neurons

1 Introduction

Synucleinopathies, including Parkinson's disease (PD) and dementia with Lewy bodies (DLB), are the second-most common progressive neurodegenerative disorders after Alzheimer's disease (AD). These disorders are associated with substantial loss of midbrain dopaminergic (mDA) and cortical glutamatergic neurons, albeit with different time course and severity. Current treatments only alleviate signs and symptoms, with dire need for drug candidates and other therapeutics capable of reversing or halting the progressive neuropathology [1]. Considerable evidence implicates the protein alpha-synuclein (αSyn), a small 14 kDa protein associated with phospholipids in membranes and synaptic vesicles in

Ricardo Sanz and Patrick Ovando-Roche contributed equally to this work.

Tim Bartels (ed.), *Alpha-Synuclein: Methods and Protocols*, Methods in Molecular Biology, vol. 1948,
https://doi.org/10.1007/978-1-4939-9124-2_18,

neurons. Misfolding and mistrafficking of αSyn within specific subsets of neuronal cells lead to neurodegeneration but the specific etiology and protein interactome maps this misfolding affects remain unknown. Understanding the role of αSyn from its localization to its native interactions may provide essential clues on disease mechanisms. So far, several methods have been used to identify the αSyn-interacting proteins. Low-throughput co-immunoprecipitation (co-IP) followed by biochemical validation identified a handful of interacting proteins in a candidate-based manner [2]. As an unbiased and high-throughput approach, co-IP was combined with mass spectrometry (MS)-based proteomics [3]. The robustness of co-IP in general relies heavily on the efficiency and quality of the antibody against the protein of interest. Immunoprecipitated protein complexes may only detect stable strong interactions, leading to false-negative results [4, 5]. In addition, IP is performed after the lysis of tissue or cells, making it impossible to probe the nature of interactions in the native cellular context.

Recently, in situ enzymatic biotinylation (so-called proximity labeling) approaches were adapted to shed light on the protein neighborhood of αSyn. Ting and colleagues initially developed this approach using an engineered ascorbate peroxidase, APEX, to define a proteome of subcellular organelles and distinct compartment by fusing APEX to a known protein with specific localization to subcellular organelles or compartment [6–9]. Upon exposure to H_2O_2, APEX oxidizes biotin-phenol (BP) to form short-lived (<1 ms) biotin-phenoxyl radicals that can conjugate to proteins that are in the vicinity of the APEX fused protein [6–8]. This biotinylation labeling creates covalent bonds on electron-rich amino acids such as Tyrosine with an estimated reaction diameter of less than 10 nm. When fused to a protein of interest, APEX can capture native cellular compartments in which the protein and its closest neighboring proteins function together. Combined with mass spectrometry, this method is unbiased and of high throughput. In contrast to the co-IP approach which requires cell lysis, the APEX approach preserves protein interactions in native cellular context without being dependent on the presence and quality of antibodies. Nevertheless, despite its many merits, APEX does not distinguish between proximity and direct (or indirect) protein interactions in a complex. To confirm the latter, further exploration of hits, such as with candidate-based co-IP, is necessary.

In this chapter, we describe in detail how to harness this approach to investigate the protein interactome of αSyn [10]. By fusing αSyn to APEX2 (an evolved form of APEX that has activity in neurons [9]), we have defined the cellular compartments in which αSyn resides and the proteins in its immediate vicinity in rat primary cortical neurons. This method demonstrated the known interacting partners of αSyn and correctly identified physiological compartments in which αSyn is known to be enriched. In addition, the approach elucidated novel interactors of αSyn in novel functional compartments, providing an opportunity to understand its lesser known physiological roles.

2 Materials

2.1 Animals

1. Timed pregnant Sprague-Dawley rat.

2.2 Dissection Tools

1. Stereomicroscope.
2. Surgical scissors.
3. Two microdissection forceps.
4. Sharp-pointed dissecting scissors.
5. Disposable transfer pipets.
6. Petri dishes.
7. Pasteur pipets (5 and 10 mL).
8. 1000 μL Pipet and pipet tips.

2.3 Solutions for Cortical Dissection and Neuron Dissociation

1. Leibovitz's L-15 Media (L-15, Thermo Fisher Scientific).
2. Accumax cell detachment solution (Stem Cell Technologies).
3. Heat-inactivated fetal bovine serum (HI-FBS) (Thermo Fisher Scientific).
4. Dulbecco's modified Eagle media, DMEM 1× (Thermo Fisher Scientific) with 10% FBS.

2.4 Cell Culture

1. Growth medium: Neurobasal Media (Thermo Fisher Scientific), 2% B27 supplement, 1% N2 supplement, 25 mM ß-mercaptoethanol, Glutamax (100×, Life Technologies), 100 IU/mL penicillin, 100 mg/mL streptomycin. Equilibrate the growth medium at 37 °C and 5% CO_2 prior to use.
2. PBS: 145 mM NaCl, 2.7 mM KCl, 8.1 mM Na_2HPO_4, 1.5 mM KH_2PO_4, pH 7.4.
3. Poly-L-ornithine: Dissolve poly-L-ornithine (Sigma Aldrich) in distilled water to a final concentration of 100 μg/mL (equivalent to 0.1% w/v).
4. Laminin: Dilute laminin (BD Biosciences) in PBS solution to a final concentration of 10 μg/mL.
5. 5% CO_2 incubator at 37 °C.
6. Hemocytometer.

2.5 Fixation and Immunofluorescence

1. Fixative: 4% Paraformaldehyde (PFA) in PBS containing 20% sucrose. Under a fume hood, heat 1× PBS to approximately 60 °C while stirring. Add PFA powder to the stirring PBS

solution to a final concentration of 4% w/v. For 500 mL, add 20 g PFA to 450 mL of PBS. Add 1 N NaOH dropwise until the solution clears (approximately 1/2 drop per gram of PFA). Verify pH with Litmus pH test strips and adjust if necessary, using small amounts of diluted HCL to pH 7.4. Filter the solution through 0.45 μm Whatman membrane filter, aliquot, and cover with aluminum foil to protect from light. 4% PFA can be frozen or stored at 2–8 °C for up to 1 month. Prepare fixative solution by dissolving sucrose to a final concentration of 20% w/v in 4% PFA.

2. Primary antibodies (βIII tubulin, Millipore).
3. Secondary antibody (Molecular Probes, Life Technologies).
4. Fluorescent-conjugated streptavidin (Molecular Probes, Life Technologies).
5. Goat serum (Jackson ImmunoResearch).
6. Hoechst 33342 (Life Technologies).

2.6 Virus Production

1. HEK 293 cells.
2. Lipofectamine 2000.
3. Lentivirus expression and packaging plasmids.
4. 0.45 μm Low-protein binding filter.
5. Lenti-X Maxi Purification kit (Clontech).
6. Lenti-X Concentrator (Clontech).
7. QuickTiter Lentivirus titer kit (Lentivirus-Associated HIV p24, Cell Biolabs).

2.7 Ascorbate Peroxidase (APEX) Labeling

1. Labeling reagents: 500 mM Biotin tyramide in neurobasal media and 1 mM hydrogen peroxide (H_2O_2).
2. Quenching reagent: 5 mM Trolox and 10 mM ascorbic acid in PBS.
3. Lysis buffer: Radioimmunoprecipitation assay buffer (RIPA, 50 mM Tris–HCl pH 7.4, 150 mM NaCl, 1% Triton X-100, 0.5% sodium deoxycholate, 0.1% SDS, and 1 mM EDTA) with protease inhibitor cocktail (Sigma Aldrich).
4. BCA assay for protein concentration measurement.
5. Streptavidin magnetic beads (Pierce).

2.8 On-Bead Digestion

1. 400 mM NaCl.
2. 1 M KCl.
3. 0.1 M Na carbonate (pH 11.5).
4. 2 M urea.

5. RIPA: 50 mM Tris–HCl pH 7.4, 150 mM NaCl, 1% triton X-100, 0.5% sodium deoxycholate, 0.1% SDS, and 1 mM EDTA.
6. 4 mM DTT.
7. 10 mM Iodoacetamide.
8. Trypsin.
9. Formic acid.
10. C18 stage tips.

2.9 iTRAQ Labeling Kit

1. iTRAQ (4-plex Sciex, Foster City, CA).
2. Ethanol.
3. 1 M Tris–HCl (pH 8).
4. Vacuum concentrator (SpeedVac).
5. C18 stage tips.

2.10 Mass Spectrometry Analysis

1. 0.1% Formic acid (FA).
2. 3% MeCN/0.1% FA.
3. A microcapillary column (360 mm outer diameter × 75 mm inner diameter), which contains an integrated electrospray emitter tip (10 mm), packed to approximately 22 cm with ReproSil-Pur C18-AQ 1.9 mm beads (Dr. Maisch GmbH) and heated to 50 °C.
4. Q Exactive mass spectrometer (Thermo Fisher Scientific).
5. Proxeon Easy-nLC 1000 (Thermo Fisher Scientific).

2.11 Data Analysis

1. Spectrum Mill software package pre-release v5.0 (Agilent Technologies).

3 Methods

3.1 Substrate Preparation

Substrates should be prepared prior to the dissection. Basal neurite outgrowth and survival from dissociated cortical cultures are promoted when culture dishes are coated with poly-L-ornithine and laminin.

1. One day before the dissection, coat tissue culture plates with poly-L-ornithine solution overnight at 4 °C. Keep plates sterile by sealing them with Parafilm and Saran wrap.
2. The following day, remove poly-L-ornithine and wash three times with sterile 1× PBS (*see* **Note 1**).
3. Coat cell culture plates with 10 μg/mL laminin diluted in PBS for 1 h at 37 °C (*see* **Note 2**). Remove laminin before seeding cortical neurons.

3.2 Isolation of Rat Cortical Neurons

1. Sterilize dissection instruments and dissecting hood with 70% ethanol.
2. Euthanize pregnant rat (E18-19) following the Animal Care Facility and University guidelines, such as CO_2 asphyxiation, in conjunction with cervical dislocation.
3. Sterilize the rat's abdomen by spraying the fur with 70% ethanol and use surgical scissors to make a cesarean section. Extract embryos using forceps and place into a petri dish containing cold L-15. You should expect to recover 7–12 embryos.
4. Remove the embryos from their individual amniotic sacs, decapitate at the head/neck junction, and transfer heads to a petri dish containing cold L-15.
5. With one hand, use forceps to stabilize the head by piercing the eyes and use your other hand to gently cut the skin along the midline, advancing toward the nose to expose the skull.
6. Fold back the skull with forceps to expose the brain.
7. Make two lateral cuts on the posterior region of the brain hemispheres to separate the cerebellum from the rest of the brain.
8. Scoop out brain and place into a petri dish containing cold L-15.
9. Divide the cerebrum along the midline into two cerebral hemispheres. Place each cerebral hemisphere medial side up and carefully remove the olfactory bulbs, diencephalon, meninges, and hippocampus without damaging the cortical region.
10. Transfer cortical tissue to a 15 mL conical tube and centrifuge for 3 min at 500 × *g*.
11. Remove the medium using a Pasteur pipet and add 2 mL of pre-warmed Accumax. Incubate cortical tissue in Accumax for 30 min in a 37 °C and 5% CO_2 incubator.
12. Centrifuge the digested cortical tissue for 3 min at 500 × *g*, remove the Accumax, and suspend the cortical tissue with 5 mL of DMEM medium containing 10% FBS to neutralize residual Accumax.
13. Centrifuge the digested cortical tissue for 3 min at 500 × *g* and carefully remove the supernatant.
14. Add 1 mL growth medium and triturate the cortical tissue by gently pipetting up and down with a P1000 pipet tip until the solution becomes homogeneously turbid (about 10–15 times) (*see* **Note 3**).
15. Dilute the cell suspension in growth medium to a final volume of 5 mL, allow debris to settle to the bottom of the tube, and remove the supernatant containing dissociated cortical neurons.

16. Count the cells with a hemocytometer and plate at a density appropriate to the experimental application. Approximately, 5–10 million cells can be obtained from one single brain. Cell densities around 30×10^3 cells/cm^2 typically generate healthy cultures. For cortical neurons, the plate requires to be previously coated with poly-L-ornithine (100 μg/mL) and laminin (10 μg/mL).
17. After seeding the neurons, place plate in a humidified incubator at 37 °C and 5% CO_2 and age neurons for 3 weeks. Half media change should be performed every 3–4 days.

3.3 Viral Mediated Delivery in Cortical Neurons

Viral transduction is an efficient way to stably integrate a genetic construct into the genome, targeting 90–100% of the cellular population. Postmitotic neurons can also be targeted using a lipid-based method or nucleofector to deliver plasmids or shRNAs. However, these approaches have shown limited success in targeting large cellular populations, and hence are discouraged for biochemical assays. To isolate alpha-synuclein interactors in living neurons, we generated a virus encoding a genetically engineered αSyn tagged with APEX2 at the C-terminus and under the control of a synapsin promoter.

1. Lentiviral particles are initially generated in human cell lines (HEK 293) via liposome-based transfection of the lentivirus expression and packaging plasmids (Pmd2.G and psPAX2).
2. Collect medium from transfected cells after 24, 36, and 48 h and replace with fresh medium. Store collected supernatants at 4 °C while you are waiting to collect all the other time points.
3. Filter collected supernatants through 0.45 μm low-protein-binding filter.
4. Purify and concentrate lentivirus particles from the media using Lenti-X Maxi Purification kit and Lenti-X Concentrator, respectively, following the manufacturer's instructions.
5. Determine virus titer using QuickTiter Lentivirus titer kit, as described in the manufacturer's protocol.
6. After 10 days in culture, dissociated cortical neurons are exposed to the purified lentivirus (*see* **Note 4**).
7. The following day, completely replace media with fresh growth media and age neurons for 3 days.

3.4 Peroxidase-Labeling Assay of Rat Primary Cortical Neurons

1. Following aging of cultured neurons and viral transduction (13 days in total), wash neurons twice with neurobasal media to remove any cellular debris and residual growth factors.
2. Incubate neuronal cultures with neurobasal media containing 500 mM biotin tyramide for 30 min at 37 °C.

3. Add hydrogen peroxide to the media at a final concentration of 1 mM for exactly 1 min at room temperature (*see* **Note 5**).
4. Wash neurons with PBS-containing quenching reagents (5 mM Trolox and 10 mM ascorbic acid) for 30 s, followed by two washes with PBS.
5. Lyse neurons with radioimmunoprecipitation assay (RIPA buffer) with a protease inhibitor cocktail.
6. Incubate neuronal lysates on ice for 20 min and centrifuge for 20 min at 10,000 × *g* at 4 °C.
7. Collect supernatant and perform a BCA assay to measure protein concentration.
8. For mass spectrometry, incubate 1–2 mg of neuronal lysates with 200 μL of streptavidin magnetic bead slurry for 1 h at room temperature.

We recommend validating the presence of biotinylated samples by staining fixed neuronal cultures with a fluorescently conjugated streptavidin (*see* **Note 6**), or by western blotting using HRP-conjugated streptavidin, prior to continuing with the proteomic analysis after washing (*see* Subheading 3.5, **step 1**).

3.5 On-Bead Digestion of Rat Primary Cortical Neuron Protein Samples

As the starting material, we recommend combining two biological replicates of neuronal pulldown with magnetic streptavidin beads. We recommend collecting and storing supernatants to determine the efficiency of pulldown (*see* **Note 7**).

1. Wash magnetic beads with rocking in 1 mL of RIPA containing 400 mM NaCl two times, with 1 M KCl once, with 0.1 M Na carbonate (pH 11.5) once, with 2 M urea twice, and with RIPA twice.
2. Spin down at 5000 × *g* for 1 min. Under constant agitation, incubate samples with 4 mM DTT for 30 min at room temperature.
3. Following reduction, alkylate samples by applying 10 mM iodoacetamide for 45 min while shaking in the dark at room temperature.
4. Digest samples by adding 0.5 μg of trypsin and shake overnight at room temperature.
5. The following day, acidify samples (pH < 3) by applying neat formic acid to a final concentration of 1% (w/v).
6. Spin down and desalt samples on C18 stage tips as previously described [11].

3.6 Isobaric Tags for Relative and Absolute Quantification (iTRAQ) Labeling of Peptide Samples for Quantitative Proteomics (See Note 8)

Desalted peptides were labeled with iTRAQ (4Plex) reagents in the following manner. We recommend the labeling be done in duplicate for each sample.

1. Allow each iTRAQ reagent to reach room temperature.
2. Spin each reagent solution to the bottom of the tube and add 70 μL of ethanol.
3. Resuspend samples in 30 μL of dissolution buffer.
4. Transfer the contents of each reagent to each sample tube (i.e., 114-Control, 115-NES, and 116-αSyn) (*see* **Note 9**).
5. Vortex each sample, spin down, and incubate at room temperature for 1 h with gentle shaking.
6. Following the incubation, take 3% of sample by volume and store it for testing label incorporation.
7. Quench the remaining sample with 10 μL 1 M Tris–HCl (pH 8) for 10 min with gentle shaking at room temperature.
8. Dry the labeled peptides using a SpeedVac and desalt on C18 stage tips as previously described [11].

3.7 Mass Spectrometry Analysis

1. Resuspend desalted sample peptides in 9 μL of 3% MeCN/0.1% FA.
2. Analyze samples by nanoflow liquid chromatography-tandem mass spectrometry (LC-MS/MS) using a Q Exactive mass spectrometer coupled online to Proxeon Easy-nLC 1000.
3. Four microliters of resuspended peptides are loaded onto a microcapillary column, which contains an integrated electrospray emitter tip, packed to approximately 22 cm with ReproSil-Pur C18-AQ 1.9 mm beads (Dr. Maisch GmbH) and heated to 50 °C. The HPLC Solvent A was 0.1% FA, and Solvent B was 90% MeCN/0.1% FA.
4. Sample peptides are eluted into the mass spectrometer at a flow rate of 200 nL/min. A 150-min inject-to-inject with a linear gradient of 0.3%B/min, followed by a ramp to 60% B (10% B/min). The Q Exactive is operated in the data-dependent mode acquiring HCD MS/MS scans ($R = 17{,}500$) after each MS1 scan ($R = 70{,}000$) on the top 12 most abundant ions using an MS1 target of 3×10^6 and an MS2 target of 5×10^4. The maximum ion time utilized for MS/MS scans is 120 ms; the HCD-normalized collision energy is set to 27; the dynamic exclusion time is set to 20 s, and the peptide match and isotope exclusion functions are enabled (*see* **Note 10**).

3.8 Data Analysis

Analyze collected data using a software package capable of performing database searches and rapidly identifying matched proteins and peptides. We use the Spectrum Mill software package from Agilent Technologies (*see* **Note 11**). Here, we describe data analysis utilizing Spectrum Mill software package pre-release v5.0:

1. Set criteria for the analysis. We recommend including in the search parameters MS/MS spectra within 750–4000 Da precursor MH+ range and a resulting sequence tag length less than zero. It is possible to merge MS scans with similar precursor mass-to-charge ratio (m/z) if they are within a 60-s retention time and 1.4 m/z tolerance.
2. Set the MS/MS spectra search against UniProt Rat complete isoform database.
3. For fixed modifications, include carbamidomethylation and iTRAQ labeling of the N-terminus and internal lysines. For variable modifications, include N-terminal acetylation and methionine oxidation.
4. Set the maximum peak intensity to 30%, with precursor mass tolerance of ±20 ppm.
5. The ratio of signal intensity of the iTRAQ reporter ions is an indication of the relative abundance of peptides from different labeled samples. Calculate the median ratio from all MS/MS spectra contributing to a protein subgroup.
6. To account for differences between samples, iTRAQ ratios are median normalized [10].
7. Protein hits identified by APEX can be further validated by immunoprecipitation or membrane-2 hybrid (MYTH) [12], as in our original publication [10].

4 Notes

1. Soluble poly-L-ornithine can be toxic, and extensive washes are necessary to prevent poor neuronal survival.
2. Do not allow the laminin coating to dry, as this may reduce the adhesion of the cortical neurons. Long-term storage of pre-coated dishes is not advised, as this may reduce cell adhesion.
3. Excessive trituration of cortical tissue will reduce neurite outgrowth and cell viability. It is advised to triturate from 10 to 15 times and continue with the isolation process.
4. Titers of lentivirus can vary between preparation. Since we have been interested in physiological, not pathological, state of αSyn, it was important to titrate to the lowest MOI that does not cause toxicity (overexpression of αSyn in neurons causes toxicity) but that gives sufficient biotin labeling. To filter out false-positive hits that result from random cytosolic labeling

and enrich the relevant labeling for αSyn APEX2, a control is critically important. In our previous paper, we used a nuclear exclusion signal (NES) APEX2 as a control, which labels cytosolic proteins. Depending on the question being asked, different controls might be appropriate. For example, a study investigating the local nuclear proteome of αSyn might add NLS-APEX2 as an additional control for the nuclear compartment. If one were comparing the effect of αSyn mutations on αSyn biology then αSyn^{WT}-APEX2 would become the control, and so forth.

5. It is critical to keep the labeling time to exactly 1 min for all the conditions. To do this, we recommend staggering the labeling in a timed manner along with washing. Since labeling is staggered, washing (*see* Subheading 3.4, **step 4**) is also staggered.
6. Fix cortical neurons with 4%PFA/20%sucrose in PBS for 10 min. Wash neuronal cultures three times with PBS and block for 1 h with 10% goat serum and 0.2%Triton x-100 in PBS. After 1 h, stain for a neuronal marker (ßIII tubulin) in 5% blocking buffer. The next day, perform three PBS washes and stain for a fluorophore-conjugated secondary antibody, a fluorophore-conjugated streptavidin, and a nuclear stain (Hoechst) diluted in 5% blocking buffer for 1 h at 37 °C and 5% CO_2 incubator. Wash three times with PBS and image with an epifluorescence or confocal microscope.
7. We recommend collecting and storing supernatant from all the washes to validate the pulldown efficiency.
8. For our previous work, we used iTRAQ4 for quantification. TMT10 quantification can also be used and increases the throughput threefold relative to iTRAQ4, at similar proteome coverage and high reproducibility (Mertins et al., Accepted Nature Protocols, 2018).
9. As a control for the pulldown, we recommend including NES-APEX2 and mito-APEX2 to target different cellular components.
10. The methods described are specific for analysis on a Q Exactive MS. However, these experiments can be run on other MS systems such as an Orbitrap Lumos MS, or new and improved QE systems (QE+ or QEHFx). If another system is used, the method parameters will vary.
11. We recommend and describe Spectrum Mill software from Agilent Technologies, but there are many other software packages which can be used, including MaxQuant.

Acknowledgments

We thank the MSA Coalition and National Ataxia Foundation for funding.

References

1. Goedert M, Spillantini MG, Del Tredici K, Braak H (2013) 100 years of Lewy pathology. Nat Rev Neurol 9(1):13–24. https://doi.org/10.1038/nrneurol.2012.242
2. Sharma M, Burre J, Bronk P, Zhang Y, Xu W, Sudhof TC (2012) CSPalpha knockout causes neurodegeneration by impairing SNAP-25 function. EMBO J 31(4):829–841. https://doi.org/10.1038/emboj.2011.467
3. McFarland MA, Ellis CE, Markey SP, Nussbaum RL (2008) Proteomics analysis identifies phosphorylation-dependent alpha-synuclein protein interactions. Mol Cell Proteomics 7 (11):2123–2137. https://doi.org/10.1074/mcp.M800116-MCP200
4. Brunner Y, Schvartz D, Coute Y, Sanchez JC (2009) Proteomics of regulated secretory organelles. Mass Spectrom Rev 28(5):844–867. https://doi.org/10.1002/mas.20211
5. ten Have S, Boulon S, Ahmad Y, Lamond AI (2011) Mass spectrometry-based immuno-precipitation proteomics—the user's guide. Proteomics 11(6):1153–1159. https://doi.org/10.1002/pmic.201000548
6. Hung V, Zou P, Rhee HW, Udeshi ND, Cracan V, Svinkina T, Carr SA, Mootha VK, Ting AY (2014) Proteomic mapping of the human mitochondrial intermembrane space in live cells via ratiometric APEX tagging. Mol Cell 55(2):332–341. https://doi.org/10.1016/j.molcel.2014.06.003
7. Rhee HW, Zou P, Udeshi ND, Martell JD, Mootha VK, Carr SA, Ting AY (2013) Proteomic mapping of mitochondria in living cells via spatially restricted enzymatic tagging. Science 339(6125):1328–1331. https://doi.org/10.1126/science.1230593
8. Hung V, Udeshi ND, Lam SS, Loh KH, Cox KJ, Pedram K, Carr SA, Ting AY (2016) Spatially resolved proteomic mapping in living cells with the engineered peroxidase APEX2. Nat Protoc 11(3):456–475. https://doi.org/10.1038/nprot.2016.018
9. Lam SS, Martell JD, Kamer KJ, Deerinck TJ, Ellisman MH, Mootha VK, Ting AY (2015) Directed evolution of APEX2 for electron microscopy and proximity labeling. Nat Methods 12(1):51–54. https://doi.org/10.1038/nmeth.3179
10. Chung CY, Khurana V, Yi S, Sahni N, Loh KH, Auluck PK, Baru V, Udeshi ND, Freyzon Y, Carr SA, Hill DE, Vidal M, Ting AY, Lindquist S (2017) In situ peroxidase labeling and mass-spectrometry connects alpha-synuclein directly to endocytic trafficking and mRNA metabolism in neurons. Cell Syst 4(2):242–250 e244. https://doi.org/10.1016/j.cels.2017.01.002
11. Udeshi ND, Mertins P, Svinkina T, Carr SA (2013) Large-scale identification of ubiquitination sites by mass spectrometry. Nat Protoc 8 (10):1950–1960. https://doi.org/10.1038/nprot.2013.120
12. Stagljar I, Korostensky C, Johnsson N, te Heesen S (1998) A genetic system based on split-ubiquitin for the analysis of interactions between membrane proteins in vivo. Proc Natl Acad Sci U S A 95(9):5187–5192

Chapter 19

Probing Alpha-Synuclein Conformations by Electron Paramagnetic Resonance (EPR) Spectroscopy

Julia Cattani, Theresa Braun, and Malte Drescher

Abstract

Electron paramagnetic resonance (EPR) spectroscopy in combination with site-directed spin labeling is ideally suited to study structure, dynamics, and interactions of intrinsically disordered proteins as alpha-synuclein.

Here we describe all steps required for a corresponding study: the spin labeling procedure, sample preparation, spectroscopic experimental procedure, and data analysis.

Key words Electron paramagnetic resonance (EPR) spectroscopy, Protein structure, Protein dynamics, Site-directed spin labeling, Alpha-synuclein, Intrinsically disordered protein

1 Introduction

Alpha-synuclein (aS) is an intrinsically disordered protein that adopts a variety of secondary structures upon macromolecular interaction. Typical interaction partners are other aS molecules, other proteins, or lipids. Its physiological function is still unclear, even though it is highly abundant in the human brain. Studying aS is of particular interest, because the protein is a key player in Parkinson's disease, both its wild-type form and the disease variants bearing point mutations. Electron paramagnetic resonance (EPR) spectroscopy is a highly suitable spectroscopy technique to investigate dynamics and conformations of aS [1–11]. Using a site-directed spin labeling (SDSL, *see* Fig. 1) site scan, interactions with membranes or other macromolecules and involved residues can be monitored [7, 9, 11]. The general approach is not limited to aS [12–16]. In order to use the SDSL site scan approach, a set of aS mutants with cysteines introduced at the desired sites are expressed, purified, and subsequently labeled with nitroxide spin labels (*see* Fig. 2). The corresponding EPR spectra (*see* Fig. 3) are obtained and analyzed (*see* Fig. 4).

Tim Bartels (ed.), *Alpha-Synuclein: Methods and Protocols*, Methods in Molecular Biology, vol. 1948,
https://doi.org/10.1007/978-1-4939-9124-2_19, © Springer Science+Business Media, LLC, part of Springer Nature 2019

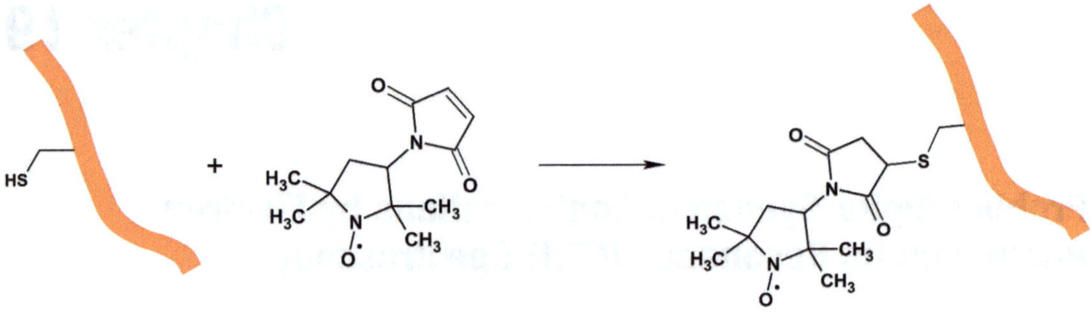

Fig. 1 aS (orange) with an introduced cysteine residue is spin labeled with 3-maleimido-PROXYL

A)

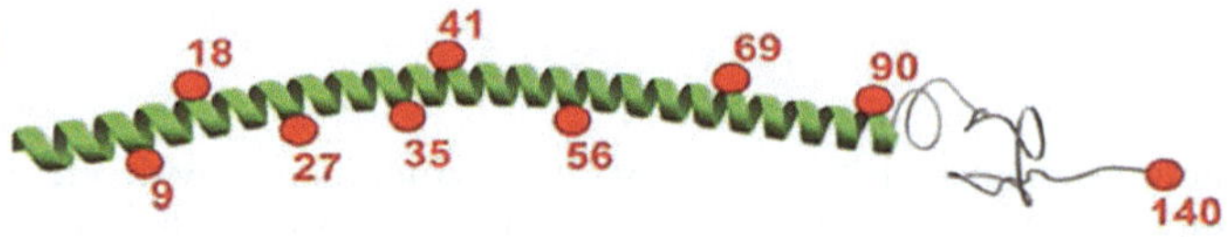

B)

10	20	30	40
MDVFMKGLSK	AKEGVVAAAE	KTKQGVAEAA	GKTKEGVLYV
50	60	70	80
GSKTKEGVVH	GVATVAEKTK	EQVTNVGGAV	VTGVTAVAQK
90	100	110	120
TVEGAGSIAA	ATGFVKKDQL	GKNEEGAPQE	GILEDMPVDP
130	140		
DNEAYEMPSE	EGYQDYEPEA		

Fig. 2 (**a**) Model of an extended α-helical aS structure and (**b**) amino acid sequence of αS with positions of spin-labeled residues indicated in red. The amino acids Ser9, Ala18, Ala27, Glu35, Gly41, Ala56, Ala69, Ala90, and Ala140 were each substituted with cysteine [8]

2 Materials

Prepare all solutions using ultrapure water (prepared by using a Milli-Q Academic Ultrapure Water System (Merck KGaA)). Prepare and store all reagents at room temperature (unless indicated otherwise). Diligently follow all applicable safety regulations and waste disposal regulations when disposing waste materials.

1. Template plasmid: pT7-7 aS WT, (Plasmid #36046, Addgene).
2. Phusion® High-Fidelity DNA Polymerase, 2000 units/mL.
3. Phusion HF Buffer (5×), dNTPs (10 mM) (#M0530, New England BioLabs Inc.).
4. Thermocycler (e.g., peqSTAR 96X Universal Gradient, PEQLAB).
5. Water bath (e.g., MyBath™ Laboratory Water Baths).

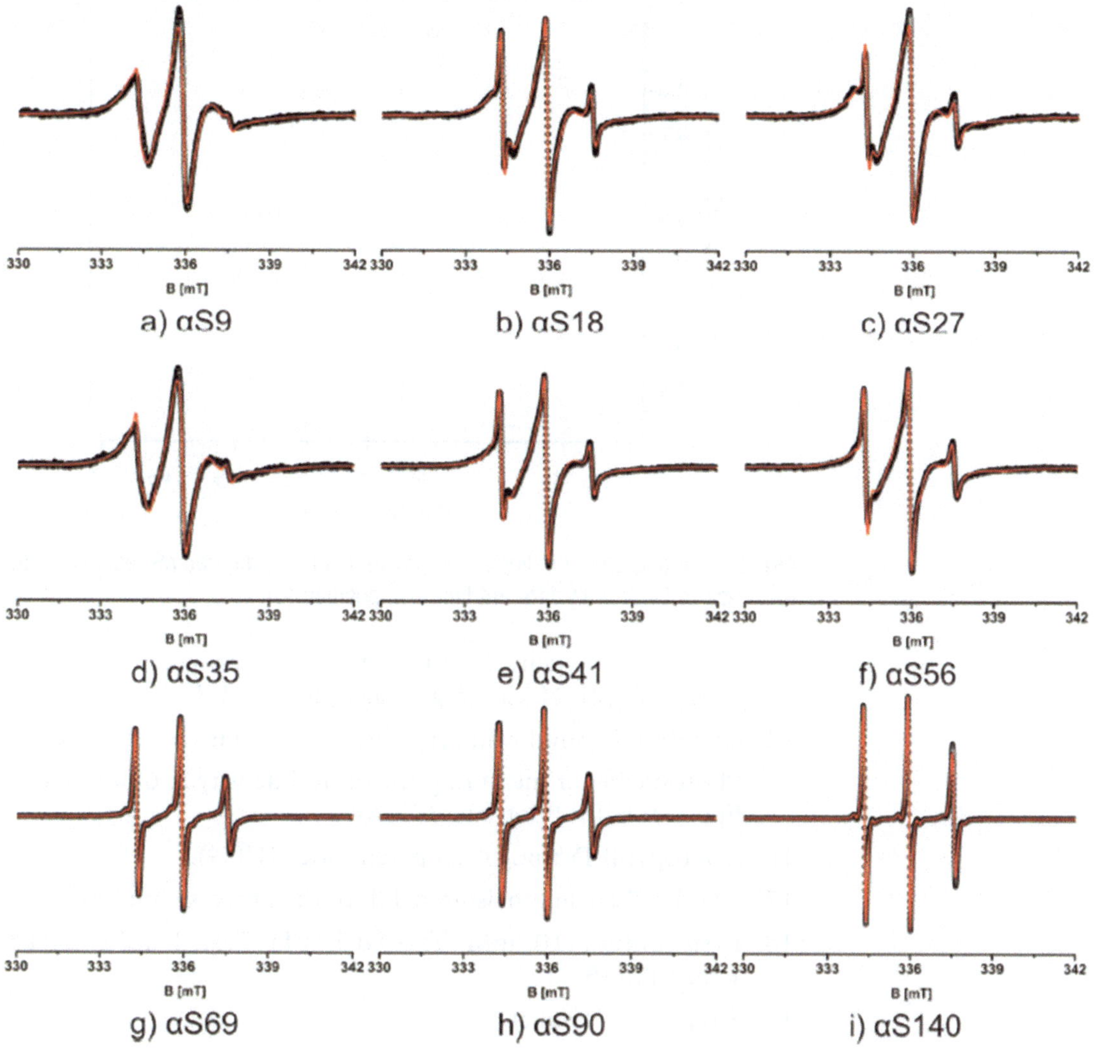

Fig. 3 Experimental spectra (black) and corresponding spectral simulations (red) of spin-labeled αS single-cysteine mutants in the presence of LUVs (POPC:POPE:CL:Ch lipids in 2:1.3:1:0.6 molar ratio) [8]

6. DpnI, methylation-sensitive restriction enzyme, 20,000 units/mL, e.g., #R0176, New England BioLabs Inc.
7. GH371 cells, iGEM.
8. Super Optimal Broth with Catabolic repressor (SOC): 0.5% Yeast extract, 2% tryptone, 10 mM NaCl, 2.5 mM KCl, 10 mM $MgCl_2$, 10 mM $MgSO_4$.
9. BL21-Gold(DE3) Competent Cells, Agilent Technologies.
10. Incubation shaker, e.g., Ecotron, INFORS HT.
11. Carbenicillin (stock solution): 50 mg/mL in 50% ethanol.
12. LB medium (Lysogeny Broth): 10 g/L Tryptone, 5 g/L yeast extract, 5 g/L NaCl, pH 7.

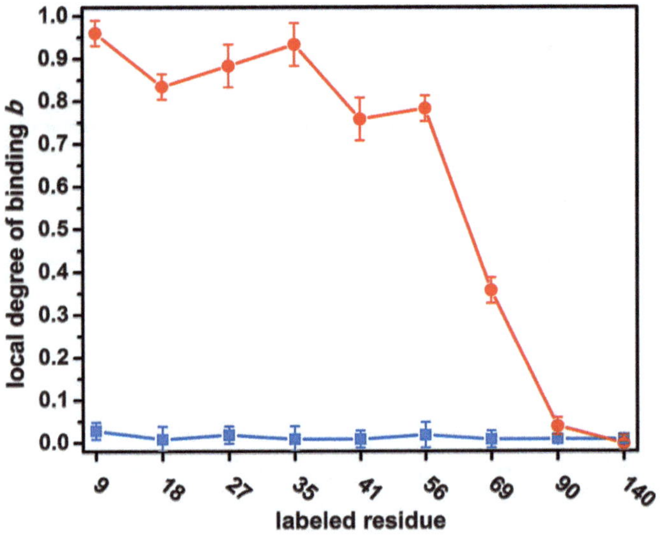

Fig. 4 Local degree of binding in the proximity of investigated aS residues in the presence of LUVs with different lipid composition [8]

13. LB-agar (Lysogeny Broth): 10 g/L Tryptone, 5 g/L yeast extract, 5 g/L NaCl, 15 g/L agar-agar, pH 7.
14. GeneJET Plasmid Miniprep Kit, Thermo Fisher Scientific.
15. Photometer for measuring the optical density at 600 nm, e.g., Eppendorf BioPhotometer® D30.
16. Isopropyl-β-D-thiogalactopyranoside (IPTG).
17. PMSF: Phenylmethylsulfonyl fluoride, protease inhibitor.
18. Lysis buffer: 10 mM Tris–HCl, pH 7.4, 1 mM EDTA, 1 mM PMSF.
19. Streptomycin sulfate.
20. Ammonium sulfate.
21. Äkta start buffer: 10 mM Tris–HCl, pH 7.4.
22. ÄKTAprime plus chromatography system (GE Healthcare Biosciences) equipped with Resource Q anion exchange chromatography column.
23. Äkta elution buffer: 10 mM Tris–HCl, pH 7.4, 1 M NaCl.
24. Electrophoresis chamber (e.g. Mini-PROTEAN® Tetra Vertical Electrophoresis Cell), 15% SDS-Gel, Laemmli sample buffer (Tris-HCl, 6×) (*see* **Note 1**).
25. Coomassie Brilliant Blue staining solution: Dissolve 1 g of Coomassie Brilliant Blue (Bio-Rad) in 500 mL methanol, 100 mL glacial acetic acid, and 400 mL H_2O.
26. Destaining solution: 200 mL methanol, 100 mL glacial acetic acid, and 700 mL H_2O.

27. Storage buffer: 10 mM HEPES, 50 mM NaCl, pH 7.4.
28. Amicon Ultra 0.5 mL centrifugal filters MWCO 3 kDa (Sigma-Aldrich).
29. Water Bath, e.g., 2 L Digital Laboratory Water Bath, MyBath™ Laboratory Water Baths.
30. Dithiothreitol (DTT) solution: Dissolve 10 mg DTT in 65 μL ultrapure water (*see* **Note 2**).
31. Spin labeling solution: Dissolve 18.9 mg 3-Maleimido-PROXYL (e.g., Merck KGaA) in 200 μL dimethyl sulfoxide (DMSO). Store at −75 °C.
32. Labeling buffer: 10 mM Tris–HCl, pH 7.4. Dissolve 12.1 g of Tris base in 80 mL ultrapure water. Adjust the pH to 7.4 while consecutively adding HCl (*see* **Note 3**). Make up to 100 mL with ultrapure water. Filter Tris–HCl with 0.22 μM sterile filter. Dilute 1:100 with ultrapure water to get 10 mM buffer.
33. Variable-speed centrifuge with holder for 15 mL conical tubes and temperature control (4 °C).
34. Zeba Spin Desalting columns, 7 K MWCO, 2 mL. Store at 4 °C.
35. Spectrophotometer (e.g., NanoDrop 1000 Thermo Scientific).
36. EMXnano EPR spectrometer (Bruker Biospin).

Software: Matlab (MathWorks) and EPR analysis software EasySpin [17] implemented in Matlab.

3 Methods

3.1 Molecular Cloning

Site-directed exchange of a single amino acid for cysteines is achieved by mutation of the respective nucleic acid code. Therefore, the aS gene of the template plasmid is modified by QuikChange mutagenesis site-directed mutagenesis.

Design primers as follows: Both oligonucleotides must contain the TGT mutation encoding for the cysteine, and anneal to the same sequence on opposite strands of the plasmid. Choose a design with a length between 25 and 45 bases and an estimated melting temperature T_m above 78 °C which can be predicted as follows:

$$T_m/^\circ C = 81.5 + 0.41(\%\text{GC}) - (675/N) - \%\text{mismatch},$$

where N is the primer length in bases, %GC the portion of guanosine and cytosine, and %mismatch the portion of nucleic acids to be exchanged.

Place the desired mutation into the center of the primer and allow for a 1 to 3 nucleic bases overhang at the primer ends. Choose a GC content higher than 40% and one or more G or C bases as primer terminus (*see* **Note 4**).

Table 1
Oligonucleotides that can be used as primers in QuikChange PCR

aS mutant	Forward primer sequence 5′ to 3′	Forward primer sequence 5′ to 3′
S9C	CAACTCCCTC CTTGGCCTTA CAAAGTCCTT TCATGAATAC ATCC	GTATTCATGA AAGGACTTTG TAAGGCCAAG GAGGGAGTTG TGG
A18C	CCCTGTTTGG TTTTCTCAGC ACAAGCCACA ACTCCCTCCT TGG	GGAGGGAGTT GTGGCTTGTG CTGAGAAAAC CAAACAGGGT GTGGC
A27C	GTCTTTCCTG CTGCTTCACA CACACCCTGT TTGGTTTTCT CAGC	GAGAAAACCA AACAGGGTGT GTGTGAAGCA GCAGGAAAGA CAAAAG
E35C	GCAGAAGCAG CAGGAAAGAC AAAATGTGGT GTTCTCTATG TAGGC	GGAGCCTACA TAGAGAACAC CACATTTTGT CTTTCCTGCT GCTTC
G41C	CATGCACCAC TCCCTCCTTG GTTTTGGAAC ATACATAGAG AACACCC	GTTCTCTATG TATGTTCCAA AACCAAGGAG GGAGTGGTGC ATGGTG
A56C	GGTGCATGGT GTGGCAACAG TGTGTGAGAA GACCAAAGAG C	CTCTTTGGTC TTCTCACACA CTGTTGCCAC ACCATGCACC AC
A69C	ACACTGCTGT CACACCCGTC ACCACACATC CTCCAACATT TGTCAC	CAAATGTTGG AGGATGTGTG GTGACGGGTG TGACAGCAGT GTCC
A90C	GGGAGCAGGG AGCATTGCAT GTGCCACTGG CTTTGTCAAA AAGG	CTTTTTGACA AAGCCAGTGG CACATGCAAT GCTCCCTGCT CCCTC
A140C	GGGAGCAAAG ATATTTCTTA ACATTCAGGT TCGTAGTCTT GATACCCTTC C	GGGTATCAAG ACTACGAACC TGAATGTTAA GAAATATCTT TGCTCCCAGT TTC

Oligonucleotides for a set of aS cysteine mutants are listed in Table 1.

Order oligonucleotides as 100 μM aqueous solutions at Sigma-Aldrich/Merck and store them at −20 °C. For usage, dilute primers 1:10 in ultrapure water.

1. Prepare solution for polymerase chain reaction (PCR) by mixing the components listed in Table 2 on ice.
2. Perform QuikChange PCR reaction in 25 μL total volume using the thermocycler and the procedure shown in Table 3.
3. Digest the methylated, nonmutated parental DNA template by incubating the reaction 1–6 h with 0.5 μL DpnI (*see* **Note** 7).
4. Amplify plasmids in *E. coli* cells. Therefore, transform plasmid DNA into chemically competent GH371 cells by mixing 4 μL DpnI digest solution with a 50 μL cell aliquot followed by 30 min incubation on ice. Heat shock the cells at 42 °C for 30 s in a pre-warmed water bath (e.g., MyBath™ Laboratory Water Baths), incubate again on ice for 2 min, and rescue in 1 mL pre-warmed SOC (37 °C). Incubate the cells at 37 °C

Table 2
Components of the PCR reaction to be mixed on ice

Component	Volume (μL)	Final concentration
Ultrapure water	17	
HF Phusion buffer (5×)	5	1×
Template plasmid (10–100 ng/mL)	1	10–100 ng
dNTPs (10 mM)	0.5	200 nM
Forward primer (10 μM)	0.5	200 nM
Reverse primer (10 μM)	0.5	200 nM
Phusion® High-fidelity DNA Polymerase (*see* **Note 5**)	0.5	1.25 U

Table 3
PCR program (*see* Note 6)

Cycle	Temperature (°C)	Time
Initial denaturation	95	30 s
Denaturation	95	30 s
Annealing	55	60 s
Elongation	68	8 min (16×)
Final elongation	68	7 min
Storage	4	∞

with 1400 rpm shaking for 1 h, put onto LB-agar plates containing 50 μg/mL carbenicillin, and incubate at 37 °C overnight. Pick a single colony and grow in 10 mL LB medium containing 50 μg/mL carbenicillin overnight. Isolate the plasmids using GeneJet Miniprep Kit following the manufacturer's protocol. Successful cloning is verified by sequencing, e.g., at GATC Biotech, using a T7 promoter forward primer (TAATACGACT CACTATAGGG) to reveal the sequence of the aS gene.

3.2 Gene Expression

1. For gene expression, transform BL21-Gold(DE3) competent cells with the purified plasmid via heat shock as described in **step 4** of Subheading 3.1.
2. Inoculate 10 mL LB medium supplemented with 50 mg/mL carbenicillin with a single colony from the LB-agar plate. Incubate at 37 °C, 180 rpm, overnight.
3. Dilute overnight culture in 1 L freshly prepared LB medium supplemented with 50 μg/mL carbenicillin and incubated at

37 °C, 180 rpm. Monitor OD_{600nm} photometrically (e.g., Eppendorf BioPhotometer® D30) (*see* **Note 8**).

4. As soon as OD_{600nm} reaches 0.5–0.7 induce the expression by adding 1 mM IPTG (238 mg) into the medium. Incubate the culture at 37 °C, 180 rpm, for 6 h and then harvest by centrifugation (3320 × *g*, 10 min). The supernatant is discarded and the pellets are stored at −20 °C.

3.3 Protein Purification

1. Resuspend the cell pellets in 50 mL lysis buffer. Cell lysis is carried out by sonication and followed by centrifugation at 10,000 × *g* for 30 min at 4 ° C. Transfer supernatant to a clean bottle.
2. Perform DNA precipitation. Therefore, add 1% (w/v) streptomycin sulfate (0.5 g), stir for 15 min at 4 °C, and remove DNA via centrifugation at 13,500 × *g* for 30 min at 4 ° C. Transfer supernatant to a clean bottle.
3. Perform ammonium sulfate precipitation. Therefore add 30% (w/v) ammonium sulfate (15 g) (*see* **Note 9**), incubate for 1 h at 4 ° C, and spin down at 13,500 × *g* for 30 min at 4 °C. Discard the supernatant and dissolve the pellet in Äkta buffer.
4. Load the solution onto a Resource Q anion-exchange column on an ÄKTAprime plus chromatography system, collect the fractions during elution with a salt gradient according to the manufacturer's manual, and assay for the presence of aS protein by SDS-PAGE (*see* **Note 1**) followed by Coomassie staining (*see* **Note 10**). Pool fractions containing aS, and transfer into storage buffer. Concentrate to approximately 250 μM using Amicon ultra 0.5 mL centrifugal filters MWCO 3 kDa (sigma-Aldrich, *see* **Note 11**) according to the manufacturer's manual.
5. Determine protein concentrations by measuring the absorbance of 1 μL at 275 nm using an extinction coefficient of 5600 M^{-1} cm^{-1} using a NanoDrop 1000 spectrophotometer (Thermo Scientific).
6. Store purified protein at −75 °C in 0.5 mL aliquots in low-binding tubes.

3.4 Spin Labeling

1. Prior to labeling, reduce aS mutant proteins with sixfold molar excess of DTT for 30 min at room temperature (*see* **Note 12**).
2. Remove DTT from protein samples with Zeba Spin Desalting Columns, 7 K MWCO, 2 mL (*see* **Note 13**). Prepare the desalting columns by putting them into Falcon 15 mL conical centrifuge tubes and by three washing steps: Remove the storage buffer by centrifugation at 1000 × *g* for 2 min. Load 1 mL Tris buffer onto the column and centrifuge again at 1000 × *g* for 2 min. Repeat two times (*see* **Note 14**). Exchange the Falcon 15 mL conical tube by a fresh one and apply the protein sample on top of the resin. Centrifuge at 1000 × *g* for 2 min.

3. Immediately add a sixfold molar excess of 3-maleimido-PROXYL (3-maleimido-2,2,5,5-tetramethyl-1-pyrrolidinyloxy, Sigma-Aldrich) (*see* **Note 15**) spin label to the protein and incubate the sample overnight in the dark at 4 °C (*see* **Note 16**).
4. Remove free label by using two additional desalting steps with Pierce Zeba 2 mL desalting columns (*see* **Note 17**). Prepare the columns as described in 2 (*see* **Note 18**). Replace the Falcon 15 mL conical tube by a fresh one and apply the protein sample on top of the resin. Centrifuge at 1000 × *g* for 2 min at 4 °C. Replace again the Falcon 15 mL conical tube by a fresh one and apply the protein sample on top of the resin. Centrifuge at 1000 × *g* for 2 min at 4 °C.
5. Optional control experiment: Determine the labeling degree. Prepare samples as described in Subheading 3.2, **step 2** and record the spectra as described in Subheading 3.4. Load the baseline-corrected spectrum and click the *Quantitative EPR* button. Click *Absolute Number of Spins*. Define the integration region using *Define Region* and the cursors popping up. The left and right cursors are adjusted to the left and right baseline of the spectrum. The middle cursors defined the region of integration. Click *Double Integration* and *Calculate.* A new window appears where you enter 1.05 mm as diameter, and 125 mm as Center (*see* **Note 19**) and the length of your sample (*see* **Note 20**) and the electron spin quantum number, which is ½ for nitroxide spin labels. After clicking the *OK* button a results report shows up. In the fourth column named *M* the molar spin concentration of the sample is listed.

Determine protein concentration by measuring the absorbance of 1 μL at 275 nm using an extinction coefficient of 5600 M^{-1} cm^{-1} using a NanoDrop 1000 spectrophotometer (Thermo Scientific), which should be around 200–250 μM.

Calculate the labeling efficiency by dividing the spin label concentration by the protein concentration (*see* **Note 21**).

3.5 Sample Preparation for EPR Measurements

1. Incubate singly labeled aS under specific conditions for the respective study, e.g., in the presence of lipid vesicles, organelles, or other interaction partners (*see* **Note 22**) [8, 9].
2. Load 20 μL singly labeled aS in solution and the incubated aS samples, respectively, into glass capillaries (ringcaps 50 μL, Hirschmann Laborgeräte) by using a pipette controller micro-classic (Brand). Seal the capillary with a sealing compound (Cha seal) at the bottom (*see* **Note 23**).
3. Place the capillary into a 3 mm EPR guiding tube (Bruker Biospin) (*see* **Note 24**).

3.6 Continuous-Wave (cw)-EPR Measurement

1. Find the correct position of the sample volume within the resonator by using the cavity template with the room-temperature adapter to adjust the sample holder onto the guiding tube (*see* **Note 25**).
2. Insert the adjusted sample holder into the resonator of the EPR spectrometer.
3. Click the *Lock Search* button to start the automatic tuning of the spectrometer.

   Following the status *Operate*, adjust the acquisition parameters as listed in Table 4.
4. Switch the AFC fine tuning button to *on* and wait till fine-tuning is finished.
5. Acquire the cw-EPR spectra by pressing the *Start* button.
6. Save the finished measurement.
7. For measuring further samples, repeat **steps 1** and **2**. Click the *Lock Search* button to start the automatic tuning of the spectrometer. Switch to *Operate*, and switch the microwave attenuation back to 20 dB and follow with **step 4**.

3.7 Data Postprocessing

Use the spectrometer Xenon software for data postprocessing. Load the spectrum. Click the *baseline correction* in the menu. Continue with *Polynomial* and choose *second order*. By clicking the *subtract* button the baseline-corrected spectrum is obtained and needs to be saved afterwards.

3.8 Data Analysis

Analyze the spectra using Matlab R2016b (The MatWorks, Inc. 3 Apple Hill Drive, Natick, MA 01760-2098, USA) and the

Table 4
Parameters for cw EPR measurement at the EMX Nano spectrometer

Center field (G)	3430.8
Sweep width (G)	200
Sweep time (s)	84.52
Modulation frequency (kHz)	100
Modulation amplitude (G)	0.6
Points/modulation amplitude	6
Conversion time (ns)	42.34
Time constant (ns)	40.92
Number of points	2000
Receiver gain (dB)	60
Attenuation (dB)	20

toolbox EasySpin 5.2.8 [17]. Perform least-squares fits with varying simulation parameters to the experimental data by use of EasySpin's fitting function esfit with the core function *chili*. Take for all simulations $A_{xx} = A_{yy} = 13$ MHz and $g = [g_x\, g_y\, g_z] = [2.00906\ 2.00687\ 2.003]$ and linewidth lw $= 0.13$ mT as fixed. Introduce an offset of the experimental microwave frequency as an additional fitting parameter to correct for systematic errors of the magnetic field measurement.

1. Simulate the spectra in solution (component S_A) (*see* **Note 26**) and the spectrum of aS labeled at position 140 upon incubation with the chosen system (*see* **Note 27**) by fitting the hyperfine interaction A_{zz} and the rotational correlation time τ_r.
2. Simulate the spectra recorded upon incubation of the protein with the chosen system (e.g., lipid vesicles, organelles, or interaction partners) with two-component fits where the spectrum $S = (1 - b)S_A + bS_B$ [7]. Take the fit parameters individually obtained for the single mutant in solution for component S_A and simulate the hyperfine interaction Azz, the rotational correlation time τ_r, and fraction b of the second spectral component S_B (*see* **Note 28**). Figure 3 shows spectra of aS upon interaction with negatively charged large unilamellar vesicles (LUVs) and the corresponding simulation using the two-component fit.
3. Vary the parameter b systematically and monitor rmsd to test in which range acceptable simulations of the data can be obtained to determine the error margins.
4. The parameter b reflects the local degree of interaction and can be compared for the different labeling positions [8]. Plot b against the labeled residues (*see* Fig. 4).

4 Notes

1. A detailed instruction on the performance of SDS-PAGE including gel casting, sample preparation, and electrophoresis performance is provided by Biorad [18].
2. The DTT solution needs to be prepared freshly each day.
3. The addition of concentrated HCl increases the temperature of the buffer solution and affects the pH. Allow the solution to cool down to room temperature to finally adjust the pH correctly.
4. For individual primer design, Agilent's online tool (www.agilent.com/genomics/qcpd) might be helpful.

5. It is important to add the polymerase last in order to prevent any primer degradation caused by the $3' \rightarrow 5'$ exonuclease activity.
6. Heat the lid of the PCR cycler to 110 °C if possible to prevent condensation in the lid of the tube.
7. The reaction can be performed in the thermocycler overnight but should be stopped after 6 by lowering the temperature to 8 °C.
8. According to our experience, BL21-Gold(DE3) doubling time is around 30 min.
9. Avoid high local ammonium concentration by stirring the solution while slowly adding the salt.
10. Best Coomassie staining results are achieved by incubating the gel with staining solution overnight followed by destaining.
11. Filter systems provided by other suppliers or dialysis might drastically decrease protein yield, since aS tends to adhere to many surfaces.
12. During the incubation time start with preparing the Zeba Spin Desalting columns needed for the next step.
13. Zeba Spin Desalting columns, 7 K MWCO, 2 mL, are suited for 200–700 μL protein solution. With larger protein volumes we advise to use columns with a volume of 5 mL. For low-volume samples (<350 μL), apply a stacker of buffer to the resin bed after the sample has fully absorbed.
14. In the instructions of the Zeba Spin Desalting columns it says to repeat the washing step two to three times. We found that repeating it twice is enough.
15. PROXYL features a higher stability of its thioether bond against a reducing environment compared to MTSL ((1-oxyl-2,2,5,5-tetramethylpyrroline-3-methyl)methanethiosulfonate) and is therefore better suited for experiments in, e.g., the cellular environment.
16. Alternatively, incubation times of 2 h at room temperature can be used with the risk of protein aggregation due to the higher incubation temperature.
17. Using two columns ensures a negligible amount of free label left in the protein solution.
18. To save time it is possible to do the washing steps for the two columns needed per sample simultaneously.
19. This is a standard value for the EMX Nano.
20. Measure the sample length by using the cavity template of the spectrometer.

21. Owing to the high reactivity of the label and the fact that the cysteine residues are freely accessible in the poorly folded structure, near-quantitative labeling can be achieved under these conditions.
22. For samples with lipid vesicles keep a molar protein-to-lipid ratio of 1:250 to prevent disruption of the vesicles and incubate for 30 min.

    For samples with organelles like mitochondria it is necessary to use higher protein concentrations (~1 mM) and short incubation times of 3 min with subsequent washing steps for achieving a recordable signal. Use Amicon Ultra centrifugal filters (Millipore) to achieve protein concentrations of about 1 mM.
23. Allow an air bubble between the seal and the sample solution. This is necessary to exclude artifacts due to the EPR signal of the seal compound.
24. At two positions the guiding tube was narrowed by a gaffer to fix the sample capillary into place.
25. For room-temperature operation the EMXnano requires the Protection Sleeve to be installed and the blueRoomTemp adapter must be inserted into the Cavity Template for adjustment of the right sample positioning in the cavity.
26. Component S_A represents the protein in solution with its spectrum in a fast motional regime.
27. The C-terminal end of aS shows commonly no interaction. Therefore, the spectrum of aS labeled at position140 can be described by only using the component S_A.
28. Component S_B represents the bound protein with its spectrum in a slow motional regime.

Acknowledgments

This work was supported by the DFG within SFB 969, project C03.

References

1. Drescher M, Veldhuis G, van Rooijen BD et al (2008) Antiparallel arrangement of the helices of vesicle-bound α-synuclein. J Am Chem Soc 130:7796–7797. https://doi.org/10.1021/ja801594s
2. Kumar P, Segers-Nolten IMJ, Schilderink N et al (2015) Parkinson's protein α-synuclein binds efficiently and with a novel conformation to two natural membrane mimics. PLoS One 10:e0142795. https://doi.org/10.1371/journal.pone.0142795
3. Robotta M, Braun P, van Rooijen B et al (2011) Direct evidence of coexisting horseshoe and extended helix conformations of membrane-bound alpha-synuclein. ChemPhysChem 12:267–269. https://doi.org/10.1002/cphc.201000815
4. Jao CC, Hegde BG, Chen J et al (2008) Structure of membrane-bound α-synuclein from site-directed spin labeling and computational refinement. Proc Natl Acad Sci 105:19666–19671. https://doi.org/10.1073/pnas.0807826105

5. Georgieva ER, Ramlall TF, Borbat PP et al (2008) Membrane-bound α-Synuclein forms an extended helix: long-distance pulsed ESR measurements using vesicles, bicelles, and rod-like micelles. J Am Chem Soc 130:12856–12857. https://doi.org/10.1021/ja804517m
6. Trexler AJ, Rhoades E (2009) α-Synuclein binds large unilamellar vesicles as an extended helix. Biochemistry (Mosc) 48:2304–2306. https://doi.org/10.1021/bi900114z
7. Robotta M, Hintze C, Schildknecht S et al (2012) Locally resolved membrane binding affinity of the N-terminus of α-synuclein. Biochemistry (Mosc) 51:3960–3962. https://doi.org/10.1021/bi300357a
8. Robotta M, Gerding HR, Vogel A et al (2014) Alpha-synuclein binds to the inner membrane of mitochondria in an α-helical conformation. Chembiochem 15:2499–2502. https://doi.org/10.1002/cbic.201402281
9. Robotta M, Cattani J, Martins JC et al (2017) Alpha-synuclein disease mutations are structurally defective and locally affect membrane binding. J Am Chem Soc 139:4254–4257. https://doi.org/10.1021/jacs.6b05335
10. Theillet F-X, Binolfi A, Bekei B et al (2016) Structural disorder of monomeric α-synuclein persists in mammalian cells. Nature 530:45–50
11. Cattani J, Subramaniam V, Drescher M (2017) Room-temperature in-cell EPR spectroscopy: alpha-synuclein disease variants remain intrinsically disordered in the cell. Phys Chem Chem Phys 19:18147. https://doi.org/10.1039/C7CP03432F
12. Li Q, Fung LW-M (2009) Structural and dynamic study of the tetramerization region of non-erythroid α-spectrin: a frayed Helix revealed by site-directed spin labeling electron paramagnetic resonance. Biochemistry (Mosc) 48:206–215. https://doi.org/10.1021/bi8013032
13. Watanabe Y, Inanami O, Horiuchi M et al (2006) Identification of pH-sensitive regions in the mouse prion by the cysteine-scanning spin-labeling ESR technique. Biochem Biophys Res Commun 350:549–556. https://doi.org/10.1016/j.bbrc.2006.09.082
14. Sahu ID, Craig AF, Dunagan MM et al (2015) Probing structural dynamics and topology of the KCNE1 membrane protein in lipid bilayers via site-directed spin labeling and electron paramagnetic resonance spectroscopy. Biochemistry (Mosc) 54:6402. https://doi.org/10.1021/acs.biochem.5b00505
15. Koteiche HA, Reeves MD, Mchaourab HS (2003) Structure of the substrate binding pocket of the multidrug transporter EmrE: site-directed spin labeling of transmembrane segment 1. Biochemistry (Mosc) 42:6099–6105. https://doi.org/10.1021/bi0342867
16. Isas JM, Langen R, Haigler HT, Hubbell WL (2002) Structure and dynamics of a helical hairpin and loop region in Annexin 12: a site-directed spin labeling study. Biochemistry (Mosc) 41:1464–1473. https://doi.org/10.1021/bi011856z
17. Stoll S, Schweiger A (2006) EasySpin, a comprehensive software package for spectral simulation and analysis in EPR. J Magn Reson 178:42–55. https://doi.org/10.1016/j.jmr.2005.08.013
18. http://www.bio-rad.com/webroot/web/pdf/lsr/literature/Bulletin_6040.pdf

Chapter 20

Expression and Purification of Untagged α-Synuclein

Alex E. Powers and Dushyant S. Patel

Abstract

α-Synuclein (αS) is an abundant neuronal protein which has been implicated, among others, in the pathogenesis of neurodegenerative diseases like Parkinson's disease (PD) and dementia with Lewy bodies (DLB). In fact, αS is the major constituent of Lewy bodies, the primarily proteinaceous inclusions found in the nervous tissue of PD and DLB patients. While its physiological role is unclear, it is believed to be involved in the regulation of synaptic vesicle exocytosis. However, in a disease state, αS will "misfold" and aggregate, leading to neuronal dysfunction and death. The study of the molecular events underlying pathogenesis, especially with biophysical and biochemical approaches, requires highly pure untagged αS. In this protocol we describe a method to purify untagged recombinant αS, which can be used for binding, folding, and aggregation studies. The purification method includes a cell lysis step, followed by two chromatography steps: ion-exchange chromatography first, and size-exclusion chromatography for polishing.

Key words α-Synuclein, Parkinson's disease, Dementia with Lewy bodies, Ion-exchange chromatography, Size-exclusion chromatography, Protein expression, Protein purification

1 Introduction

Biophysical studies of changes in the physical chemistry of α-synuclein (αS) and many other biochemical assays require the expression of highly pure, untagged recombinant αS.

The αS purification protocol described here (adapted from [1]) is composed of three steps: (1) cell lysis, (2) ion-exchange chromatography, and (3) size-exclusion chromatography (Fig. 1). The vector used as the expression system, pET21a, contains the human αS gene (SNCA) under the control of a lac-repressed T7 promoter [2] (Fig. 2). The T7 polymerase is expressed constitutively (through a lacUV5 promoter) in BL21(DE3) *E. coli*, but the lacI repressor, binding to both the T7-lac and the lacUV5 promoters in the absence of inducer, maintains a low basal expression level of αS. Upon induction with isopropyl β-D-1-thiogalactopyranoside

Tim Bartels (ed.), *Alpha-Synuclein: Methods and Protocols*, Methods in Molecular Biology, vol. 1948,
https://doi.org/10.1007/978-1-4939-9124-2_20,

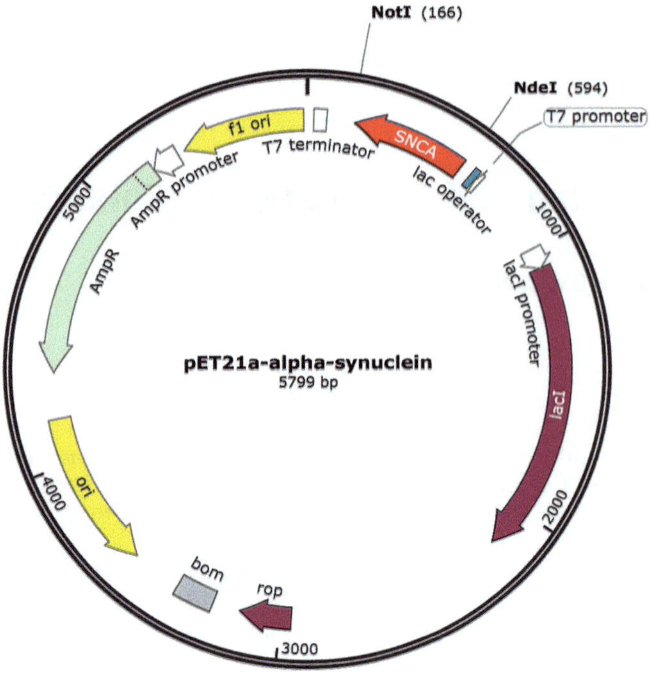

Fig. 1 Plasmid map of pET21-a-alpha-synuclein. The α-synuclein gene is under the control of a LacUV5-T7 promoter

(IPTG), the bacteria express high levels of αS. It is important to note that highly expressed αS is not toxic to *E. coli* because it is soluble in the bacterial cytosol up to very high concentrations and it is also likely translocated to the periplasmic space [3, 4]. After αS expression, cell lysis is achieved by boiling. Since αS has no stable secondary structure in *E. coli* it is unaffected by high temperatures, while proteases and many other protein contaminants will be denatured and precipitated out of solution (Fig. 1). Following cell lysis, ion-exchange chromatography (IEX) is performed. IEX columns contain a positively charged Sepharose resin [5]. When run on the column, αS and other proteins from the boiled lysate will bind to the resin depending on their charge and will then be eluted by using a high-salt buffer gradient, which will displace them and release them sequentially (Fig. 3). IEX is followed by a size-exclusion chromatography (SEC) "polishing" step. SEC, also known as gel-filtration chromatography, separates molecules based on their sizes. The SEC resin is composed of porous beads and smaller molecules will readily diffuse through these pores. Larger molecules, on the other hand, will simply pass through the column without entering the beads' pores, thus shortening their path through the resin. This results in larger molecules eluting from the column earlier than smaller molecules. After SEC, >95% αS purity is usually achieved (Fig. 4).

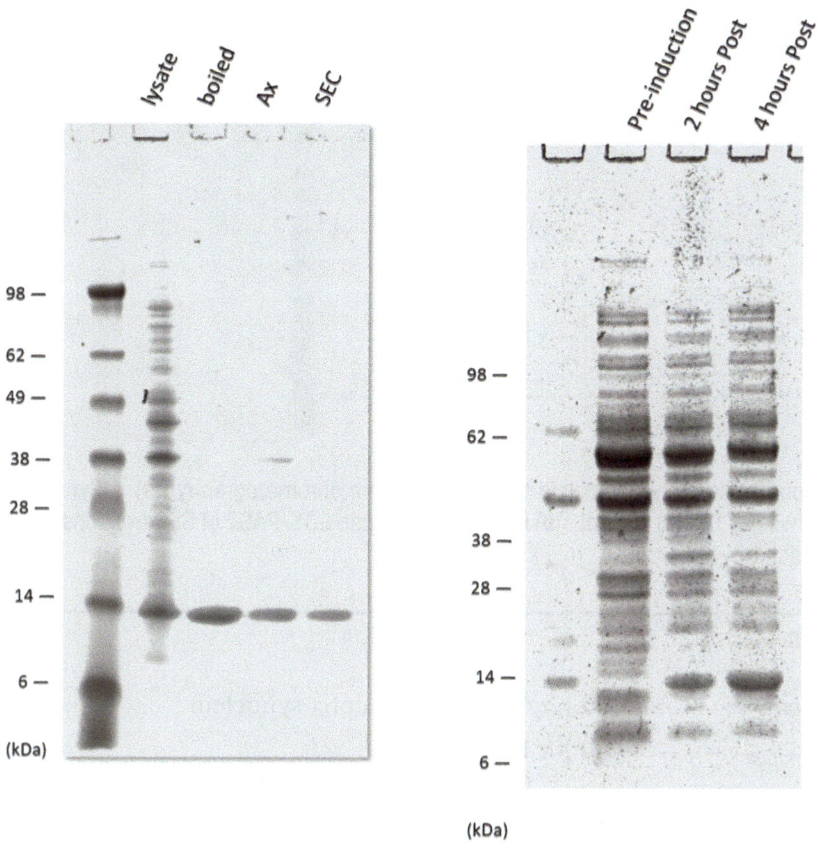

Fig. 2 Overview of the purification stages and induction times. (**a**) Lane 1 contains the lysate after freezing overnight. Lane 2 contains the lysate after the boiling step. Lane 3 contains the boiled lysate after purification via IEX. Lane 4 contains the pooled fractions for IEX after "polishing" with SEC. (**b**) Lane 1 contains the lysate pre-induction. Lane 2 contains the lysate 2 h post-induction. Lane 3 contains the lysate 4 h post-induction

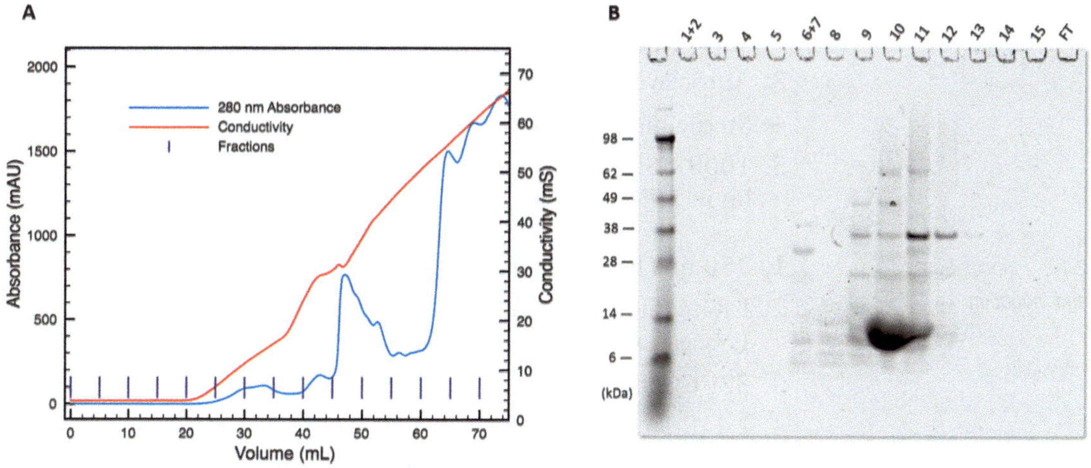

Fig. 3 Purification of α-synuclein by ion-exchange chromatography. (**a**) α-Synuclein elutes off the IEX column around 25–35 mS. (**b**) Coomassie-stained SDS-PAGE of IEX fractions show greatest αS concentration in fractions 9 + 10

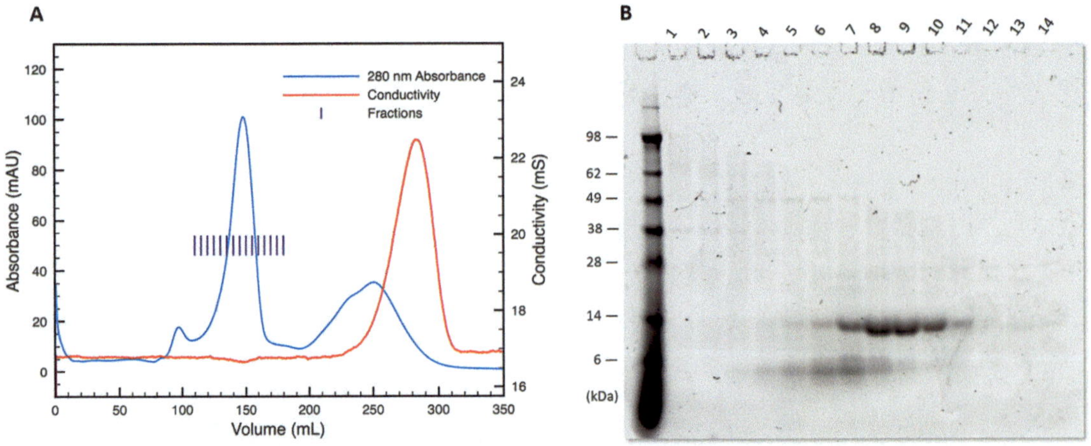

Fig. 4 Further purification of α-synuclein via size-exclusion chromatography. (**a**) α-Synuclein elutes off the SEC column between 110 and 180 mL. (**b**) Coomassie-stained SDS-PAGE of SEC fractions show highest purity (>95%) in fractions 7–10

2 Materials

2.1 Expression

2.1.1 Transformation Reagents

1. 50 ng/μL pET21a-alpha-synuclein.
2. 50 μL BL21(DE3) *E. coli*.
3. 1 mL SOC medium: 20 g/L Tryptone, 5 g/L yeast extract, 4.8 g/L $MgSO_4$, 3.6 g/L dextrose, 0.5 g/L NaCl, 0.186 g/L KCl.
4. Incubator.
5. Water bath.

2.1.2 Selection Plates

1. 37 g of mixed LB/agar powder (5 g peptone, 10 g peptone from casein, 10 g NaCl, 12 g agar).
2. 1 L of Milli-Q water.
3. Autoclave.
4. Sterile dishes (100 × 15 mm).
5. 1000× Stock of antibiotic (in our case 100 mg/mL of ampicillin in 50% ethanol).

2.1.3 Inoculation and Induction

1. 250 mL Erlenmeyer flask.
2. LB medium (10 g/L casein digest peptone, 10 g/L NaCl, 5 g/L yeast extract, 1.5 g/L Tris–HCl).
3. 37 °C incubator.
4. 2 L Erlenmeyer flask.
5. 1 mM Isopropyl β-D-1-thiogalactopyranoside.
6. 1000× Stock of antibiotic.

2.2 Purification

2.2.1 Cell Lysis

1. Centrifuge capable of handling 1 L total volume at 9000 × *g*.
2. 2× 250 mL centrifuge tubes.
3. Hot plate.
4. 2 L Erlenmeyer flask.

2.2.2 Ion-Exchange Chromatography (IEX)

1. Ion-exchange Buffer A: 20 mM Tris-base HCl, 25 mM NaCl, 1 mM EDTA, pH 8.00.
2. Ion-exchange Buffer B: 20 mM Tris, 1 M NaCl, 1 mM EDTA, pH 8.00.
3. 5 mL HiTrap Q HP columns (GE Life Sciences, 17516301).
4. ÄKTAprime plus FPLC system.

2.2.3 Size-Exclusion (SEC)

1. 50 mM ammonium acetate, pH 7.40.
2. 13 mL HiPrep 26/60 Sephacryl S-200 HR (GE Life Sciences, 17119501).
3. ÄKTAprime plus FPLC system.

3 Methods

3.1 Expression

1. Thaw 20–50 μL of BL21(DE3) competent *E. coli* on ice for ~10 min or until melted (*see* **Note 1**).
2. Once the cells are completely thawed, add 1–5 μL of 10 pg–100 ng of pET21a-alpha-synuclein (*see* **Note 2**) and gently mix by inverting the tube.
3. Following the addition of the plasmid, incubate the cell and plasmid mixture on ice for about 30 min (*see* **Note 3**).
4. Then, heat-shock the cell and plasmid mixture for exactly 10 s in a 42 °C water bath.
5. Remove the mixture from the water bath, place it on ice for 5 min, and then add 950–980 μL of SOC medium to the tube (*see* **Note 4**).
6. Incubate the bacteria at 37 °C for 60 min while shaking at ~250 rpm. After incubation, repeatedly invert the tube to mix the culture well and then plate 100 μL onto selection plates (*see* Subheading 2.1). Incubate the plate(s) upside-down overnight at 37 °C (*see* **Notes 5** and **6**).
7. On the following day, choose a single colony to inoculate in ~180 mL of LB medium at 37 °C overnight (pre-culture).
8. The next day, add a 1:20 dilution of the pre-culture to LB medium and grow at 37 °C until OD_{600} = 0.5–0.6. Once an OD_{600} = 0.5–0.6 is reached, induce the expression of αS with

1 mM IPTG and let the culture grow at 37 °C for ~4 h (*see* **Note 7**).

9. Then, spin the culture at 9000 × *g* for 20 min (RT). After the spin, remove the supernatant and freeze the pellet overnight at −20 °C. Freezing the pellet in this fashion already lyses most of the bacteria.

3.2 Purification

This protocol is written for an ÄKTAprime plus FPLC system equipped with a 280 nm UV and conductivity detectors. Although other FPLC systems will largely follow the same protocol, it is advised to always refer to the manufacturer's specifications.

3.2.1 Cell Lysis

1. Preheat a hot plate and a 2 L Erlenmeyer flask to the minimum temperature to initiate boiling. While the hot plate and flask are reaching the right temperature (enough to bring the culture to a boil but not too high that it will cause charring), submerge the frozen pellet in IEX buffer A (20 mM Tris, 25 mM NaCl, 1 mM EDTA, pH 8.00) and vortex until the pellet is completely resuspended, taking extra care not to leave any pellet in solution (*see* **Note 8**).
2. Pour the resuspended cell suspension into the preheated 2 L Erlenmeyer flask. Increase the hot plate temperature and let the mixture rise to an even boil (*see* **Note 9**).
3. After the first visible signs of boiling, allow the mixture to boil for an additional 15 min to denature and precipitate proteases and other protein contaminants.
4. After boiling, let the lysate cool down to 4 °C (*see* **Note 10**), transfer it to 250 mL centrifuge bottles, and spin it at 20,000 *g* for 45 min (4 °C). Filtrate the supernatant through a 0.22 μm or 0.45 μm filter. Bring the supernatant up to approximately 300 mL with IEX buffer A.

3.2.2 Ion-Exchange Chromatography (IEX)

1. Wash the FPLC system with IEX buffer A at 5 mL/min.
2. Connect the columns (2× 5 mL HiTrap Q HP columns) in the presence of a flow rate of 0.3 mL/min to ensure that no air enters the column (*see* **Note 11**). After the columns are attached, equilibrate them with IEX buffer A at 1 mL/min for 40 min or until the conductivity reaches a steady measurement for ~20 min.
3. Following equilibration, load the cell lysate mixture onto the columns overnight at a max flow rate of 1 mL/min.
4. Wash the column with IEX buffer A until a steady (around 40 mL) plateau in the 280 nm absorbance is achieved, in order to remove any weakly bound contaminants (*see* **Note 12**).

5. After the UV absorbance plateaus, elute the protein with a gradient, from 0% to 100%, of IEX buffer B (20 mM Tris, 1 M NaCl, 1 mM EDTA, pH 8.00) over 75 mL and collect fractions every 5 mL.
6. Analyze all the fractions and the flow-through (the unbound lysate fraction which passes through the column and is collected in the FPLC waste during the lysate loading step) via Coomassie-stained SDS-PAGE or Western blot (αS usually elutes between 25 and 35 mS) to determine the fractions with the highest concentration of αS (*see* **Note 13**).
7. The fractions (usually around a conductivity of ~25 mSi when adopting this protocol and purification system) containing the greatest amount of αS can be pooled and (eventually) concentrated down to 10 mL, which will then be loaded on a size-exclusion column.

3.2.3 Size-Exclusion Chromatography (SEC)

1. Wash the FPLC system with 50 mM ammonium acetate, pH 7.40 (5 mL/min). Attach the SEC column (HiPrep 26/60 Sephacryl S-200 HR) in the presence of a buffer flow rate of 0.2 mL/min, to ensure that no air enters the column.
2. Attach a 10 mL loop to the six-way valve of the FPLC system and flush it with 20 mL of 50 mM ammonium acetate to remove any aggregated protein or built-up waste.
3. Inject the pooled IEX fractions in the loop, making sure not to inject air bubbles into the system or column. The valve position must be set to "Load" when injecting the sample onto the loop (*see* **Note 14**).
4. At the beginning of the run, switch the valve position to "Inject." At this point, the pooled IEX fractions will be injected onto the SEC column. After 25 mL of 50 mM ammonium acetate is run through the loop, switch the valve position back to "Load."
5. Flush a total of 350 mL of ammonium acetate buffer through the FPLC, collecting 14× 5 mL fractions between 110 and 180 mL corresponding to a molecular weight of 60 kDa.
6. Analyze the fraction purity via Coomassie-stained SDS-PAGE and pool 5–6 of the purest fractions. Measure the absorbance of these fractions at 280 nm ($\varepsilon(\alpha S) = 0.412$ mL mg^{-1} cm^{-1}). Usually purity of >95% can be obtained.
7. Aliquot the pooled fractions into 1 mg aliquots, then flash-freeze them using liquid nitrogen or a dry ice/ethanol bath, and lyophilize (or store frozen at −80 °C) the samples.
8. After lyophilization, seal the tubes with Parafilm, to prevent moisture from entering, and store at 4 °C for short-term storage (1–2 weeks) or −20 °C for long-term storage.

4 Notes

1. If more than 50 μL of cells are to be thawed, the remaining aliquot can be frozen using a 100% ethanol and dry ice bath.
2. Our lab has had consistent success using ~50 ng of pET21a-alpha-synuclein DNA for transformation.
3. This time can be reduced to a minimum of 2 min, but the transformation efficiency significantly decreases with each halving of the time.
4. SOC medium is ideal; however LB medium can also be used but will result in a roughly twofold loss of transformation efficiency.
5. When first plating the transformants, it is recommended that different volumes of the culture (e.g., 50, 100, 200 μL) are plated to ensure that at least 5–6 single colonies grow on each plate, but not so many that picking one becomes challenging.
6. If the plasmid contains resistance for ampicillin, the outgrowth step is not necessary [6]. However, if different resistance genes are present in the expression vector, it is necessary to perform this step to allow for cell recovery and expression of the antibiotic resistance gene(s).
7. The amount of IPTG to be used for the induction (typically 0.1–1 mM) depends on the cells, construct, and culture conditions. It is recommended to perform a quick optimization, when setting up the expression protocol, in order to determine the optimal amount of IPTG.
8. If cells are spun in a 250 mL centrifuge tube, approximately 60 mL of IEX buffer A should be enough to fully submerge the pellet.
9. If the pellet is not fully resuspended after generous vortexing, a 5 mL pipette can be used to break up the remaining clumps.
10. It is highly recommended to incubate the lysate mixture at 4 °C after cooling on the benchtop to avoid damaging centrifuge tubes upon high-speed centrifugation.
11. If columns are stored in 20% ethanol it is recommended to wash the columns with Milli-Q before equilibrating the column with buffer. This will ensure that salts will not precipitate in the column.
12. The wash step can also be performed using low percentages of IEX buffer B (e.g., 5–10%), in order to have more stringent wash conditions. However, this can also wash off small amounts of αS and should thus be tested by checking the waste of the washing step by Coomassie-stained SDS-PAGE

or WB. In total, the volume of IEX buffer A used during the wash step should be between 150 and 200 mL.

13. For a lab-scale expression (typically 1 or 2 L expression), 2× 5 mL HiTrap columns should be enough to allow for complete αS binding. If a larger expression is required, it is recommended to increase the number of HiTrap columns in the system or switch to a column(s) with a larger bed volume(s). Always analyze the flow-through fraction to ensure complete binding of αS to the columns. If αS is detected in the flow-through fraction, increase the number of columns in the system to increase the overall binding capacity.
14. Injecting small amounts of air into the columns is unavoidable (especially when working with a 10 mL sample loop). Air bubbles, when eluting, will give off characteristic spikes in the UV absorbance trace [7]. On the other hand, if large amounts of air are injected onto the column (i.e., resin discoloration from the presence of air is visible in the bed), generously flush the system and column with buffer. This applies to both IEX columns and SEC columns, but SEC columns are much more sensitive to the presence of air in their resin because of their resin packing being critical for resolution (and cracks or lacunae can disrupt it completely and require column re-packing).

References

1. Giasson BI, Uryu K, Trojanowski JQ, Lee VMY (1999) Mutant and wild type human - α-synucleins assemble into elongated filaments with distinct morphologies in vitro. J Biol Chem 274:7619–7622. https://doi.org/10.1074/jbc.274.12.7619
2. Mierendorf RC, Novy RE, Morris BB, Hammer B (1998) Expression and purification of recombinant proteins using the pET system. Methods Mol Med 13:257–292
3. Ren G, Wang X, Hao S et al (2007) Translocation of alpha-synuclein expressed in Escherichia coli. J Bacteriol 189:2777–2786. https://doi.org/10.1128/JB.01406-06
4. Volles MJ, Lansbury PT (2007) Relationships between the sequence of α-synuclein and its membrane affinity, fibrillization propensity, and yeast toxicity. J Mol Biol 366:1510–1522. https://doi.org/10.1016/j.jmb.2006.12.044
5. GE (2016) Size exclusion chromatography columns and media—selection guide. https://doi.org/10.1111/j.1467-8330.1974.tb00606.x
6. Pope B, Kent HM (1996) High efficiency 5 min transformation of Escherichia coli. Nucleic Acids Res 24:536–537. https://doi.org/10.1093/nar/24.3.536
7. GE Healthcare Bio-Sciences (2009) ÄKTAprime™ plus Operating Instructions. p 59

Chapter 21

Generation and Analysis of Viral Vector-Mediated Rodent Models for Parkinson's Disease

Akua A. Karikari, James B. Koprich, and Chi Wang Ip

Abstract

The use of viral vector-mediated α-synuclein-overexpressing rodent models for Parkinson's disease has become increasingly accepted to study associated pathology and to use as a platform for therapeutic efficacy testing. We here describe methods to generate such models and how to analyze them by behavioral assessments, histological investigations, and examination of nigrostriatal dopaminergic function.

Key words Parkinson's disease, Rat model, Mouse model, Dopamine, Substantia nigra, Neurodegeneration

1 Introduction

In Parkinson's disease (PD) research, the selection of an appropriate animal model is critical to testing your hypothesis. PD is an α-synucleinopathy, and thus the animal model of choice, in particular for efficacy testing, should have at its core, pathology related to mechanisms that involve α-synuclein (aSyn) [1]. The major advantage of PD models with a viral vector-mediated overexpression of aSyn is the pathophysiologic resemblance to the human PD condition with loss of dopaminergic neurons in the substantia nigra (SN), aSyn accumulation, and development of Lewy bodies and neurites in the dopamine neurons that remain [2]. Additionally, by implementing a unilateral viral vector injection, we can mimic the predominant occurrence of unilateral symptoms in PD patients, therefore allowing us to apply more sophisticated behavioral analyses like the cylinder test. We here present methods on how to generate a PD mouse and rat model by unilateral viral vector-mediated aSyn overexpression. Moreover, we give a step-by-step description on how to perform behavioral analyses, histological assessments of neurodegeneration and Lewy body aggregation, as well as analysis of the dopaminergic metabolism in the PD models.

Tim Bartels (ed.), *Alpha-Synuclein: Methods and Protocols*, Methods in Molecular Biology, vol. 1948,
https://doi.org/10.1007/978-1-4939-9124-2_21, © Springer Science+Business Media, LLC, part of Springer Nature 2019

2 Materials

2.1 General Materials

1. RotaRod Advanced (TSE systems, Bad Homburg, Germany).
2. Transparent plexiglass cylinder of 12 cm (mice) or 20 cm (rats) diameter and 30/40 cm height.
3. Two mirrors.
4. JVC GC-Px 100 video camera (JVC Kenwood Inc., Yokohama, Japan).
5. Benchtop shaker.
6. NIH ImageJ software.

2.2 Materials for Stereotaxic Injection

Unless stated otherwise, the equipment was all purchased from Stoelting (Wood Dale, IL, USA).

1. Mouse/rat stereotaxic frame.
2. Mouse/rat anesthesia stereotaxic mask apparatus.
3. Motorized microinjection pump.
4. Isoflurane vaporizer (Dräger AG., Lübeck, Germany).
5. Induction box (custom made).
6. Drill with handpiece (Foredom Co., Bethel, CT, USA).
7. Ocular lubricant.
8. 70% ethanol S.
9. Scalpel #10 (Feather Co., Osaka, Japan).
10. Tissue forceps.
11. Gauze.
12. Cotton swabs.
13. Suture materials (V97D9) (Ethicon Inc., Somerville, NJ, USA).
14. 75N 5 μL microinjection syringe (Hamilton Co., Reno, NV, USA).
15. Oxygen tank.
16. Electric razor.
17. Human AAV1/2-A53T- aSyn or empty AAV1/2- viral vector (EV).
18. C57BL/6 wild-type mice.
19. Pain medication.
20. Isoflurane.

2.3 Materials for Tissue Processing and Antibodies

1. Sodium pentobarbital.
2. Bone scissors.

3. Heparin.
4. 25 G needle.
5. Tissue pincers.
6. Brain matrix for mice and rats.
7. Aluminum foil.
8. Dry ice.
9. Optimal cutting temperature (OCT) cryomold.
10. Methylbutane.
11. Object slides.
12. Coverslips.
13. Aquatex (aqueous mounting media).
14. Distilled water.
15. Leica CM 1950 cryostat (Leica Co., Wetzlar, Germany).

2.3.1 Solution Composition and Preparation

1. 10× (0.1 M) phosphate-buffered saline (PBS): 80 g NaCl, 14.2 g $NaHPO_4 \times 2H_2O$, and 2 g KCl dissolved in distilled water. Bring pH to 6.8 with HCl. Dilute to 1× (0.01 M) PBS with distilled water and adjust pH to 7.4.
2. PBS with heparin solution: To 150 mL of 1× PBS, add 0.5 mL of heparin for perfusion.
3. Antifreeze cryoprotectant: 30% glycerol, 30% ethanol, 40% 0.01 M PBS.
4. 4% paraformaldehyde (PFA): Dissolve 10 g PFA in 250 mL 0.01 M PBS. Add two drops of 5 M NaOH and keep stirring solution at 60 °C. When solution is clear, cool on ice and bring pH to 7.4 with HCl. Filter the solution.
5. 30% sucrose solution: Dissolve 30 g of sucrose in 100 mL 1× PBS.
6. Blocking solution: 10% Normal goat serum (NGS), 2% bovine serum albumin (BSA), 0.5% Triton X (Tx) in 0.01 M PBS.
7. Antibody diluent: 2% NGS, 2% BSA, 0.5% Tx in 0.01 M PBS.
8. Proteinase K: Prepared as 1:100 in dH_2O (Enzo Life Sciences, Farmingdale, NY).
9. Rabbit anti-mouse TH (1:1000, Abcam cat # 112, Cambridge, UK).
10. Biotinylated goat anti-rabbit (1:100, Vector Laboratories cat # BA-1000, Burlingame, CA, USA).
11. Mouse anti-NeuN (1:500, Merck Millipore cat # MAB377, CA, USA).
12. Biotinylated goat anti-mouse (1:100, Vector Laboratories cat # BA-9200, Burlingame, CA, USA).

13. Ultra-sensitive avidin-biotin complex (ABC) peroxidase standard staining kit (Thermo Fisher Scientific cat # 32050, Waltham, MA, USA).
14. 3,3-Diaminobenzidine-tetrahydrochloride (DAB) and peroxidase kit (Vector Laboratories cat # SK-4100, Burlingame, CA, USA).
15. Chicken anti-mouse TH (1:500, Abcam cat # 76442, Cambridge, UK).
16. Rabbit anti-human aSyn (1:30,000, Sigma-Aldrich cat # S3062, St. Louis, MO, USA).
17. Goat anti-rabbit Cy3 (1:300, Jackson ImmunoResearch Inc. cat # 111-165-003, West Grove, PA, USA).
18. Goat anti-chicken Alexa Fluor 488 (1:300, Invitrogen cat # A-11039, Carlsbad, CA, USA).
19. 4′,6-Diamidino-2-phenylindole dihydrochloride (DAPI) nuclear stain (1:500, 000, Sigma-Aldrich cat # D8417, MO, USA).
20. Mouse anti-human aSyn primary antibody (Thermo Fisher Scientific, cat # 180215 (LB509; 1:500), Waltham, MA, USA).
21. Mouse anti-α- phosphorylated at serine 129 (Ser129, 81A; 1:8000) (BioLegend, cat # 825701, San Diego, CA, USA).

2.4 Materials for Stereological Analysis

1. Stereo Investigator software package version 11.07 (MicroBrightField Biosciences, Williston, VT, USA).
2. BX53 microscope (Olympus Co., Tokyo, Japan).

2.5 Dopamine Analysis

1. Binding buffer: 50 mM Tris, 120 mM NaCl, and 5 mM KCl at RT.
2. [^{125}I]-RTI-121 (specific activity 2200 Ci/mmol).
3. GBR 12909 dihydrochloride.
4. [^{125}I]- microscale standards.
5. [^{14}C] microscale standards.
6. HPLC binding buffer: 0.1 M Trichloroacetic acid, 10^{-2} M sodium acetate, 10^{-4} M EDTA, 10.5% methanol.
7. HPLC with autosampler (Waters 717 HPLC with 515 pump).
8. Phenomenex Nucleosil (5u, 100A) C18 HPLC column (150 × 4.60 mm).
9. Elution buffer HPLC: 89.5% 0.1 M trichloroacetic acid, 10^{-2} M sodium acetate, 10^{-4} M EDTA, 10.5% methanol.
10. Waters Empower Software.
11. Bicinchoninic acid assay kit (Pierce).

3 Methods

3.1 Stereotaxic Injection

1. Switch on oxygen tank to 2 L/min and set the isoflurane vaporizer to 3–4% (*see* **Note 1**).
2. Place the rodent in an isoflurane induction box (close after opening) until it is fully anesthetized.
3. The rodent will not respond to nociceptive stimuli, such as pinching the toes when deeply anesthetized.
4. Next, remove the hair from the scalp using an electric razor.
5. Set isoflurane vaporizer between 1.5% and 2% mixed with 2 L/min of O_2. Keep monitoring mouse throughout the procedure and readjust the vaporizer when necessary.
6. Place rodent on the stereotactic frame, secure teeth on the mouthpiece of the mask, and tighten the nose piece.
7. Fasten the head with ear bars such that it is not tilted and there is no movement horizontally.
8. Apply an ocular lubricant to prevent drying of the eyes.
9. Disinfect the shaved skin with 70% alcohol.
10. With a scalpel, make an incision in the rostro-caudal direction (enough to expose both Bregma and Lambda), midline of the shaved scalp.
11. Gently push away the periosteal connective tissue that adheres to the bone with cotton swabs exposing clearly the cranial sutures of Bregma and Lambda.
12. Load the microinjection syringe with 2.1 μL of the AAV1/2-A53T- aSyn or empty AAV1/2- viral vector (EV) ensuring that there are no bubbles.
13. Attach the microsyringe to the injection apparatus.
14. By moving the microsyringe from Bregma to Lambda and from the left to right side of the sagittal suture, ensure that the head of the mouse is leveled in the rostro-caudal and horizontal directions. In this case, the needle should be close enough to the skull but not touching it to avoid bending of the tip.
15. The nose and ear bars can always be readjusted if needed, to ensure that the head is level.
16. When the head is correctly positioned, proceed to identify the coordinates on the skull surface. For male C57BL/6 wild-type mice, at the age of 12 weeks, the coordinates for right-sided, unilateral injection are from Bregma: AP—3.1 mm; ML—1.4 mm; and DV—4.2 mm according to Ip et al. [2]. For male rats with the weight of 280 g the coordinates for a right-sided, unilateral injection are from Bregma: AP—

5.2 mm; ML—2.0 mm; and DV—7.5 mm according to Koprich et al. [3] (*see* **Note 2**).

17. With the handheld drill, carefully drill hole in the skull at the AP-ML coordinate identified.
18. To avoid damage to brain parenchyma, be careful when applying pressure during drilling.
19. Stop drilling immediately when breaking through the skull and remove any bone residue and dura that may interfere with the insertion of needle.
20. Insert needle into hole and with the microinjector controller to ensure that the fluid in the microsyringe flows to the tip of the needle. This is to be certain that the fluid will flow from the syringe without any hindrance in the tissue.
21. Insert needle till the tip touches the tissue and move the syringe in the DV direction, moving slowly to the DV target point.
22. The injection of viral vector takes 2 min, at a rate of 0.5 μL/min, controlled by the motorized microinjection pump.
23. After injection, leave the needle in place for 5 min; afterwards gently and slowly remove the needle from the brain (*see* **Note 3**).
24. Close the skin incision with suture material.
25. Apply pain medication to the rodent.

3.2 Rotarod Performance Test

1. For rotarod performance test, rodents are trained for 2 days prior to their measurements.
2. Rodents are placed on a rotarod with an accelerating speed from 5 rpm and maximum of 50 rpm for a duration of 300 s (*see* **Note 4**).
3. During the 2-day training session, animals are placed back on the rod when they fall too quickly.
4. Each day of training consists of five runs of 300 s each.
5. The measurements are recorded on the third day.
6. All measurements after the 2-day training consist of five runs lasting 300 s each.
7. Rodent falls and latency to falls are automatically recorded by light beam sensors in the rotarod system.
8. For repetitive data acquisition, data is represented relative to the preoperative measurement.

3.3 Cylinder Test

1. Rodents are placed into a transparent plexiglass cylinder for mice or rats in front of two mirrors.
2. The mirrors are placed at an angle such that the entire diameter of the cylinder can be viewed in the video recording.

3. When animal is placed in the cylinder, video record a card labeled with mouse number, date, etc. for about 2 s.
4. Record animal movement for 12 min (*see* **Note 5**).
5. The videos are subsequently scored by an observer for the first 10 min.
6. For each rearing, the first limb of the mouse whether the right (ipsilateral) or left (contralateral) or both limbs to touch the inner surface of the cylinder is scored as a score for that limb or for both limbs.
7. For another limb use score, the mouse should completely remove both limbs from the wall. If mouse uses both paws to explore the walls of the cylinder laterally, the movement is scored as a use of both paws.
8. The data is presented as a percentage of the ipsilateral forepaw (corresponding to the site of injection) use and calculated as follows: [(ipsilateral paw + 0.5 both paws)/(ipsilateral paw + contralateral paw + both paws)] × 100.
9. The observed forepaw use post-AAV1/2 injection is represented in relation to the preoperative score.
10. Thus, calculated percentage shows the preference of the forepaw use as follows: 100% = symmetric use of both forepaws; <100% = preference of the left forepaw; and >100% preference of the right forepaw.

3.4 Tissue Processing: Brain Dissection

1. Sedate animals with sodium pentobarbital or isoflurane (or as has been approved) followed by sacrifice by way of exsanguination.
2. Perfuse the animal transcardially with 15 mL/100 mL of 0.01 M PBS with heparin solution using a 25 G needle.
3. Carefully remove brain from the skull with bone scissors and tissue pincers.
4. Put brain in a matrix with the ventral surface (region where optic chiasm is seen) upward and leveled.
5. Dissect brain in the coronal plane at +0.14 mm from Bregma (mice) and −0.12 mm from Bregma (rats).

3.5 Tissue Processing for Unbiased Stereology of SN Neuron Number

1. Put the caudal part of the dissected brain, which includes the SN, in 4% PFA (*see* **Note 6**) for 2 days at 4 °C.
2. Next, transfer brain into 30% sucrose solution for another 2 days (mouse brain) or 4 days (rat brain) at 4 °C.
3. After incubation with sucrose solution, seat the brain into a cryomold filled with optimal cutting temperature (OCT) compound and slowly freeze the tissue in liquid dry ice-cooled methylbutane (*see* **Note 7**).

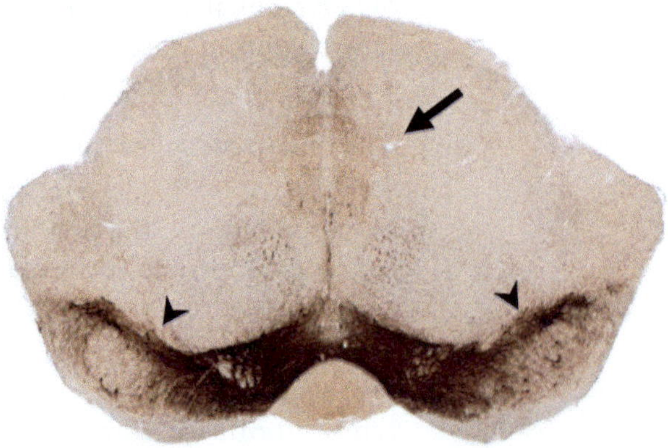

Fig. 1 Immunohistochemical staining for TH-positive dopaminergic neurons of a 40 μm mouse SN section. Arrowheads point toward the SN pars compacta on both sides. Note the tiny hole that was placed during the cryosectioning of the tissue (arrow). The hole should be placed away from the SN. The size of the hole should be limited; otherwise the section might tear during staining procedure

4. For free-floating sections, cut 40 μm cryosections of the PFA-fixed brain in the coronal plane in four series (sections in the same series are separated by 160 μm) for mice and six series (240 μm distance) for rats collecting sections into an antifreeze cryoprotectant and store at −20 °C (*see* **Note 8**).
5. Mark one hemisphere to be able to distinguish the right from the left SN by placing a tiny hole at the upper part of the brain stem (*see* **Note 9**, Fig. 1).
6. Sections starting at −2.46 mm from Bregma and ending at −4.04 mm from Bregma for mice and from −4.36 to −6.72 from Bregma (rats) are collected. This will cover the whole SN.

3.6 Tissue Processing for Striatal HPLC and Dopamine Transporter Autoradiography

1. If the rostral part is not to be processed directly for HPLC or dopamine transporter (DAT) binding assay, freeze it immediately after dissection in liquid dry ice-cooled methylbutane.
2. Quickly wrap tissue in aluminum foil and keep on dry ice until it is stored at −80 °C (*see* **Note 10**).

3.7 Tissue Processing for Analysis of Striatal Dopaminergic Terminals

1. With tissues meant for striatal dopaminergic fibers and aSyn staining, immediately immerse rostral region into cryomold filled with OCT compound, slowly freeze the tissue in liquid dry ice-cooled methylbutane, and store at −20 °C.
2. Fresh frozen tissues are cryosectioned at 10 μm, mounted on object slides, and stored at −20 °C. Sections are collected from 0.26 to 1.10 mm from Bregma.

3.8 Immunohistochemical Stainings: Tyrosine Hydroxylase (TH) Free-Floating Sections for Stereological Analysis of SN Neuron Number

1. Select one series of SN from the 4/6 series described above.
2. Drain free-floating sections in antifreeze cryoprotectant into petri dish containing 0.01 M PBS for 3 min.
3. Move sections into a well containing ¾ full 0.01 M PBS and wash sections in PBS 3× for 5 min (*see* **Note 11**).
4. Incubate sections in blocking solutions at room temperature (RT) for 1 h on a shaker.
5. After blocking, transfer sections into a well with primary rabbit anti-mouse TH dissolved in diluent and incubate overnight at RT on shaker (*see* **Note 12**).
6. Wash sections 3 × 5 min in 0.01 M PBS.
7. Next, incubate sections with secondary antibody solution of biotinylated goat anti-rabbit and incubate at RT on shaker for 2 h.
8. Wash sections 3 × 5 min in 0.01 M PBS.
9. Transfer sections into avidin/biotin reagent solution and incubate for 2 h on shaker. Avidin/biotin solution is prepared 30 min before use.
10. Stain with DAB activated with H_2O_2 under visual sight (time may vary depending on how quickly the staining develops). Stop reaction with distilled water (*see* **Note 13**).
11. Wash sections 3 × 5 min in 0.01 M PBS.
12. Mount sections on object slides with Aquatex and put on coverslips (Fig. 1, *see* **Note 14**).

3.9 TH Staining on Fresh Frozen Sections for Analysis of Striatal Dopaminergic Terminals

1. Quickly dry sections at RT and encircle with liquid blocker.
2. Postfix sections with 4% PFA for 25 min at RT.
3. Wash sections 3 × 5 min in 0.01 M PBS.
4. Block for 30 min at RT.
5. Add primary antibody rabbit anti-mouse TH dissolved in diluent.
6. Incubate overnight at 4 °C.
7. Wash sections 3 × 5 min in 0.01 M × PBS.
8. Add secondary antibody, biotinylated goat anti-rabbit, for 1 h at RT.
9. Wash sections 3 × 5 min in 0.01 M PBS.
10. Incubate sections with avidin/biotin solution for 30 min.
11. Wash sections 3 × 5 min in 0.01 M PBS.
12. Add DAB-peroxidase substrate solution under visual sight. Stop reaction with distilled water.
13. Wash 2 × 3 min.
14. Add a drop of Aquatex and put coverslips on sections (Fig. 2).

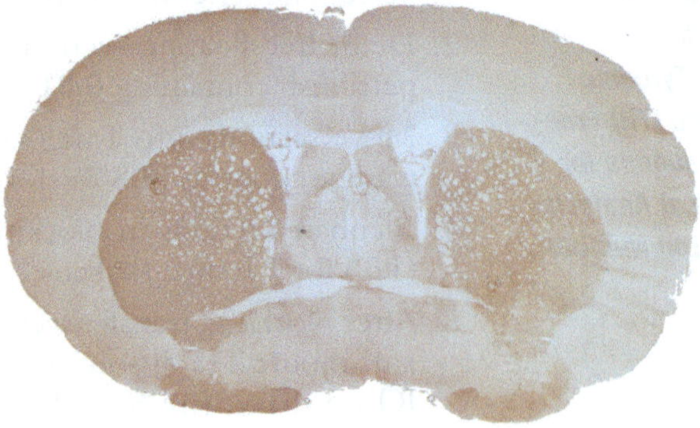

Fig. 2 Immunohistochemical staining for TH-positive dopaminergic terminals in a 10 μm mouse striatum section at the region of +0.26 mm from Bregma

3.10 Anti-NeuN Free-Floating Sections

1. Thick sections (40 μm) are washed and blocked as stated for free-floating sections above.
2. Incubate sections with primary antibody mouse anti-NeuN in diluent overnight at 4 °C.
3. Wash sections as stated above and apply biotinylated goat anti-mouse secondary antibody for 2 h.
4. Wash sections followed by avidin/biotin reagent incubation and finally stain with DAB activated with H_2O_2.

3.11 TH/aSyn Free-Floating Sections

1. For immunofluorescence double staining, 40 μm sections are blocked with blocking solution for 1 h at RT.
2. Sections are simultaneously incubated overnight at 4 °C with chicken anti-mouse TH and rabbit anti-human aSyn in diluent.
3. Wash sections 3 × 5 min in 0.01 M PBS.
4. Stain sections with goat anti-chicken and goat anti-rabbit secondary antibodies for 2 h at RT (*see* **Note 15**).
5. Wash sections 3 × 5 min in 0.01 M PBS.
6. Next, apply DAPI nuclear stain for 20 min at RT.
7. Wash sections 3 × 5 min in 0.01 M PBS.
8. Mount sections on object slides with Aquapolymount and coverslips.

3.12 TH/aSyn Fresh Frozen Sections

1. Postfix sections with 4% PFA for 25 min at RT.
2. Wash sections 3 × 5 min in 0.01 M PBS.
3. Block for 45 min at RT same as for free floating.
4. Incubate with primary antibodies (as stated for free-floating sections) overnight at 4 °C.
5. Wash sections 3 × 5 min in 0.01 M PBS.

6. Incubate with secondary antibody, for 1 h at RT.
7. Wash sections 3 × 5 min in 0.01 M PBS.
8. Incubate sections with DAPI for 10 min at RT.
9. Wash sections 3 × 5 min in 0.01 M PBS.
10. Use coverslips with Aquapolymount.

3.13 aSyn Aggregate/pSer129 Staining

1. 40 μm PFA-fixed mouse brain sections are brought down to water through graded alcohols (70%, 90%, 95%).
2. Sections are then treated with proteinase K for 15 min at 37 °C water followed by 5-min treatment at RT.
3. Perform antigen retrieval using citrate buffer (98 °C for 20 min and let it cool to RT for 20 min).
4. Quench endogenous peroxidase activity with 3% hydrogen peroxide.
5. Next, incubate sections with primary antibody LB509 at a dilution of 1:500 overnight at 4 °C.
6. To detect the phosphorylated form of aSyn at serine 129 (pSer129) sections are incubated in formic acid for 5 min followed by proteinase K digestion as described above.
7. Incubate sections with primary antibody 81A, at a dilution of 1:8000 for 1 h at RT.
8. Apply secondary antibodies and use an alkaline phosphatase polymer kit to detect staining (Vector Laboratories, Burlington, ON).
9. Develop color with Warp Red (Biocare Medical, Concord, CA) and counterstain sections with Mayer's hematoxylin (Fig. 3).
10. Dehydrate in alcohols (70%, 90%, 95%), clear in xylene (100%), and mount sections (*see* **Note 16**).

3.14 Dopamine Transporter (DAT) Binding Assay

1. Fresh frozen sections are incubated in binding buffer at RT.
2. Sections are then placed in buffer **step 2** with additional 50 pM [^{125}I]-RTI-121 for 2 h at 25 °C.
3. Nonspecific binding is described as that observed in the presence of 100 mM GBR 12909.
4. Next, slides are washed in ice-cold binding buffer, followed by rinsing in ice-cold distilled water, and dried.
5. Juxtapose slides with an autoradiographic film with [^{125}I]-microscale standards and leave for 2 days at 4 °C before development.
6. Autoradiograms are evaluated with MCID software (Image Research Inc.) (Fig. 4).

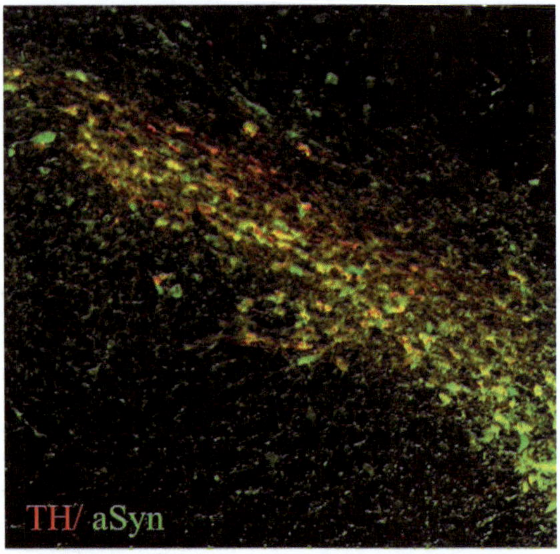

Fig. 3 Double-label immunofluorescence to label aSyn (LB509) in substantia nigra dopamine neurons (tyrosine hydroxylase)

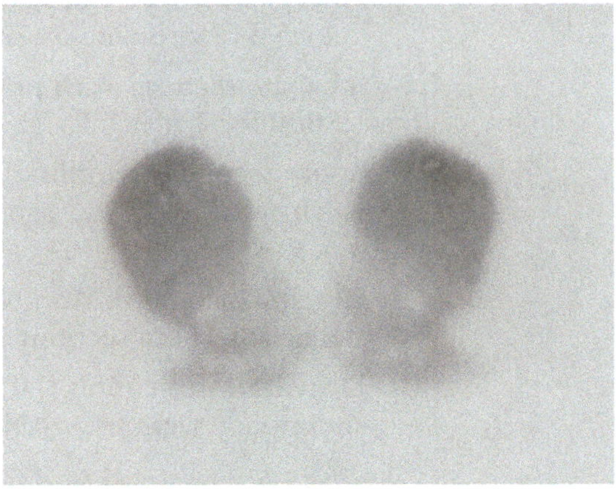

Fig. 4 Dopamine transporter autoradiography in the striatum of a rat derived from 10 μm thin sections cut on a cryostat

7. Densitometric analysis of three striata slices per mouse is done with a reference curve of c.p.m. against optical density and calculated from b-emitting [^{14}C] microscale standards and used to quantify the intensity of signal as nCi/g.
8. The background intensity is also subtracted from each reading.
9. Finally, the data is shown as mean ± SEM signal intensity for the treatment groups.

3.15 Catecholamine Quantification by High-Performance Liquid Chromatography (HPLC)

1. For HPLC analysis, brain tissues are homogenized in HPLC binding buffer.
2. Centrifuge samples at 10,000 *g* for 20 min.
3. Remove supernatant and save pellet for protein analysis.
4. For catecholamines, analysis is done with a specific HPLC assay with an Antec Decade II (oxidation: 0.5) electrochemical detector operated at 33 °C.
5. Samples are injected by using a Water 717+ autosampler onto a Phenomenex Nucleosil (5u, 100A) C18 HPLC column (150 × 4.60 mm).
6. Elution buffer delivery is at a rate of 0.8 mL/min with a Waters 515 HPLC pump.
7. Analytes are examined as follows: 3,4-dihydroxyphenylacetic acid (DOPAC), dopamine (DA), and homovanillic acid (HVA).
8. Data acquisition and control of HPLC are done with Waters Empower software.
9. Protein concentration for each sample is performed by Pierce BCA protein assay.
10. Catecholamine values are expressed as ng analyte/mg total protein.

3.16 Analysis of Dopaminergic Fibers in Striatum

1. The TH+ fiber density in striatum is measured by the optical density assessment using the NIH ImageJ software.
2. Transform image color into 8-bit grayscale images.
3. With the freehand tool, measure both left and right striata as well as the nonspecifically stained corpus callosum.
4. For each section analyzed, subtract the value of left and right striata from that of the corpus callosum.
5. Striatal optical density is represented as a percentage of the right in relation to the left striatum.
6. At least four striatal slices per mouse are analyzed.

3.17 Unbiased Stereological Estimation of TH+ Neurons in SN

1. The unbiased estimation of dopaminergic cells in the SN is done with the optical fractionator method of the Stereo Investigator software package.
2. Open software and click on Probes.
3. Select Optical Fractionator to open the workflow window.
4. In the pop-up window, choose Start a new subject.
5. Next, insert the investigator's name under "Name" and SN in the "Subject's name."
6. Enter the thickness of the cut tissue sections, in this case 40 μm.

7. When unsure of the number of sections per series to count, leave this option blank as sections can always be added in the last step.
8. For the next step, set the microscope to low magnification of 2× and choose 2× from the drop-down menu of the workflow window.
9. With the microscope still set at ×2, trace your region of interest with the left mouse button. TH+ neurons in the SN pars compacta and reticulata are included as described by Baquet et al. [4].
10. For the next step, set the microscope objective to ×100 and select ×100 from the drop-down menu of the workflow window.
11. The mounted thickness can either be determined before or during counting. To do this, focus to the top and bottom of the section and thickness will be automatically determined.
12. Next, define the counting frame size as 50 × 50 μm (size of the optical dissector) for mice and 60 × 60 μm for rats.
13. The grid size is defined as 110 × 110 μm for mice and for rats: 130 × 130 μm for TH and 200 × 200 μm for NeuN staining.
14. Set the guard zone at 1.5–2.0 μm.
15. Review sampling parameter settings and when necessary save the settings.
16. Ensure that the cursor is inside the traced region in the "Macro view" by right clicking "go to."
17. In the next step, select "count objects" to count TH+ neurons in each grid at every counting site.
18. Select a preferred marker and ensure that this is used for all sections in a series.
19. After counting is done for one section, click on the "Add New Section" button to begin the Optical Fractionator workflow for the next SN section of the same series.
20. After the last section of a series is quantified, select "I've Finished Counting."
21. Next, choose the "View Results" button to display the sampling data.
22. Enter the section interval, which will be 4 (mice) or 6 (rats), because cutting was done in a series of 4 or 6 in these cases.
23. The Gundersen coefficient of error for $m = 1$ should be less than or equal to 0.1 for series counted.

4 Notes

1. The surgery room should be well ventilated to prevent accidental intoxication with isoflurane (*see* Subheading 3.1, **step 1**).
2. The determination of the Bregma is crucial. If the Bregma is difficult to find, try to look for other anatomical hallmarks (like the Lambda) to orientate yourself (*see* Subheading 3.1, **step 16**).
3. Slow removal of the Hamilton syringe after injection is important, since viral vector might be drawn out of the SN by a fast-moving syringe (*see* Subheading 3.1, **step 23**).
4. For rotarod performance test the 2-day training session is important to get valid results. In case a rodent does not perform well enough (threshold 200–250 s), one should consider omitting this rodent from the rotarod performance test (*see* Subheading 3.2, **step 2**).
5. For the analysis of the cylinder test video recordings, one needs a high-quality video recorder and a video player that is able to process the movie at slow motion as one suggested under *see* Subheading 2.1.
6. When handling PFA, note that PFA is toxic, carcinogenic, and allergenic (*see* Subheading 3.4, **step 2**).
7. When handling dry ice, note that dry ice is extremely cold with a temperature around −78 °C. Therefore, use personal protective equipment (PPE) like gloves, when working with dry ice. Additionally, since dry ice evaporates in $C0_2$, one should work under a fume hood (*see* Subheading 3.5, **step 3**).
8. Stereology starts with tissue processing. Therefore, the serial sections of the SN must be obtained very carefully to prevent loss of sections (*see* Subheading 3.5, **step 4**).
9. To be able to perform stereological cell number estimation it is important to mark one hemisphere in order to distinguish the right from the left SN. This step is crucial. Check each section for the placement of the hole at the upper part of the brain stem (*see* Subheading 3.5, **step 5**).
10. For HPLC analysis the striatal tissue has to be processed as fast as possible after brain dissection. To prevent degradation of the tissue, store it at −80 °C if not processed immediately (*see* Subheading 3.6, **step 2**).
11. Handling the 40 μm free-floating brain sections can be tricky and there are many possibilities for how one can damage the tissue by accident. Try to use very fine brushes if you need to handle the sections. Otherwise try to touch the sections as less as possible (*see* Subheading 3.8, **step 3**).

12. Although we give recommendations for the dilution we used for the primary antibodies, the optimal working concentration should be determined by titration experiments (*see* Subheading 3.8, **step 5**).
13. When handling DAB, note that DAB is suspected to be carcinogenic. Therefore, use PPE when working with DAB (*see* Subheading 3.8, **step 10**).
14. Avoid air bubbles under the coverslip by lowering the coverslip very slowly from one end to the other. This will apply the Aquatex evenly along the bottom of the coverslip to cover the section (*see* Subheading 3.8, **step 12**).
15. In case you find some fluorescent precipitation on the section after performing the immunofluorescence staining, try to centrifuge the secondary antibody before application to get rid of unbound fluorescent dye during the staining process (*see* Subheading 3.11, **step 4**).
16. When handling xylene, note that xylene is an aromatic hydrocarbon that is moderately toxic and inflammable. Work under a fume hood and wear PPE (*see* Subheading 3.13, **step 10**).

Acknowledgments

This project was supported by the Interdisciplinary Center for Clinical Research (IZKF) at the University of Würzburg (A-303).

References

1. Koprich JB, Kalia LV, Brotchie JM (2017) Animal models of alpha-synucleinopathy for Parkinson disease drug development. Nat Rev Neurosci 18(9):515–529
2. Ip CW, Klaus LC, Karikari AA, Visanji NP, Brotchie JM, Lang AE et al (2017) AAV1/2-induced overexpression of A53T-alpha-synuclein in the substantia nigra results in degeneration of the nigrostriatal system with Lewy-like pathology and motor impairment: a new mouse model for Parkinson's disease. Acta Neuropathol Commun 5(1):11
3. Koprich JB, Johnston TH, Reyes MG, Sun X, Brotchie JM (2010) Expression of human A53T alpha-synuclein in the rat substantia nigra using a novel AAV1/2 vector produces a rapidly evolving pathology with protein aggregation, dystrophic neurite architecture and nigrostriatal degeneration with potential to model the pathology of Parkinson's disease. Mol Neurodegener 5:43
4. Baquet ZC, Williams D, Brody J, Smeyne RJ (2009) A comparison of model-based (2D) and design-based (3D) stereological methods for estimating cell number in the substantia nigra pars compacta (SNpc) of the C57BL/6J mouse. Neuroscience 161(4):1082–1090

INDEX

A

B

C

D

E

F

G

H

I

Tim Bartels (ed.), *Alpha-Synuclein: Methods and Protocols*, Methods in Molecular Biology, vol. 1948,
https://doi.org/10.1007/978-1-4939-9124-2,

T

V

W

Y

Z

MIX
Papier aus verantwortungsvollen Quellen
Paper from responsible sources
FSC® C105338

If you have any concerns about our products,
you can contact us on
ProductSafety@springernature.com

In case Publisher is established outside the EU,
the EU authorized representative is:
Springer Nature Customer Service Center GmbH
Europaplatz 3, 69115 Heidelberg, Germany

Printed by Libri Plureos GmbH
in Hamburg, Germany